**Production**

Michelle Bloom, WB1ENT

Jodi Morin, KA1JPA

David F. Pingree, N1NAS

Printed in USA



ISBN: 978-1-62595-135-9

Second Edition
First Printing

We strive to produce books without errors. Sometimes mistakes do occur, however. When we become aware of problems in our books (other than obvious typographical errors), we post corrections on the ARRL website. If you think you have found an error, please check **www.arrl.org/notes** for corrections. If you don't find a correction there, please let us know by sending e-mail to **pubsfdbk@arrl.org**.

# CONTENTS

# PREFACE

Welcome to the 2nd edition of *Antenna Physics, an Introduction*. The 1st edition was remarkably successful, with four printings. The text was written for amateurs, engineering students, and professionals alike, and many copies were sold to all groups. Many university-level antenna theory courses are using the text as a supplement to the more traditional academic treatments. With each subsequent printing, typos and errors were corrected, and these corrected segments of the book remain. I would like to thank the multitude of readers who contributed to finding errors and taking the time to document and report them to me.

The largest addition to this 2nd edition is Chapter 4, concerning transmission lines. This chapter was designed to supplement the discussion in *The ARRL Antenna Book* with considerably more depth and with a comprehensive treatment of open wire lines. Also, Chapters 2, 3, 5, 8, and 9 have additions and/or modifications.

## Why write a book about Antenna Physics?

After nearly 50 years of studying the subject, designing, building, and using antennas for both amateur and professional applications, I had concluded that there here are many very fine books available on this fascinating subject, which usually fall within one of two categories: (1) relatively simple treatments that provide "how-to" material as well as basic theory to support some level of fundamental understanding and (2) highly advanced theoretical graduate-level engineering texts.

There appeared to be a gap between these two extremes. Thanks to the well-written and comprehensive "how-to" publications of the ARRL, I was able to master much of the basic theory of antennas by the time I graduated from high school. These books provide enough theory to gain a sufficient background to design, build, measure, and use antennas. However, these texts only instilled a craving for deeper understanding of the underlying physics of antennas. The leap from these introductory texts to the masterpieces of Kraus, Jasik, Balanis, and others necessarily had to wait for competence in advanced mathematics, including the indispensable calculus. After many years of using these books as references and thus providing ever deeper insights into the workings of antennas, I concluded that many of the more advanced treatments could be explained, at least in part, using a simpler approach. This book is an attempt, therefore, to "bridge the gap" between the introductory and "how-to" books and the advanced texts. I hope some readers will use this introductory text as a springboard to the more advanced treatments of this fascinating subject, indispensable to all forms of radio applications. Although many of the antenna applications will be most useful to amateur radio operators, about 2/3 of the text should also find more widespread use. I hope that it will serve electrical engineering students and professional radio engineers, as well as the amateur community.

Advanced mathematics is an essential tool for solving complex antenna problems. However, it is not necessary for understanding the deeper principles. This is one of the key motivations and challenges to writing this book: you can understand the physics without being able to solve a calculus problem, but it is of great advantage to understand what the equations are saying. Indeed, many of the complex-looking equations found in this text are actually only special forms of much simpler equations that are second nature to anyone with a competence in electronics. In this text I often point out the simpler forms as the calculus is presented. For example, many of the complex equations are merely special forms of well-understood relationships like Ohm's Law or the power law.

Some competence in mathematics is required for this text, particularly algebra. Also, being able to deal with some trigonometry as well as logarithms is essential to grasping the fundamental concepts. When more advanced equations are presented, explanations are often offered to explain what the reader is looking at, and what it means, particularly for equations involving calculus and linear algebra. Of course, a mastery of calculus and other forms of advanced mathematics will afford an even deeper insight when combined with dedicated study of antenna physics proper.

Physics is classified as one of the natural sciences, together with mathematics, chemistry, biology, and geology. Students of these disciplines (purists) are only interested in describing nature. In contrast, the engineering sciences, concern themselves with designing and building something. The engineering sciences are completely dependent upon the natural sciences for the laws and theories needed to do or build something useful. On the other hand, engineering, which utilizes theories based in natural science, provides excellent vindication of the laws and theories rooted in natural science. The business of natural science is to provide dependable repeatable experiments with similar results. Engineers need repeatable results to design and build something that will be manufactured. For example, the esoteric theoretical subject of sub-atomic physics was validated with the first atomic explosion. For the pragmatic person, this is the ultimate test of a theory: what are the useful results? Or, what are the possible useful results armed with this new theory? What can we do with this new knowledge?

In this book we will explore how a mathematician (Maxwell) first predicted that radio waves exist. Then a physicist (Hertz) actually built an experimental apparatus to prove radio waves exist, and finally an engineer (Marconi) actually developed radio as something useful. Of course, it wasn't that straightforward. Thousands of scientists and engineers have made contributions, large and small to develop "wireless" as we know it today, but the sequence of theory, proof, and finally "let's do something with this" is clear. Therefore, an antenna designer, builder, or even just a user can become more capable if the basic underlying principles (the physics) are understood.

After a lifetime of learning, experimenting, mentoring and teaching antenna theory, I found several topics to be commonly misunderstood. Having to repeat the same ideas multiple times over the years, extra attention is given to topics that reflect these misunderstandings among not only amateurs, but also professionals. These topics, understandably, are the more complex and less intuitive. Some repetition of the basic "how-to" books is essential in writing such a text. However, I attempted to minimize repetition assuming the reader already understands these basics or has access to basic texts such as *The ARRL Antenna Book*.

"Applied" examples were chosen to gain insights into common antenna types, which can then lead to aids in engineering antennas. Also, many of these applied examples, indeed chapters, reflect my serious interest in amateur radio and the publisher, ARRL. Amateur radio has been an avocation since my childhood, and participation in this wonderful hobby led me ultimately into a lifelong career in electronic engineering with a specialty in "RF". ARRL's publications have been and remain indispensable for the radio amateur and many engineers and technicians alike. In this vein I hope that this text provides value to amateurs, students, and professionals.

## Preface to the Text

The organization of this text is different from both the more traditional introductory and advanced texts. The beginning chapters weave actual electricity and magnetism (E&M) material, i.e. Maxwell's equations taken from pure classical physics with how these concepts relate directly to antennas. The reasoning is simple. In order to develop a deeper understanding of antennas, it is necessary to gain a deeper understanding of the fields with which they interplay and the characteristics of free space, the medium in which they perform.

Many of the terms in common use regarding antennas are redefined in relation to terms typically quantified in the more advanced texts but ignored or only briefly encountered in the basic texts. Hopefully these discussions will shed new light and deeper insights on these basic terms used in discussions on antennas. This is the essence of the first three chapters.

It is common practice for authors to simply reference equations already presented in previous chapters. I have always found this practice awkward, having often to thumb back through the chapters to find the applicable equation. I would prefer readers to find the text a bit repetitive rather than assuming equations and other points made earlier are now embedded in the readers' memory. Also, this book is written somewhat in a "narrative" form. The intent is for chapters to flow from one to the next. Hopefully, however, the chapters will also stand by themselves as references to the applicable topics. Many important topics are presented in an abbreviated form, especially when these topics are very well covered by introductory and/or specialized texts commonly available.

## Part 1: Antenna Physics

**Chapter 1:** ***Development of Antenna Physics***: We first present an introduction of both the theoretical development of radio and then show how antenna theory was co-developed as an essential element to "radio." Taken in a historical context, it also offers an acknowledgement of some of the great indispensable contributors to the art.

**Chapter 2:** ***Fundamentals***: Building upon the introductions in Chapter 1, we begin by introducing some of the basic physical parameters of antennas. The approach quickly diverges from the "how-to" books and concentrates on developing standards of measurements, the concept of aperture, analogy to light, relationships among directivity, gain and aperture, the concept of an isotropic antenna, path loss, and radiation resistance. In this edition, more detail is presented on the effects of aperture on link budgets.

**Chapter 3:** ***Radiation of Radio Waves***: Again, building from the previous chapter, we develop, in detail, how radio waves "radiate" and how antennas perform this task. An introduction is made to Maxwell's equations, the wave equation, relationship between the magnetic and electric fields, the impedance of free space, power, energy and fields, detailed definition of radiation resistance, and how fields and antenna patterns are formed. In order to clarify some concepts, the dipole antenna is also introduced. This chapter, in particular, diverges from most antenna texts in that it roots antenna performance and characteristics in electromagnetic physics. In this edition considerably more material and explanation is provided on different types of electric and magnetic fields and expanded discussions of radiated fields and how they combine in the far field.

**Chapter 4:** ***Transmission Lines***: This chapter has been greatly expanded in this edition. The original material is largely intact, stressing the differences between radiated and guided waves and extending previous chapters' discussions into transmission lines. New material then a very detailed treatment of reflected waves; their causes, characteristics, and consequences. Specific step-by-step descriptions are presented to illustrate how standing waves are formed, the roll conjugate match "tuners" play, and the reciprocity of tuned lines. The extensive final section offers a very comprehensive discussion of open wire transmission lines including a technique for accurate modeling of all the important characteristics of open wire lines using *EZNEC*, with extensive explanations and graphs plotting modeling results.

**Chapter 5:** ***Gain Patterns from Multiple Sources***: We explain how directive antennas can be constructed from multiple elements, including multi-element driven and parasitic arrays. The concept of "PADL" is also introduced. In this edition, the concept of phased array antennas is introduced using the familiar 4-square array as an example. Expanded performance of the 4-square is explained in detail.

## Part 2: Applied Antenna Physics

**Chapter 6:** ***Dielectric effects upon Radio Waves***: Most antennas we encounter are located close to Earth's surface. Therefore, no antenna text is (even a physics book) complete without a treatment of the physics of the Earth "geophysics" as relating to radio waves. Since the Earth is a dielectric, dielectrics are treated first in a general discussion. Then the earth is discussed as a special example of dielectrics.

**Chapter 7:** ***Vertical Antenna***: Much of the material presented in Chapter 6 becomes particularly relevant to ground-mounted vertical antennas. Amateur radio experiments may find this chapter particularly interesting stemming from the intense interest in the amateur community of vertical antennas. A mathematical format (scalar matrix) is introduced for the quantification of vertical antenna losses. Also, a simple comparison of vertical configurations is also offered with their relative major advantages and disadvantages.

**Chapter 8:** ***Yagi-Uda and Cubical Quad Parasitic Arrays***: The Yagi-Uda configuration has become one of the most commonly used antenna forms. Although volumes have been written on this antenna type, the purpose of this chapter is to relate its performance to the physical parameters already developed in this text.

**Chapter 9:** ***Specialized Antenna Configurations***: This chapter presents some of the more important, but less-known antenna configurations. Loop and Helix antennas are presented first, followed by "aperture" antennas, traveling wave antennas, and finally a detailed discussion on small antennas (relative to wavelength) The discussion on small loop antennas has been greatly expanded in this edition.

**Chapter 10:** ***Noise, Temperature, and Signals***: In this final chapter several goals are attempted: The primary importance of noise and noise temperature are presented and are shown to be intrinsically related. Then "noise temperature" is shown to be an important antenna specification, especially for radio astronomy and deep space applications. Radio link budgets are presented showing how noise sets fundamental limits. We also return to include some historical narrative culminating in great scientific achievements. The epic discovery of Penzias and Wilson is presented as an outstanding research application of an antenna.

The work of Claud Shannon and Information Theory are introduced. Although not related directly to antennas the discussion of Information Theory is presented since it links noise to fundamental physical limiting factors for both the radio link and on the limits to communication of information. Thus, the reader is left with all the known physical limits (and opportunities) for communicating information using radio links, for which the antenna forms a critical function. These are the same limits for garage door openers, cellphones, WiFi as well as establishing a hypothetical link with an extraterrestrial civilization.

## Acknowledgements

First, I would like to acknowledge the ARRL publications that provided me with the critical first level of understanding of radio technology and to the point of this text, antenna theory. It is my hope that this text will offer yet another useful volume to ARRL's library. Some of my mentors dating from the 1960s also deserve mention: Martin Hornak, W8MAE (SK) and Orlando O. Okleshen, W9RX (SK) provided needed guidance and help through those difficult teenage years. Later my professors at the University of Oregon Physics Department, especially Dr. K. J. Park, deserve mention. Although I never met him, Dr. John Kraus', W8JK, famous text, remains my "bible" for antenna engineering. My friends and fellow authors Wes Hayward, W7ZOI, and Cornell Drentea, KW7CD, showed me through example how hard work and dedication can result in the production of fine technical books. The narrative style of Cornell's book provided the inspiration to also present this text in a similar vein. Gary Breed, K9AY, provided an excellent editorial review of the entire text and numerous useful suggestions and corrections. Gary also wrote the introduction to this book. Finally, Dr. Nuri Celik performed a detailed check of the most difficult math, particularly from Chapter 3, and yes, he found some errors! However, the errors were minor and few and I hope this will ring true for the entire text.

Bob Zavrel, W7SX
February 21, 2020
Spokane, Washington

# ABOUT THE AUTHOR

Robert J. Zavrel Jr, W7SX, was born in Chicago in 1951. His interest in radio developed very early having dissected his mother's radio at age four, much to his mother's horror seeing him playing with an opened tube radio on the floor.

He was raised in Chicago Heights and passed his Novice exam at age 14. He earned his DXCC certificate at age 17. He graduated with his first degree at Roosevelt University (majoring in Geography) in 1974 and worked as a navigator (Raydist system) on oil exploration ships in the Gulf of Mexico, and then as a radio broadcast engineer.

In 1979 he returned to school at the University of Oregon and received his second degree in Physics in 1983. He moved to Silicon Valley upon graduation and spent 12 years there working mostly with RF semiconductors. Subsequent assignments were at IBM in RTP, NC. and ATMEL in Colorado Springs working at director level on SiGe for RF applications, Trimble Navigation in Corvallis, Oregon, and a total of 10 years in RF/business consulting for a wide range of customers. He has earned DXCC and CW DXCC Honor Roll status using only tree-supported wire antennas, plus his DXCC on all bands 160-10 meters. He recently delivered a 2-hour lecture on low band antennas to the Bordeaux DX Groupe (in French).

Bob is a Life Member of ARRL, an ARRL Technical Advisor, an IEEE Senior Member, a member of the IEEE Antennas and Propagation Society, and a part-time adjunct professor in Electrical Engineering at Gonzaga University with access to their advanced RF lab and anechoic chamber.

He has seven patents, published the first block diagram of a software defined radio, wrote the first paper on using (DDS)M for FM and HDTV broadcast applications, which are now industry standards. Bob has published over 70 articles and papers in both professional and amateur publications. He recently helped refine the search algorithm (DSP) for a major Search for Extraterrestrial Intelligence program employing three instruments, including one at the National Radio Astronomy Observatory.

Today, Bob is doing consulting work for a variety of government and private companies all involving RF/antenna designs. He also presents seminars on antenna theory, and business/marketing research. When this book went to press, he was building an antenna farm and radio antenna installation north of Spokane, Washington using trees for wire supports.

# INTRODUCTION

I'll begin by re-stating part of my introduction to the original edition: This book was written with the goal of helping amateur radio colleagues follow their curiosity, with the intention of providing a substantial step forward, beyond the how-to articles and basic concepts presented in magazines and handbooks.

As with any good technical book, feedback and questions are received from readers (and reviewers). The author and his colleagues also decide that certain things require better explanations, a bit more detail, and maybe another chart or equation to illustrate a particular concept. And so there is a new edition, incorporating as many updates and modifications as is practical. This new edition includes expanded material on transmission lines, antenna arrays, and plenty of revised descriptions of concepts and principles that should make things clearer to many readers.

Many of the updates are improved intuitive descriptions of electromagnetic principles that are difficult to state in layman terms. The need to get a "seat of the pants" grasp of such concepts is important, since it often makes it easier to follow the mathematical definitions. Those attempts to find the right way to say things will certainly continue. Let the author know if something isn't clear—he may provide you with a better explanation that will be used in a future edition!

And a final thought from the original introduction: If this book encourages readers to pursue further study to gain the additional background and find the alternate descriptions they are seeking, we will consider it a great success!

Gary Breed, K9AY

## About ARRL

We're the American Radio Relay League, Inc. — better known as ARRL. We're the largest membership association for the amateur radio hobby and service in the US. For over 100 years, we have been the primary source of information about amateur radio, offering a variety of benefits and services to our members, as well as the larger amateur radio community. We publish books on amateur radio, as well as four magazines covering a variety of radio communication interests. In addition, we provide technical advice and assistance to amateur radio enthusiasts, support several education programs, and sponsor a variety of operating events.

One of the primary benefits we offer to the ham radio community is in representing the interests of amateur radio operators before federal regulatory bodies advocating for meaningful access to the radio spectrum. ARRL also serves as the international secretariat of the International Amateur Radio Union, which performs a similar role internationally, advocating for amateur radio interests before the International Telecommunication Union and the World Administrative Radio Conferences.

Today, we proudly serve nearly 160,000 members, both in the US and internationally, through our national headquarters and flagship amateur radio station, W1AW, in Newington, Connecticut. Every year we welcome thousands of new licensees to our membership, and we hope you will join us. Let us be a part of your amateur radio journey. Visit www.arrl.org/join for more information.

225 Main Street
Newington, CT 06111-1400 USA
Tel: 860-594-0200
FAX: 860-594-0259
Email: membership@arrl.org

**www.arrl.org**

# Part 1
# Antenna Physics

# Development of Antenna Physics

## Early History

"Electricity" had probably been observed since the beginning of the human species in the form of lightning and perhaps static electricity raising hairs on skin clothing worn by early cave-dwelling experimenters. From ancient Greece, Aristotle and Thales described the effects of magnetism. Much later, in the 11th century, the Chinese scientist Shen Kuo was the inventor of the compass and employed it for navigation. Detailed rational explanations of electricity and magnetism had to wait until the 19th century. The invention of radio and thus antennas followed quickly after the fundamental breakthroughs.

## Modern History

During the Renaissance in Europe, the scientific method began to replace mystical explanations for a wide variety of unexplained phenomena. Careful, measured experiments were undertaken and documented. Following is a chronological record of some of the more important breakthroughs related to electromagnetics:

• **1590s**: William Gilbert (English) conducted the first formal experiments in electricity and magnetism. His observation that the Earth is actually a magnet directly led to refinements of the compass.

• **1746**: Jean le Rond d'Alembert (French) formulated the one-dimensional wave equation describing the motion of a vibrating string. Although his work had nothing to do with electricity or magnetism, it would prove indispensable a century later.

• **1750**: Benjamin Franklin (American) developed the law of conservation of charges, found there were two types of charges (positive and negative), and invented the lightning rod.

• **1755**: Charles Augustin de Coulomb (French) invented a very sensitive balance and used it to relate mechanical forces due to electric and magnetic fields.

• **1780**: Karl Friedrich Gauss (German) developed the divergence theorem, which later would become part of Maxwell's equations.

• **1800**: Alessandro Volta (Italian) invented the voltaic cell and the battery, thus providing, for the first time, a continuous moving charge — an electric current.

• **1819**: Hans Christian Oersted (Danish) observed that an electric current flowing on a wire deflected a compass, thus showing that electricity and magnetism were somehow intrinsically related.

• **1820**: Andre Marie Ampere (French) invented the solenoid and thus produced a significantly higher magnetic force from an electrical current. He also quantified the relationship between the current and magnetic flux.

• **1821**: Georg Simon Ohm (German) derived his famous law relating voltage, current, and resistance

• **1831**: Michael Faraday (English) discovered and quantified the phenomena of a changing magnetic field produced an electrical current, which is the reciprocal of Oersted's findings. This implied that electricity and magnetism were more closely related more than previously thought.

• **1831**: Joseph Henry (American) independently observed the same phenomena as Faraday.

## Maxwell, the Turning Point

•**1860s/70s**: James Clerk Maxwell (Scottish, **Figure 1.1**) saw that Faraday's work was the final concept needed to postulate a unified theory of electromagnetism. In effect, Maxwell combined the equations of Faraday, Gauss and Ampere into one set of relationships that precisely define the behavior of electromagnetism, and more importantly for this text, the behavior of electromagnetic waves.

**Figure 1.1 — James Clerk Maxwell.**

Maxwell also postulated (correctly) that light is an electromagnetic wave by a brilliant observation. Coulomb and Ampere, through *static* measurements, were able to determine the permittivity ($\varepsilon_0$) and permeability ($\mu_0$) of free space by measuring the physical forces of magnets and electric fields imparted on objects. Maxwell noticed that the expression

$$\frac{1}{\sqrt{\varepsilon_0 \mu_0}}$$

actually defined a *velocity*, where

$$\varepsilon_0 = 8.85\,(10)^{-12}\,\frac{(coulomb)^2}{Newton - meter^2} \qquad \text{(Equation 1.1)}$$

and

$$\mu_0 = 4\,\pi\,10^{-7}\,\frac{Newton}{(Ampere)^2} \qquad \text{(Equation 1.2)}$$

or

$$\mu_0 = 4\,\pi\,10^{-7}\,\frac{Newton}{\left(\frac{coulomb}{second}\right)^2} \qquad \text{(Equation 1.3)}$$

therefore

$$\frac{1}{\sqrt{\varepsilon_0 \mu_0}} = 299.8634 \times 10^6 \text{ m/sec} \qquad \text{(Equation 1.4)}$$

or 299,863,400 meters/second.

This is obviously a velocity, but of what? In other words the basic free space coefficients of the magnetic and electric fields were related to a velocity. Through simple algebra, Maxwell reduced these measured values (distance, mechanical force, static charge and current) to a velocity. Once he had performed these calculations, he postulated that light must be an electromagnetic wave after noticing that this velocity was close to the then-measured speed of light. Since these static (and current flow) terms could be measured with good accuracy even in the 1860s/70s, Maxwell's calculation of the speed of light was more accurate than the direct measurements of the speed of light possible in the 1870s. His calculated speed is very close to today's known measured speed of light.

Sixty years before Maxwell, Thomas Young discovered that light waves could interfere with one another, supporting the idea that light was composed of waves, and the distance between the ripples in the interfering pattern demonstrated the wavelength of light directly. If light were an electromagnetic wave, then certainly electromagnetic waves of other wavelengths must also exist! Thus in the 1870s Maxwell not only predicted the existence of radio waves, but also laid the accurate theoretical basis for their study and — more important — their application. This quantification of radio waves, even before they were discovered, still forms the basis of electromagnetic theory today. Maxwell's equations define precisely the relationships among the electric field, the magnetic field, and the direction of propagation. They also define precisely the ratio of the field strengths, and the mechanisms for producing a radiated or a directed electromagnetic wave.

The pivotal work of Maxwell was not universally accepted at the time of its formulation. Today Maxwell's equations remain the basis of all theory pertaining to electromagnetic waves, and thus also antennas. It would take decades before the validity of Maxwell's work would become difficult to refute.

The initial difficulty with Maxwell's equations was their complexity. His original formulation was 20 equations with 20 variables! In the early 1880s, Oliver Heaviside applied his understanding of both electrical circuits and advanced mathematics to "clean up" and simplify Maxwell's equations to just four equations in two variables. This work made it much easier to grasp the concepts and fundamental relationships developed by Maxwell, but we should remember that the actual equations that are familiar to us are the work of Heaviside. We will return to Maxwell's equations with more detail in Chapter 3.

• **1887**: Albert Michelson and Edward Morley (American) performed the first experiment that negated the theoretical existence of the "luminiferous ether." Previously it was accepted theory to assume that space was filled with an invisible "ether." Since light waves traveled through free space, they required some type of matter to travel in. After all, ocean waves need water, sound waves need air (or some other non-vacuum medium), and so on. This assumption was in large measure derived from Newton's Laws constituting Classical Mechanics. This experiment, together with Maxwell and a growing body of knowledge from many scientists, eventually led to Albert Einstein's Theory of Relativity. Later we shall see how Maxwell found that electromagnetic waves did not require an "ether" as a propagation medium in effect, questioning the then-assumed universality of Classical Physics.

• **1886**: Heinrich Hertz (German, **Figure 1.2**) built the first complete radio system consisting of a spark gap generator, a transmit antenna, a receive antenna, and a spark gap receiver (**Figure 1.3** and **Figure 1.4**). Since radio waves cannot be detected by any human sense, it was necessary to build a complete system to actually prove radio waves exist! Thus Hertz invented the antenna, the radio transmitter, and the radio receiver, and in so doing "discovered" the radio waves that Maxwell had predicted to exist.

**Figure 1.2 — Professor Heinrich Hertz.**

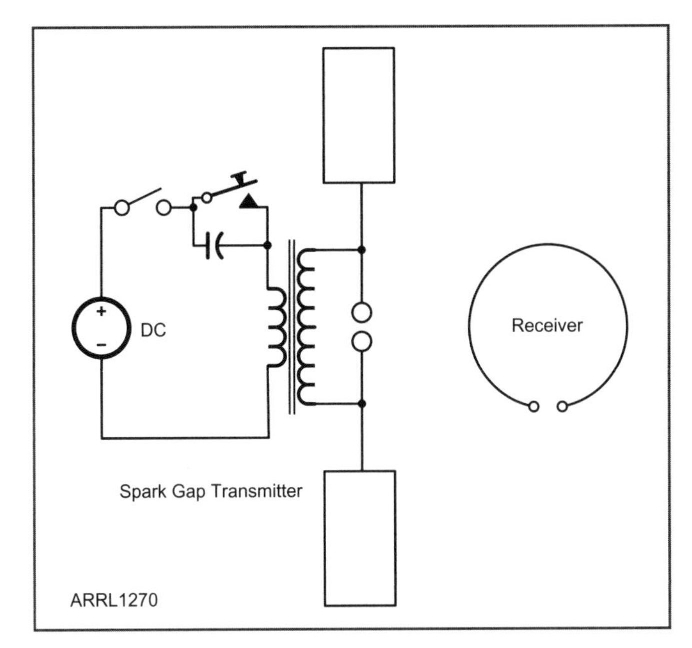


Figure 1.3 — The first radio transmit-receive system. The voltage is increased by a transformer, and a spark appears across the antenna input. The receive antenna is a single loop which receives the signal and produces its own spark.

Figure 1.5 — Guglielmo Marconi, the" father of radio," transformed "wireless" from a theoretical curiosity into a practical tool.

Figure 1.4 — An early rotary spark gap transmitter. [Joe Veras, K9OCO, photo]

Figure 1.6 — This memorial inscription for Marconi, the "inventore della radio" was placed in the Basilica of Santa Croce in Florence, Italy near the tombs of Galileo, Michelangelo, Foscolo, Machiavelli, and Rossini. [Photo by the author]

• **1895**: Guglielmo Maria Marconi (Italian, **Figure 1.5** and **Figure 1.6**) invented practical radio. Hertz was a physicist and was interested in the discovery and study of this newly-invented apparatus. Marconi was the engineer who used Hertz's inventions and findings to build something useful. If radio were to be useful, it would be necessary to significantly extend the range to provide a new and useful application — wireless communication. Successive experiments and refinements extended the range, and soon it became apparent that the antenna played a crucial role in increasing the range of the wireless devices.

In one of the experiments he greatly lengthened the wires and configured both into ground mounted "verticals." The result was a range of 2.4 kilometers. Marconi was perhaps the first to appreciate that the size of an effective antenna needed to be comparable to the wavelength being used. This was synonymous to raising the radiation resistance of an antenna, a critical specification for antennas.

After moving to England, Marconi received needed funding. After further refinements to his antennas and radios, in 1901 he shocked the world by completing a wireless trans-Atlantic link from England to Newfoundland. Modern wireless had been born.

• **1897 – 1904**: Hendrik Lorentz, George Fitzgerald, and Joseph Larmor derived the Lorentz transformation, stating that Maxwell's equations were invariant.

• **1905**: Albert Einstein used Maxwell's equations as the basis of Special Theory of Relativity.

## 1901 to Present

Since its infancy, radio technology has progressed through a continuous stream of innovations. The basic theory was derived in the 1860s and 1870s, but the technology is still evolving quickly some 150 years later. This technical evolution covers spark gap, the invention of the triode vacuum tube, discovery of the piezoelectric effect, the transistor, integrated circuits, super-miniature digital integrated circuits, miniaturization of all the passive devices, and countless other advances small and large.

Figure 1.7 — Conceived in the 1940s, Dick Tracy's two-way wrist radio was made possible today by advances in digital and RF microelectronics. The necessary antenna technology was already in place.

The first vacuum tubes had severe speed limitations and thus were usable only at relatively low frequencies. However, vacuum tubes also were far more effective in building receivers of much improved sensitivity and selectivity compared to the coherer detector. In the early 20th century, the first radio applications used the region of spectrum now used by the standard AM broadcast band, hundreds of meters in wavelength. Longer wavelengths were considered "low frequency" and shorter wavelengths were considered "high frequency."

The relationship between wavelength ($\lambda$) and frequency in Hz ($f$) is:

$$\lambda = \frac{c}{f} \quad \text{(Equation 1.5)}$$

where $c$ is the speed of light, $\lambda$ is wavelength, and $f$ is the frequency in Hertz (Hz).

As technology improved, thus pushing the useable spectrum higher in frequency, it was necessary to use progressively stronger superlatives to define these new bands as they opened up to practical use as shown in **Table 1.1**. Legacy trapped us into defining 3 MHz as a "high frequency." Perhaps frequencies above 300 GHz should be defined as *Unimaginably* High Frequencies!

Today, the opening of the microwave spectrum (approximately >1 GHz) has made possible low-cost, low-power portable devices. Microwave wavelengths are comparable to the physical dimensions of the human hand. Thus by using these wavelengths (and shorter), efficient antennas could be built into hand-held equipment. The results have been profound. For example, the personal cell phone and its numerous derivatives were made possible by a combination of two fundamental technical innovations: 1) miniaturization of microwave radio components to permit a two-way microwave radio fit in a hand-held package and be battery operated; and 2) digital integrated circuits that could permit the new miniature two-way microwave radio to behave like a consumer-friendly wide-area wireless "telephone" and/or computer link. More recently, advanced computers could also be miniaturized and included in tablets such as Apple iPads and similar hand held computer/wireless devices.

Dick Tracy had his "wrist radio" in the 1940s (**Figure 1.7**), complete fantasy at the time. The concept of "cellular radio" was conceived in the late 1940s at Bell Labs, but at that time a "portable" cellular radio would have been the size of a telephone booth. In contrast, antenna technology for portable applications at microwave frequencies is almost as old as antennas themselves. After all, for higher frequencies you simply make the antenna smaller! Antenna engineers had to wait for radio and digital circuitry to catch up. But catch up, they did, as shown in **Figure 1.8**.

We are all familiar with hand-held receivers that are quite effective at much longer wavelengths (such as AM broadcast). However, the receive antenna's inefficiency (due to very small size versus the wavelength) is overcome by the stationary transmitters' high power (kilowatts) and antennas that do approach the physical dimensions of the applied wavelength. Thus, poorly performing antennas are not out of the question when designing radio systems, as we will explore in later chapters.

The advent of personal computers such as the one shown in **Figure 1.9** was made possible by the miniaturization of the microprocessor, memory, and I/O digital circuits. The application that led

**Table 1.1**
**Frequency Band Nomenclature**

| *Frequency Range* | *Name* | *Total Bandwidth* |
|---|---|---|
| 30 – 300 kHz | Low Frequency | 270 kHz |
| 300 – 3000 kHz | Medium Frequency | 2.7 MHz |
| 3 – 30 MHz | High Frequency | 27 MHz |
| 30 – 300 MHz | *Very* High Frequency | 270 MHz |
| 300 – 3000 MHz | *Ultra* High Frequency | 2.7 GHz |
| 3 – 30 GHz | *Super* High Frequency | 27 GHz |
| 30 – 300 GHz | *Extremely* High frequency | 270 GHz |

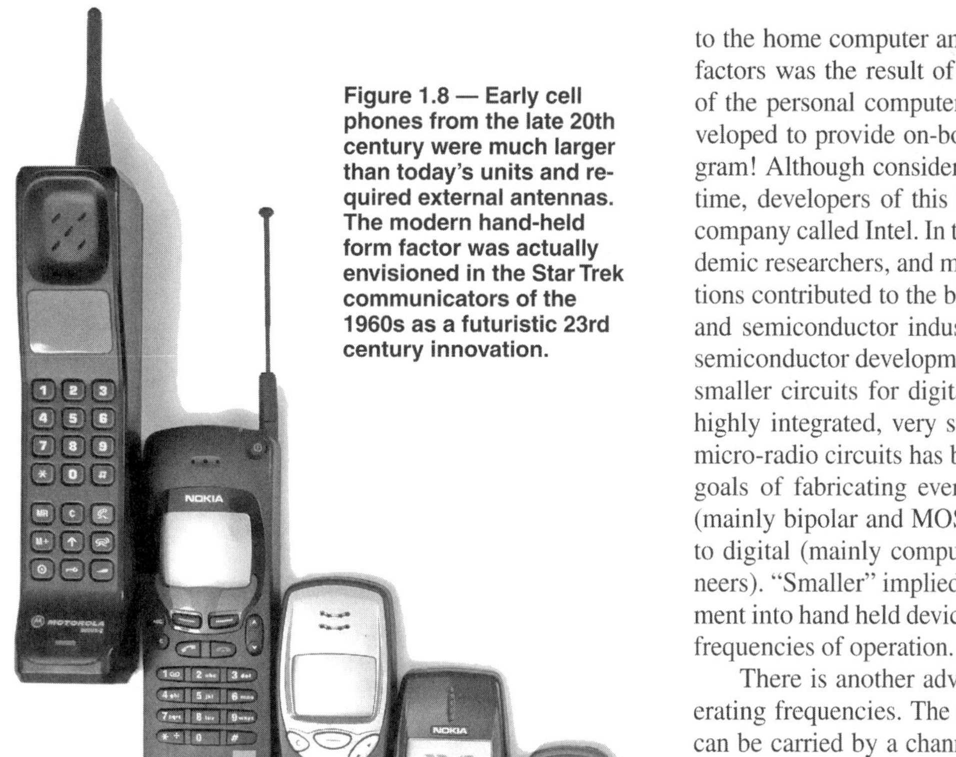

Figure 1.8 — Early cell phones from the late 20th century were much larger than today's units and required external antennas. The modern hand-held form factor was actually envisioned in the Star Trek communicators of the 1960s as a futuristic 23rd century innovation.

to the home computer and its endless applications and form factors was the result of the clever innovators and pioneers of the personal computer. The first microprocessor was developed to provide on-board navigation for the Apollo Program! Although considered an off-shoot development at the time, developers of this innovation went on to form a new company called Intel. In turn, Intel and other companies, academic researchers, and many more individuals and organizations contributed to the birth of the modern microelectronics and semiconductor industries. Although the main focus of semiconductor development remains building ever faster and smaller circuits for digital applications, the implications of highly integrated, very small and very fast components for micro-radio circuits has been equally dramatic. In effect, the goals of fabricating ever smaller and faster active devices (mainly bipolar and MOS transistors) were of equal interest to digital (mainly computer) and radio designers (RF engineers). "Smaller" implied transforming rack-mounted equipment into hand held devices, and "faster" implied ever higher frequencies of operation.

There is another advantage to applying ever higher operating frequencies. The amount of data or information that can be carried by a channel of finite bandwidth has definite limits (as we will explore in Chapter 10). Thus higher frequencies inherently permit wider operational bandwidths. For example, the entire AM broadcast band spans from 540 to 1700 kHz, a bandwidth of only 1.16 MHz. In contrast, a comparable bandwidth in the UHF region is a small fraction of the total band.

## Enter the Antenna

Antenna design has followed this upward trend in frequency and thus smaller form factors for shorter wavelengths. Unlike the radically different technologies that permitted this trend in the radio proper, many antennas remain remarkably similar in form. For example, the standard AM broadcast antenna (circa 1920s) is a ¼ wavelength, ground mounted vertical. At 1 MHz, this implies a steel tower about 75 meters high (**Figure 1.10**). In contrast, 2.4 GHz WiFi access points typically also use a ¼ wavelength (or longer) simple vertical antenna, only inches high! The differences between an AM broadcast system and a WiFi for radio technology is enormous, but the antenna design is practically identical, at least the first iterations of these antennas. There is also a continuous effort to integrate the antenna(s) into the various hand-held form factors. The external flexible "rubber duckie" antennas (**Figure 1.11**) have been gradually replaced with antennas integrated inside the unit. These internal antennas are in a constant redesign cycle to approach an optimum performance limited only by physics.

Figure 1.9 — An early desktop computer from the late 20th century. [Ruben de Rijcke, dendmedia.com, photo]

Almost all antennas, past and present, are configured of metal. Even very complex antenna arrays are built upon simpler long-known configurations but still consist of copper,

Figure 1.10 — Typical ¼ wave AM broadcast vertical monopole antenna. The steel tower itself and the ground radial system comprise the antenna. AM broadcast antennas are usually located over the best conducting soil possible to enhance the desired ground wave as we will explore in depth in Chapter 7. Wet soil enhances ground conductivity significantly. [Richard B. Johnson, photo]

Figure 1.11 — Typical 2.4 GHz monopole antenna, functionally nearly identical to the AM broadcast antenna. The only significant difference is operation frequency and thus size! As operating frequencies have increased over time, some of the simplest antenna types were simply made smaller and smaller and smaller.

aluminum, and/or other metals and alloys. Although there has been considerable progress in optimizing antenna designs for various applications, basic antenna designs remain in widespread use. More importantly for this text, they represent the focal point for discussions on antenna physics, which forms the basis for all antenna types.

In principle, virtually any type of antenna can be used at any radio frequency, but their physical sizes will dramatically change as a function of wavelength. Consequently, different antenna designs may be more or less practical as the frequency of interest changes. This is a key point in developing an intuitive understanding of antenna design: The physics remains constant, but the practical implementation and choices of design typically vary widely with frequency.

This is not to say there has not been significant progress in the development of antenna technology. Over the past several decades, enormous progress has been made by optimizing antenna parameters ever closer to physical limits. Computer modeling has permitted such optimized designs not possible only 10 or 15 years ago. Optimization of antenna design involves changing the configurations of metallic structures and often substrates upon which they are mounted.

In the early days of radio we began with Zepps, loops, dipoles, and monopoles. Today many antennas are simply manifestations of the simple dipole or monopole (as shown above). Then multielement arrays (including the Yagi-Uda design) using multiple dipoles or monopoles were designed to create more *directivity*. The helix antenna was invented in the 1940s, and the parabolic reflector was simply borrowed from optics exemplified by the reflector telescope. Patch antennas are mainly two-dimensional flat structures, but 3D structures are also in use. Diversity antennas of various forms use multiple elements of the same simple structures to improve information rate capacity and/or reduce path losses due to changing propagation conditions. These innovations often utilize sophisticated signal processing algorithms by comparing and combining signals from two or more antennas. The market wants products to be smaller, cheaper, and with better performance!

We will explore in detail the reasons why antennas, in general, require their minimum physical size to be close to the wavelength of operation, or as close as possible. This is always a critical factor in antenna design. However, this is only the first "rule of thumb" to gain an intuitive understanding of the workings of antennas. There are many fundamental terms to be mastered, but again, the basic physics is constant for all antennas. This is good news. By mastering the basics of antenna physics, any antenna at any frequency is within grasp of understanding. An antenna engineer is limited only by his/her understanding of antenna physics.

# 2 Fundamentals

## What is an Antenna?

The relevant Webster's definition of *antenna* states, "a usually metallic device (as a rod or wire) for radiating or receiving radio waves." This is indeed the most common application for antennas. Kraus provides a more technical definition: "...may be defined as the structure associated with the region of transition between a guided wave and a free-space wave, or vice-versa." In other words, it is the device that connects the guided wave (usually a one-dimensional transmission line) to three-dimensional free space and vice-versa (**Figure 2.1**). Some also refer to antennas as "transducers," converting the guided wave — represented by currents and voltages — to radiated waves and vice-versa.

Most of the practical antenna discussions in this text will be confined to linear antennas, which also constitute most of the earliest antenna designs mentioned in Chapter 1. A linear antenna is constructed from straight elements that are much smaller in diameter than their length. For example, practical linear antennas are typically constructed from wires or metal tubing. We will also discuss some nonlinear configurations in later chapters, but it is best to begin this introductory discussion by restricting ourselves to the simpler linear antenna configurations. In turn, more complex antennas build upon the basic principles derived from simpler types, but the calculations (particularly the integral calculus) become more difficult. There is also the fictional "isotropic" antenna, which is an essential reference point for all other antenna types, practical and imaginary.

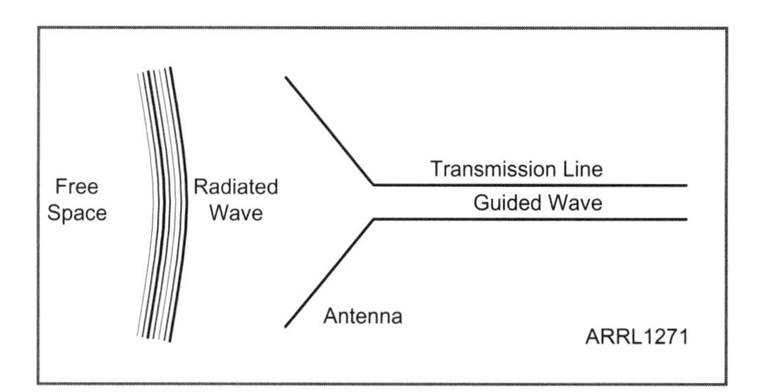


**Figure 2.1 — A conceptual diagram of a receive antenna performing the 3D to 1D transformation.**

Receive antennas capture radio frequency power from free space and concentrate it into a guided wave, and then, usually, to a receiver. Transmit antennas perform the reciprocal, receiving guided wave power from a transmitter (RF generator) and radiating — or "coupling" — that power into free space. Free space is defined as an empty three-dimensional area devoid of any matter. For most of the radio spectrum, air can also be considered free space for antenna work. While air (the atmosphere) often has profound effects on radio propagation, it typically has little effect on an antenna's characteristics *per se*.

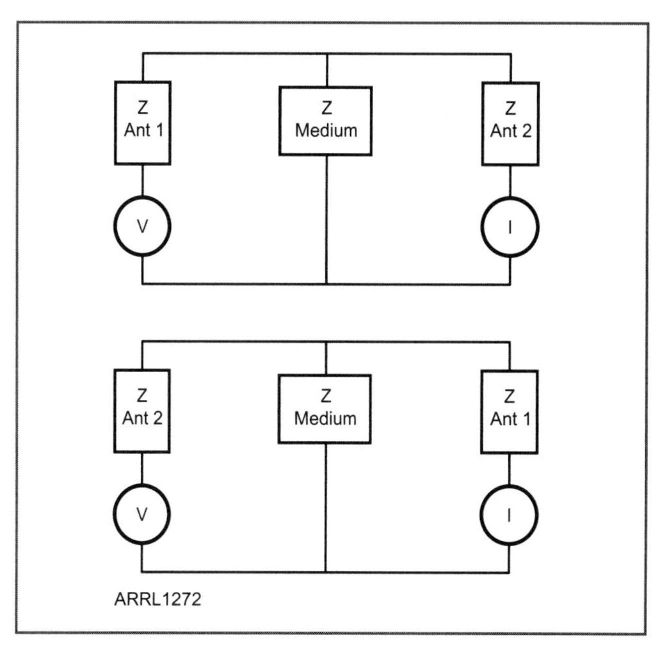


**Figure 2.2 — Reciprocity Theorem as applied to antennas: If an RF voltage (V1) is applied to the guided wave terminal of antenna 1, a Current (I1) will appear across the guided wave terminal of Antenna 2. If the same voltage (V1) is applied to Antenna 2, then the same current (I1) will appear across the guided wave terminal of Antenna 1. In the case of antennas, the intervening "medium" is usually free space with its intrinsic impedance (Z).**

Figure 2.3 — The author's home-built tower: Amateur Radio operators are frequent users of the reciprocity theorem for two-way communications around the world. Featured in this photo are three Yagi-Uda arrays, which is really a direct spin-off of the simple dipole. Notice the shorter dipole lengths on the top antenna for the VHF being covered (50 MHz) vs. the Yagi below for the 14 and 21 MHz HF bands. The antennas are used for both transmitting and receiving. Also shown is a UHF TV broadcast antenna. [Photo by the author. Tower design, fabrication, and welding also by the author.]

*The essence of Antenna Physics deals with the physical characteristics of the antenna and how it interfaces with these two ports.*

In electrical and electronic circuit theory, there are theorems that can be of great help in solving difficult problems. For example, if we have a two-port circuit (as in **Figure 2.2**) composed of linear elements and apply a source voltage (and resulting current) across Antenna 1, we will observe a resulting current in Antenna 2. Now if we apply the same observed current to Antenna 2 and remove the original source voltage from Antenna 1, we will see the same voltage across Antenna 1 that we previously applied. This is called the *Reciprocity Theorem*.

Since most antennas can be considered linear structures, we can assume that the Reciprocity Theorem also holds for an antenna (also defined as a two-port circuit described above). Notice from Figure 2.2 that both antennas reflect some impedance value, and there is also an intermediate impedance value of the medium — in this case, free space. These impedance terms are of fundamental importance to antenna physics and will be carefully developed later.

The reciprocity theorem as applied to antennas is of critical importance whenever the antenna is to be used for both transmitting and receiving (**Figure 2.3**).

## The Sphere as a Measurement Tool

Since free space is three-dimensional, 3D measurements are of fundamental importance in describing antennas' characteristics. We need some convenient construction that can serve to represent and quantify what is happening in free space relative to an antenna's performance. We will use an imaginary sphere for this measurement tool, superimposed within free space with a test antenna at its center. For antenna work, we are usually concerned with only the center of the sphere and the *surface* of the sphere upon which we can define *power density* over a portion of the entire sphere's surface. The *volume* of a sphere is usually only concerned with *energy* that is not typically used for antenna characterization. It is usually assumed that the space inside and outside the sphere is indeed free space — no matter allowed except for the antenna!

It is often impractical to perform actual measurements in actual free space. Antenna measurements are often taken on or close to the surface of the Earth. This causes difficulties in that matter (solids, liquids, and sometimes gases) can reflect and absorb power, and worse, affect the performance of the antenna proper. Reflections, in particular, are problematic in that they can add or subtract from the direct path wave between the antenna and a measurement device, thus distorting the measurement. Such reflections can cause *multipath* conditions, where a receiver receives a signal from a straight line from the transmitter and also a reflected signal. See **Figure 2.4**.

It is necessary to eliminate reflections that may distort attempts to perform accurate antenna measurements — in other words, simulate a free space test environment but not actually have to become airborne to escape the effects of the Earth. A practical solution to this problem is to use an *anechoic chamber* (**Figure 2.5**). As its name implies, enough of the radiated energy's "echoes" or reflections are absorbed by the floor, walls, and ceiling to an extent sufficient for good antenna measurements.

Since the sphere is often used in antenna measurements, a review of spherical geometry is in order with emphasis on terms used in this book. **Figure 2.6** shows a section of an imaginary sphere. The antenna is located at the center

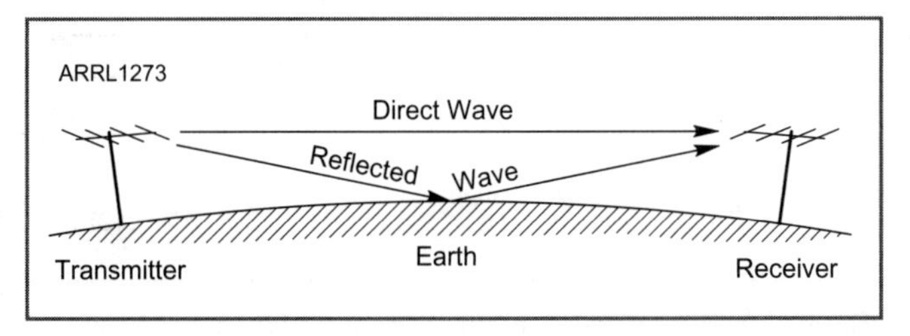

**Figure 2.4 — Example of a typical multipath condition.**

Figure 2.5 — The author inside an anechoic chamber used to simulate free space antenna testing. [Courtesy of Gonzaga University RF Lab.]

of the sphere and is usually assumed to be a "point source" in that the antenna's physical dimensions are much smaller than the radius of the sphere. We can see the traditional use of spherical coordinates analogous to latitude (θ) and longitude (ϕ). Therefore any unique point on the sphere's surface can be identified with a unique value of θ (0 – π) and ϕ (0 – 2π), where the value is given in radians. We can convert the radian measurement to degrees with this relationship: 2π radians = 360 degrees.

Of particular importance to antenna measurements is the area of a sphere's surface and also portions of that area. A sphere's surface area is $4\pi r^2$, where $r$ is the radius. Similar to a full circumference of a circle being 2π rads (360 degrees), the entire surface of a sphere is defined as 4π **Ω**, where **Ω** is the symbol for a *solid angle* measured in *steradians*. A hemisphere is defined by 2π **Ω,** a quarter of a sphere π **Ω,** and so on. In this book we use a standard bold font for (**Ω**) solid angle to differentiate the same Ω symbol for ohms.

Since the surface area of a sphere is $4\pi r^2$ and the solid angle representing a full sphere is 4π **Ω**, the actual area of a portion of a sphere is simply $\mathbf{\Omega} r^2$. Now we can use this "yardstick" to develop some basic parameters of antennas.

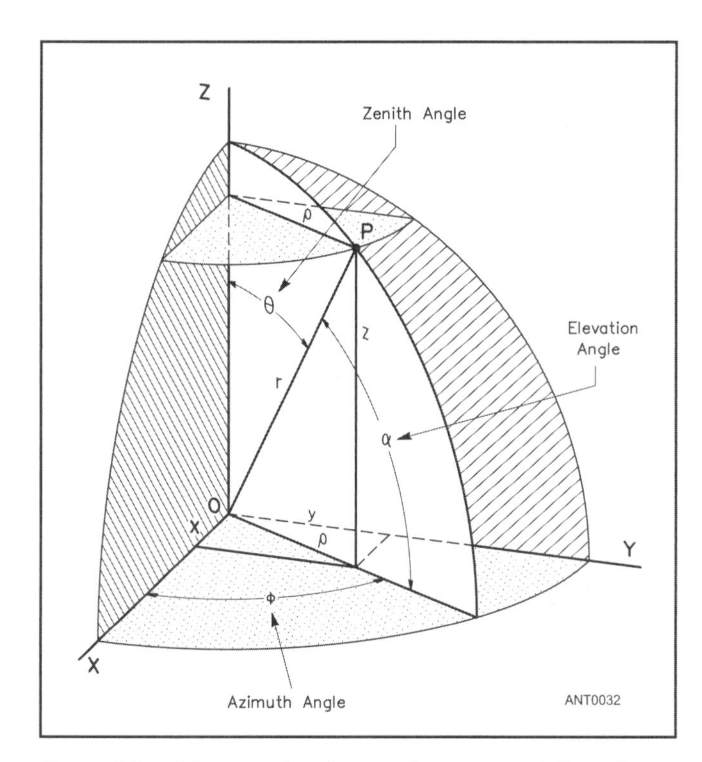


Figure 2.6 — Diagram showing a polar representation of a point P lying on an imaginary sphere surrounding a point-source antenna. The various angles associated with the coordinate system are shown referenced to the X, Y, and Z-axes.

## Analogy Between Light and Radio Waves and the Concept of Aperture

Let us place a 100 W light bulb at the center of our imaginary transparent sphere. We assume that this light source is emitting exactly 100 W of light power and ignore any heat generated through inefficiencies. Our light bulb also has a special characteristic: Its *pattern* of radiation is isotropic, which means that the light power emitted by the bulb is radiated in all directions equally. Therefore, the power radiated is distributed perfectly evenly over the surface of the sphere. If the interior of the sphere is a vacuum, then all 100 W of light power will reach the surface of the sphere. If we could calculate the total power passing through the sphere's transparent surface, it would also be 100 W evenly distributed over the surface of the sphere.

Furthermore, we can define the power of the light bulb (or any electromagnetic wave) as a power density, or *flux* over an *area,* or $W/m^2$ (watts per square meter). Another term often used is the Poynting Vector **S** (θ,ϕ). **S** is the vector defined as $Wm^{-2}$, and (θ,ϕ) represents the spherical area over which the power density is constrained, or more conveniently by a steradian value. See **Figures 2.7** and **2.8**.

On the surface of the 100 W light bulb's surrounding

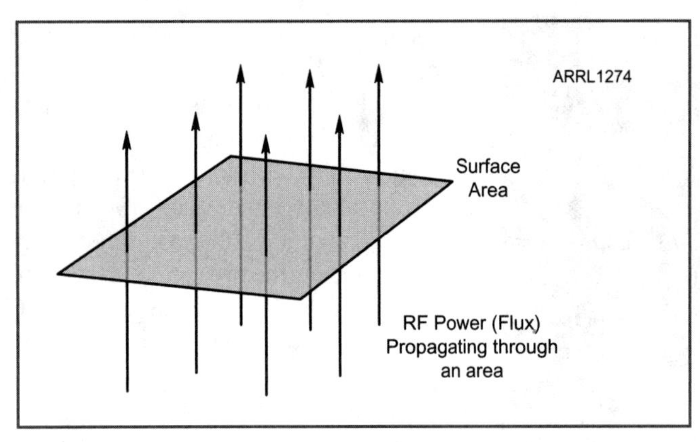


**Figure 2.7 — The Poynting Vector (S) is simply the power in W/m² propagating through a surface. In the case of $S(\theta,\phi)$ the surface is a part (or whole) surface of a sphere. The area and location of the area are defined by spherical coordinates θ and φ.**

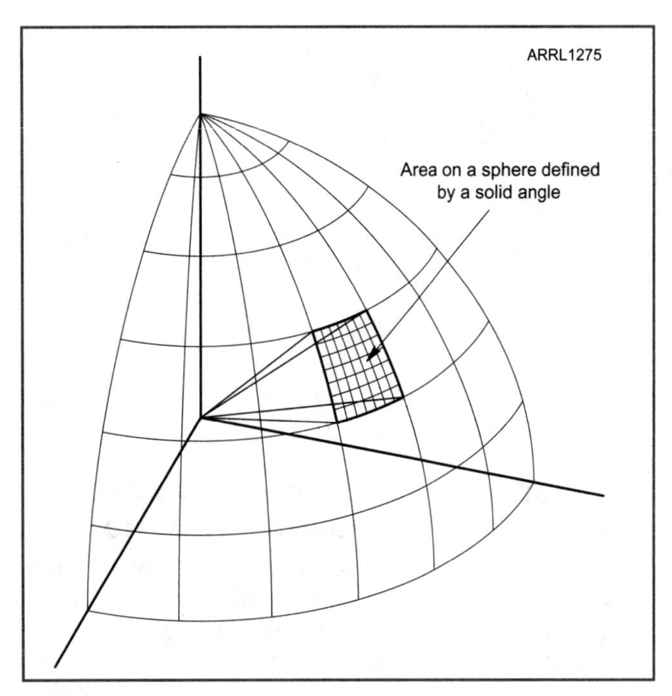


**Figure 2.8 — A solid angle projected from the sphere's center will define an area on the sphere's surface with a radius *r*. In this case, this solid angle is referred to as the Poynting Vector.**

sphere we will place a solar panel. The surface area of the sphere is taken to be 1000 m² and our solar panel is 1 m² and 100% efficient. Therefore we can capture 1/1000 of the light power, or 1/10 W, or 100 mW from the light energy falling on our solar panel. If we double the *area* of the solar panel we double the power received. Here we introduce another very fundamental relationship in antenna physics: the "area" of an antenna (which is more appropriately defined as *aperture*). As we shall soon describe, aperture and *power gain* are directly proportional. Thus we can now express how much power is received for a given distance and a given aperture of the receiver:

$$P_r = P_t\left(\frac{A_r}{4\pi r^2}\right) \quad \text{(Equation 2.1)}$$

where
$P_r$ is the power received
$P_t$ is the power transmitted in the direction of the receiver
$A_r$ is the aperture of the receive antenna
$4\pi r^2$ is the total surface area of the sphere with a radius of $r$.

Notice that the term

$$\left(\frac{A_r}{4\pi r^2}\right)$$

is simply the ratio of the antenna aperture to the area of the entire sphere, which also represents the ratio of received power to the total power transmitted.

As an aside, if we substitute $r$ with $d$ (distance) we get the coefficient $1/d^2$ by:

$$\frac{A_r}{4\pi r^2} = \frac{A_r}{4\pi}\left(\frac{1}{d^2}\right) \quad \text{(Equation 2.2)}$$

Thus we have derived the well-known "inverse-square law" from spherical trigonometry, holding for any form of electromagnetic radiation. *For a given receiver aperture the power received is proportional to the inverse-square of the distance from the transmitter.*

Of course no solar panel is 100% efficient, but this becomes a trivial calculation. If the solar panel from the above example is only 20% efficient, the electric power generated is simply 20% of the potential.

$$P_r = 100 \text{ mW} \times 0.2 = 20 \text{ mW}$$

Yet another analogy is *real* solar panels located on the Earth. In this case, the light source is the Sun, which is very close to being an isotropic radiator. The measurement sphere is defined by a radius, which is the distance between the Sun and the Earth. The Earth's atmosphere absorbs about ½ of the solar light before it reaches the Earth, and a 100% efficient 1 m² solar panel will provide about 750 W. Thus we can work in reverse and calculate the total power output of the Sun!

The Sun is about 150 billion meters from the Earth, or $150 \times 10^9$ m. Therefore the surface area of this imaginary sphere (the radius being the Earth – Sun distance) is $4\pi(150 \times 10^9 \text{ m})^2$, or about $2.8 \times 10^{21}$ square meters. Thus if the Sun's power is 1500 W per square meter at the distance of the Earth, then the total light power output of the Sun is about $4.25 \times 10^{24}$ W! Thus we can calculate any unknown variable in Equation 2.1 if the other terms are known. Also note that thus far we assume an isotropic transmitter and a *known* aperture of the receiver. Reconciling these two variables is a key task for any advanced discussion of antenna physics. The following discussion will quantify this essential relationship.

## Defining the Critical Fundamental Terms

We use the examples of a light bulb or the Sun, since they are easy to visualize. Now we can take the leap and state that there is a direct correlation between our brief discussion about light and radio waves. We *know* this is a correct correlation because Maxwell predicted the existence of radio waves using the then-known properties of permittivity and permeability of free space and their relationship to the speed of light. He then correctly postulated that both light and the yet-undiscovered radio waves were both electromagnetic radiation, in effect having identical properties except for their differing wavelengths. After the actual discovery of radio waves, endless measurements and experiments have proven the validity of his early postulations.

Solar cells remain inefficient because we remain restricted to the *photoelectric* effect that relies upon the inefficient process of a light photon dislodging electrons within the photovoltaic material, thus generating an electrical current. Building an array of receive antennas for light is possible, but converting the "guided" light wave *directly* into a dc current would require a futuristic diode. Again wavelength is the critical limitation. It is possible to build a dipole antenna for light and/or infrared (IR) wavelengths using micro-processing techniques. To convert the light power into an electrical current, however, we would need to rectify the light power into direct current. No such diode exists. If it could be developed, solar cells could be replaced with arrays of tiny dipoles and futuristic diodes. So for a *practical* comparison of light and radio waves we need some slight modifications, but remember that the physics is the same. Fortunately, antennas for radio frequencies often do approach 100% efficiency.

A transmit antenna can be compared to the light source (a light bulb, LED, or a torch) and the receive antenna compared to the solar panel. Unlike radio antennas, however, these particular choices of light sources and light "receivers" are not reciprocal. We do not normally think of a solar cell as a radiator or a light bulb as a light receiver. On the other hand, radio antennas *are* reciprocal, therefore transmit antennas are also characterized as having an aperture (the same aperture as they have when receiving).

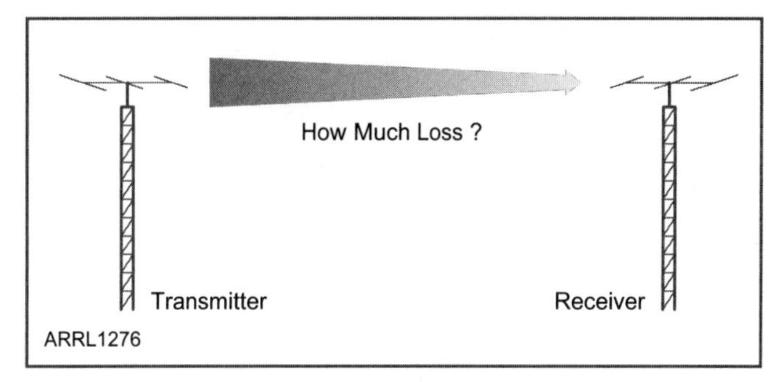


**Figure 2.10 — How can physics define a given path loss?**

A "shrinking" aperture due to *inefficiency* is analogous to a reduction in antenna gain due to power losses (usually as heat). Also, as above, if we double the aperture, we double the power gain (about 3 dB difference). This is illustrated by doubling the aperture (two rather than one solar cell) for a doubling of power capability (**Figure 2.9**). The *effect* of cutting an antenna's physical aperture in half is identical to cutting the efficiency of two solar cells in half. Therefore, the *power gain* of a solar cell or antenna follows the same rule. For example, a hypothetical 100% efficient solar cell would convert all the light power falling on it into electrical power. In a real solar cell, the absorbed power not converted into electrical power is converted into heat. This is precisely what happens in an antenna, either for transmit or receive antennas. Power applied to the antenna, but not radiated from the transmitter or not delivered to the receiver from a receive antenna, is lost as heat.

One of the advantages of using aperture instead of gain in radio link calculations is that we can *directly* calculate how much power will be received by the antenna. Indeed, *any* equation for calculating the power loss of a radio link must use aperture for the proper calculation, or a derivation thereof. The answer to the very important question posed in **Figure 2.10** requires the detailed understanding of the aperture concept which follows.

**Figure 2.9 — Solar power receiving antenna. The more panels, the more aperture, the more gain, and thus the more power recovered.**

A non-isotropic antenna will present different gain for different directions and therefore also present different apertures for different directions. Again we can use the solar cell as an example. When oriented perpendicular to the Sun (toward the light transmitter)

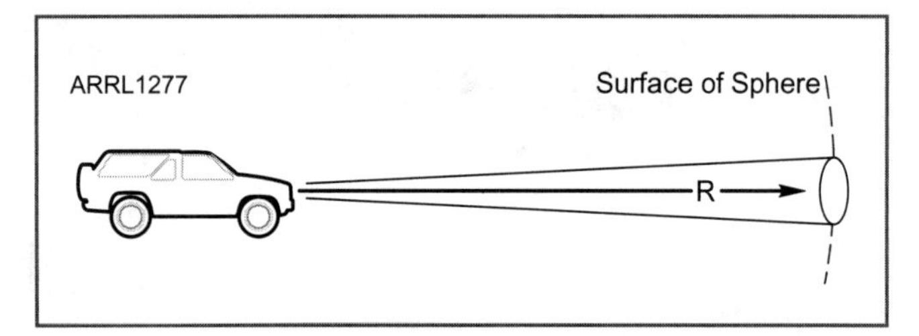


**Figure 2.11 — Automobile headlight beam uses a parabolic reflector in back of a near-isotropic light bulb to dramatically increase the effective power.**

it will receive the maximum amount of power. This is the direction of maximum gain and thus also the *direction* of maximum aperture. If you begin to rotate the panel away from the Sun's direction the effective aperture decreases, and so the power received decreases. Finally, with the panel facing 90 degrees off the Sun's direction there is no sunlight striking the panel and the received power drops to zero. An antenna's gain and aperture also depends literally on the antenna's point of "view"! Of course, those familiar with directional antennas know that as the antenna is rotated toward the direction of a transmitter, the signal power received increases and peaks where the forward gain is maximum. This is also the direction that corresponds to the antenna's maximum aperture.

We also know that any non-isotropic antenna that is 100% efficient will radiate all the power applied to it, but *not* in all directions equally. Therefore in some direction(s) a receiver will "see" the equivalent of more than 100 radiated watts and in other direction(s) a receiver will "see" less than the equivalent of 100 radiated watts. In this case the antenna has gain greater than isotropic in some favored direction(s).

Yet another optic analogy is the automobile headlamp (**Figure 2.11**). The typical headlamp bulb radiates about 10 W of light power. However, placing a parabolic mirror in back of the bulb results in a very high power beam with a small steradian value defining its narrow radiation pattern. In effect, the parabola "robs" all other directions on the sphere of light power and concentrates it within the narrow beam. If we can concentrate all the 10 W of power into a steradian of 1% of the total sphere 0.01 $(4\pi)\Omega$ we simply multiply the output power by the power *gain* (in this case 100) to derive an *effective radiated power relative to an isotropic* or 1000 W EIRP. That is why a misaligned headlight can blind an on-coming driver.

Again, the power analogy between optics and radio physics is exact. Also, here is another critical point: If we take the *average and thus total* power radiated over the entire sphere, we find that the value remains 10 W. In the case of the isotropic the *average and total* power is also 10 W. However, the *effective* radiated power (EIRP, compared to an isotropic) is 1000 W in the main beam of the headlamp, but 0 W over 99% of the sphere, thus the average is, again, 10 W.

## The ½ Wavelength Dipole Antenna

Perhaps the simplest form of a practical linear radio antenna is the ½ wavelength dipole, as seen in **Figure 2.12**. The dipole is the shortest linear antenna that is *resonant*. Resonance, in the case of antennas, implies that the source impedance (when used as a receive antenna) or load impedance (when used as a transmit antenna) is a pure resistance (no reactive part of the impedance). In Figure 2.12 we use $R_L$(load) since we are assuming (for this example) a transmit antenna. Therefore an antenna can be understood as a device that is represented by a load to a signal generator that absorbs power by radiating it into free space. Of course by reciprocity, the imaginary resistor can also form a signal source with the same impedance.

The isotropic and dipole antenna types form building blocks for both the theoretical and practical basis of antenna study. We will soon provide a detailed understanding of their behavior and their interrelationships.

## Relationship Between Aperture and Gain

Aperture is a straightforward concept in that an antenna can be defined as a simple two-dimensional area. Gain (dB) is another issue: gain relative to what? Usually antenna gain specifications reference the antenna to a *defined* gain of an isotropic (dBi) or a ½ wave dipole (dBd). Without a reference, a specification such as the antenna gain is "6 dB" is meaningless: 4 times the power of what? Since an isotropic antenna would have a power gain of 1 (0 dBi), this is a logical reference to use. The decibel measure of power is $10\log_{10}(P_1/P_2)$, where the $P$ values are linear values of power. It is essential for $P$ is be a ratio of power values, such as a ratio of some gain value to the gain value of a dipole or an isotropic.

Assuming we are comparing lossless antennas, the isotropic antenna has the minimum possible gain of any possible antenna. For this reason, antenna gain references are ultimately based on the gain of an isotropic antenna (0 dBi). As a result *it is necessary to know the aperture of an isotropic antenna in order to relate any antenna's gain to aperture.* Without equating the two, it is impossible to calculate how much power will be received by an antenna for a given power at a given distance and thus answer the critical question posed

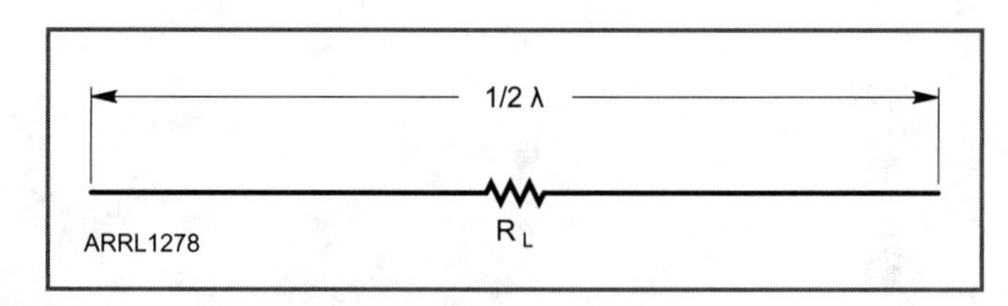


**Figure 2.12 — Simple illustration of ½ λ dipole antenna.**

in Figure 2.10. For example, you know your receive antenna has a gain of 9 dBi and that the power density is 10 microwatts per square meter at the receive location. How much power will you expect to receive? A value of 9 dBi is meaningless without equating it (or any antenna gain specification) to an equivalent area. However, we *know* that the aperture of a 9 dBi antenna will be about eight times as large as an isotropic aperture. But the trick question is: What is the aperture of an isotropic antenna?

## Empirical Calculation of the Aperture of an Isotropic Antenna

We can use the above introduction (basic physics and some simple mathematics), coupled with a set of empirical measurements, to derive the aperture of an isotropic antenna. We will also use relatively simple instruments to measure power. This calculation is critical to antenna physics, antenna engineering, and radio system engineering. But how do we calculate the aperture of the hypothetical isotropic antenna that both radiates and receives to and from all directions equally?

To begin, let us review what we now know:

1) An isotropic antenna radiates and receives in all 3D directions equally.

2) A non-isotropic antenna does not radiate in all directions equally, but radiates and receives more power in some directions and less in others (relative to the hypothetical isotropic)

3) The total and average radiated power of any antenna will be the same as an isotropic antenna (assuming 100% efficiencies). If an antenna radiates *all* the power applied to it, the power must be going *somewhere!*

4) The inverse-square law applies also to radio waves.

Now we are prepared to derive an experiment and subsequent calculations to determine the aperture of an isotropic antenna. In this experiment, we begin with two dipoles. We use dipoles because we have no idea how to design an isotropic antenna. At present we assume we do not know the gain of these dipoles, nor do we know their apertures. We make sure that the dipoles are near 100% efficient by carefully measuring the temperature of the antenna before and after we apply power to them. We find negligible difference in the temperature, so we assume near 100% efficiency. We also verify a perfect (lossless) match to our RF generator by taking a simple VSWR measurement — it is 1:1. To begin, we use 300 MHz for our first test since relatively high frequencies are needed for practical measurements inside anechoic chambers.

In the anechoic chamber we use a dipole as a source antenna and a second dipole at 10 meter distance as the receive antenna. By rotating the receive dipole in three dimensions, we can record the directivity of the dipole over the dipole's entire imaginary external sphere. We don't need a calibrated power source since we are only plotting the *relative* gain. Since our dipole is near 100% efficient, we assume the *relative gain* is the same as the relative directivity, where

$$Gain = Directivity \times Efficiency \qquad \text{(Equation 2.2)}$$

If we plot the relative gain of the ½ λ dipole in free space, we find a "doughnut" pattern (**Figure 2.13**). By our statement 3 above, we *know* that the total and average power response of this non-isotropic dipole antenna must be the same as an isotropic antenna. By adjusting both dipole antennas we find maximum power is received when the dipoles are broadside to each other and the dipole elements are in the same plane (same polarization).

We can now perform an empirical process to calculate the gain difference between the dipole and the isotropic antennas. We *know* that the gain of the dipole is constant for all points perpendicular (broadside) to the line of the dipole because we measured it. However, the gain difference can be plotted in two dimensions (a plane) that bisects the doughnut tangent to the doughnut's "hole." Since we only need to find an average of the dipole power gain, we can ignore performing a three-dimensional integration and use only a "slice" of the doughnut and perform a two-dimensional integration. We find the total area of the two sides of the slice, which we *know* corresponds to the total power output

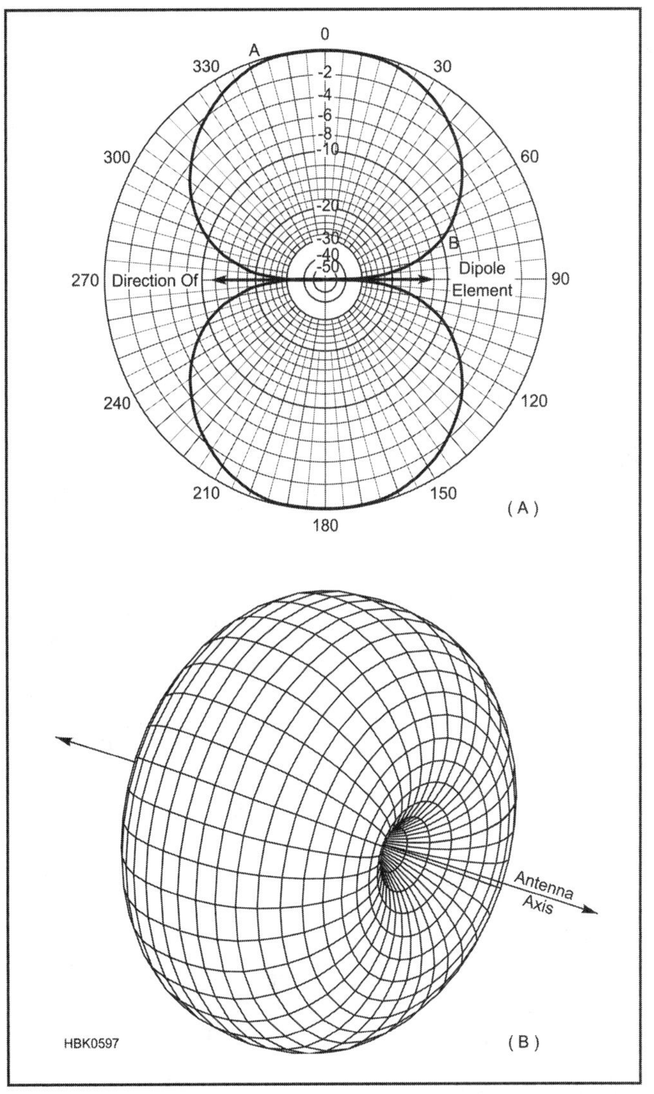


**Figure 2.13 — Doughnut pattern of a ½ wave dipole with the line of the dipole through the "center" of the doughnut.**

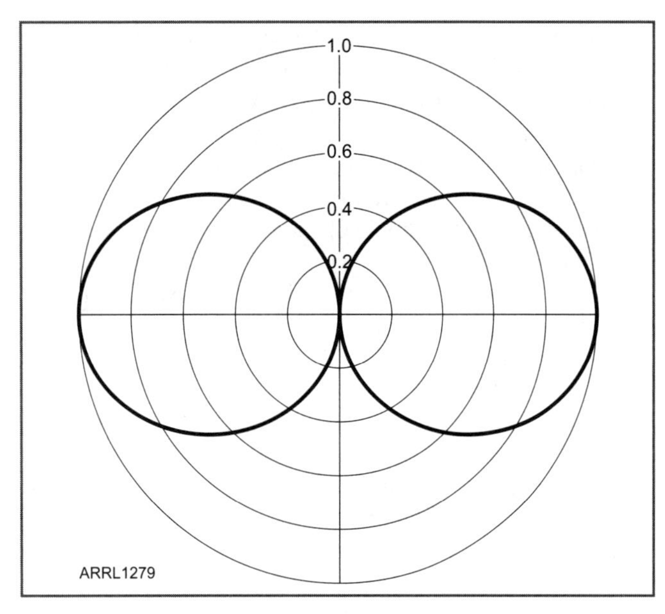


**Figure 2.14 — Linear power response of a dipole antenna in free space (or an anechoic chamber), looking at the ends of a half doughnut cut down the middle, with the dipole in a vertical configuration at the center. We can calculate the power gain of an isotropic by taking the average vector value of the dipole (or any 100% efficient antenna), which corresponds to the gain of an isotropic relative to the dipole. The average response is about 0.61, or about –2.15 dB relative to the measured dipole.**

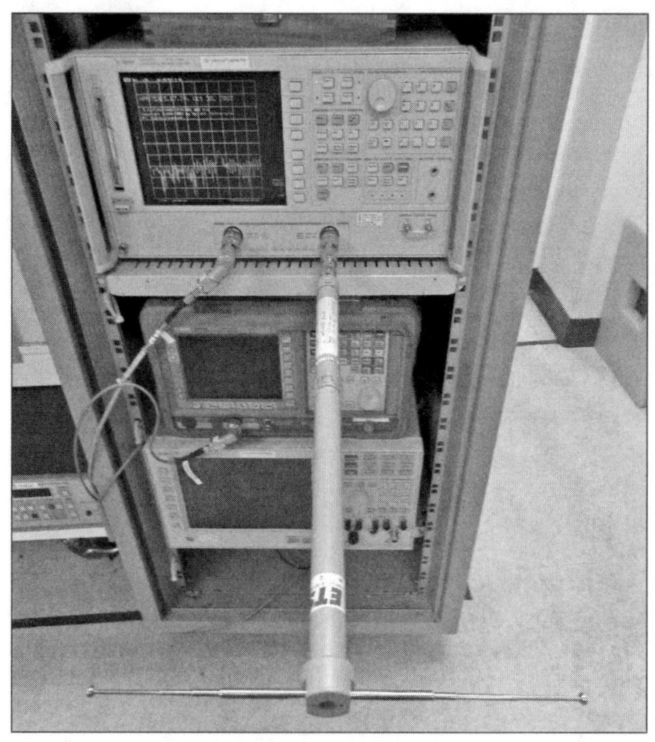

**Figure 2.15 — A simple dipole antenna for lab tests. [Courtesy of Gonzaga University RF Lab.]**

of the dipole. Then we take this total area and calculate the radius of a circle we construct with the identical area, which is the equivalent pattern of an isotropic at an identical power (circular). The radius of this circle is 0.611 of the maximum length of the dipole's doughnut in its maximum gain direction. This means the dipole-doughnut has a power gain of about 1/0.611 ≅ 1,637 over the theoretical isotropic, or 2.15 dBi. Thus we have just measured the dBi gain of a ½ wave dipole.

Since the actual sphere is 3D, we can form a conceptual understanding of the power integration in 3D by imagining that we have a doughnut shaped 3D piece of putty. We measure the radius of this doughnut to be 1 inch. Then we take the doughnut and reshape it into a perfect sphere (like making a snowball) and measure the radius of the new sphere. We find it to have a radius of 0.61 inch which is about a 2.15 dB difference.

Assuming we were very careful in our measurement techniques, **Figure 2.14** shows a linear plot of the dipole's response in linear/linear coordinates as we rotate the dipole 360 degrees perpendicular to the source.

We also *know* that the aperture of an antenna is directly proportional to its gain *in the same direction from the antenna* as in our solar cell example. Therefore we can *measure* the received power from our dipole example and calculate *directly* the aperture of the dipole and thus also the aperture of an isotropic. This is the relationship that we need to break into new territory in understanding antenna physics.

We now *know* that the power gain of our ½ wave dipole is 1.637 over an isotropic. Therefore if we apply 1 W to the transmit dipole we will have an EIRP of 1.637 W in a direction broadside to the dipole. When we place our receive di-

**Figure 2.16 — Two much longer dipole antennas, one for 3.5 MHz (in the background) and the other for 7 and 14 MHz. Notice that the dipole in the foreground uses "open wire" feed line for the radiating element, with one of the two conductors cut for 7 MHz and the other for 14 MHz. Functionally they are identical to the 950 MHz dipole in Figure 2.15 except for the wavelength of operation. [Joel Hallas, W1ZR, photo]**

pole 30 meters away from the transmit dipole (also broadside to the transmit dipole) we record a received power of just over 18.855 μW. Since we *know* that we have a *total* dBi power gain in the system of 2.6797 (product of the two dipoles' linear gain broadside to each other), we *know* the equivalent isotropic assumption would be just about 7.036 μW by equation 2.1. Our actual maximum measured broadside-to-broadside received power was 18.854 μW, or 2.6797 times the calculated isotropic received power. Now we have *normalized* the dipole to an assumed isotropic radiator radiating 1 W at the center of a 30 meter radius sphere and a *calculated* received isotropic power of 7.036 μW. We can now calculate the aperture of the isotropic antenna.

The area of the 30 meter radius sphere is about 11,310 square meters. With 1 W of isotropic radiated power at the center, the calculated power density is 88.419 μW per square meter. Since we would actually receive 7.036 μW, the aperture of our isotropic antenna is

$$\frac{7.036\ \mu\text{W}}{88.417\ \mu\text{W/m}^2}$$

or about 0.0796 square meters. This looks like the answer, but we have another observation/potential problem. We *know* that ½ wave dipoles have different lengths for different frequencies, such as the antennas shown in **Figures 2.15** and **2.16**. *Intuitively* we believe that a larger antenna will have a larger aperture, well, because it's bigger! For example, a 7 MHz dipole is considerably larger than a 144 MHz dipole. Intuitively we think that a much bigger structure will couple more power into free space, or receive more power from free space. If our intuition is correct we should arrive at different aperture values for different frequencies.

So we prepare an identical experiment, this time at 600 MHz which is ½ the wavelength and thus also ½ the dipole length in the 300 MHz experiment. If there *is* a difference in the calculated isotropic aperture, we will *know* there is a wavelength term in the isotropic aperture equation. We run the experiment again and find that the isotropic value is about 0.0199 square meters, or exactly ¼ the area of the 300 MHz dipole. Additional experiments at other frequencies show that the aperture of a dipole, and thus also an isotropic antenna, *increases at a rate equal to the square of wavelength, or the reciprocal of the square of the frequency.*

However, we have some coefficient we must also solve to get a true equation that defines the aperture of an isotropic at *any* wavelength. If we return to the original 300 MHz results, we *know* the wavelength is a nice normalized number of 1 meter, and 1 squared is 1. Thus we see that the coefficient is about 0.07958. The reciprocal of this is 12.566. Because we've been using this number so often with our spherical tool and with the insight of a James Maxwell, we immediately recognize it as 4π. We use this coefficient in our other measured results, and thus converge upon the actual definition of the aperture of an isotropic antenna as

$$A_e = \frac{\lambda^2}{4\pi} = 0\ \text{dBi} \qquad \text{(Equation 2.3)}$$

where the antenna aperture (area) is in the same units of wavelength ($\lambda$).

This is one of the most fundamental equations (and relationships) in antenna theory in that it relates antenna gain to aperture. Indeed, many misconceptions regarding antenna performance can be resolved by understanding this fundamental relationship. The most striking implication of this equation is that the aperture of an isotropic antenna (and all other antennas referenced to it) depends heavily upon the wavelength of operation. Now we have the tools to solve the problem illustrated in Figure 2.10.

## Free Space Path Loss

Path loss simply defines the ratio of power lost between the transmitter's EIRP and the power received by the receiver's antenna. In free space, we use the simple spherical model as above and assume an isotropic receive antenna with the appropriate aperture. If we receive or "capture" 1/1,000,000 of the power radiated, our path loss is $10\log_{10}(0.000001)$ or 60 dB. Therefore the power received will be 60 dB below whatever power is transmitted in the direction of the receiver. The relationship, in most situations can be considered linear and thus proportional.

Virtually all calculations for free space path loss are based upon these simple concepts: power radiated (EIRP), distance, and receive antenna aperture. These terms represent the "starting point" to the operation of a garage door opener, an interplanetary radio link, radar systems, cell phone coverage area, satellite services, and so on. The great complexity (and cost) of some radio coverage prediction software tools reflects their ability to approximate the effects of material environmental effects (such as a portion of the Earth's surface) upon the coverage. But the starting point always begins with these fundamentals. It's the refinement that is difficult. Free space calculations are simple; it's when "stuff" gets in between the transmitter and receiver that estimations become difficult. No matter how sophisticated a path loss predictor may be, the result is always one single number (the path loss, and possibly some statistical deviation around that number).

Returning to the basic situation of an isotropic antenna at the center of our sphere, we know that the power will be equally distributed over the sphere's area, $4\pi r^2$. We also know that the power received will be proportional to the aperture of the receive antenna. If our receive antenna is 100% efficient, the *portion* of the transmitted power received is simply the ratio of the antenna aperture to the area of the sphere, or,

$$\frac{A_e}{4\pi r^2} \qquad \text{(Equation 2.4)}$$

and the dB loss is simply:

$$\text{Path loss} = 10\log\frac{A_e}{4\pi r^2} \qquad \text{(Equation 2.5)}$$

Notice that Equation 2.4 is simply a ratio of two areas — the aperture of the receive antenna to the total area of the sphere on which the antenna is set.

Let's assume that our receive antenna is isotropic. After the above calculation and measurements, we know its aperture is

$$A_e = \frac{\lambda^2}{4\pi}$$

Substituting $\lambda^2/4\pi$ for $A_e$ we get

$$\text{Coefficient for isotropic path loss} = \frac{\frac{\lambda^2}{4\pi}}{4\pi r^2} = \left(\frac{\lambda}{4\pi r}\right)^2$$

Notice the beautiful mathematical simplicity of this term. This simplicity emerges because of the similarity in the equation defining the area of a sphere and the isotropic antenna, whose radiation pattern is spherical! Hopefully the elegance of using spherical geometry in calculating and defining antenna parameters is now apparent.

If we know that the transmit and receive antennas are isotropic, we can now easily calculate how much power will be received by the receive antenna.

$$P_r = P_t\left(\frac{\lambda}{4\pi r}\right)^2 \quad \text{(Equation 2.6)}$$

where $P_r$ is received power and $P_t$ is transmitted power.

As implied above, we can adjust the aperture value if we know the gain value of the antenna. For example, if we have a 6 dBi antenna, that represents about 4× power gain (calculated through using the inverse-log term), so the aperture of a 6 dBi antenna is simply

$$\frac{4\lambda^2}{4\pi} \text{ or } \frac{\lambda^2}{\pi}$$

On the transmit side, if a 6 dBi antenna is used with a 100 W transmitter, then the EIRP (effective radiated power relative to an isotropic antenna) is 400 W (in the maximum gain direction). Therefore by these simple adjustments you can calculate any free-space path loss for any power, antenna gains, frequency, and distance. The other radiated power convention is ERP (effective radiated power relative to a dipole). The difference between the two is always 2.15 dB, but caution should be used to determine which is being applied for a given application. Again, if the 6 dBi antenna has only a single lobe with a "brick wall" pattern, then the pattern — defined by steradians — is $\pi\Omega$, or ¼ of a spherical, isotropic pattern.

The received power using antennas with non-isotropic gains is:

$$P_r = P_t G_t G_r\left(\frac{\lambda}{4\pi r}\right)^2 \text{ watts} \quad \text{(Equation 2.7)}$$

where $G$ is the linear power gain coefficient related to an isotropic gain for the transmit and receive antennas with two antennas pointing at each other in free space.

Since

$$A_e = \frac{G\lambda^2}{4\pi}$$

Then

$$G = \frac{4\pi A_e}{\lambda^2}$$

By substitution we can now also use the aperture of *both* the transmit ($A_{et}$) and receive ($A_{er}$) antennas in a given link to get the path loss by substituting these terms for the isotropic in Equation 2.7, so we get

$$P_r = P_t\frac{A_{et}A_{er}}{r^2\lambda^2} \text{ watts} \quad \text{(Equation 2.8)}$$

This is the free-space Friis Transmission Formula, where again $r$ is the distance between the antennas, $\lambda$ is the wavelength, and $A_{et}$ and $A_{er}$ are the antenna apertures all using the same units. Here we can see another result of the reciprocity theorem. Notice that both the transmitter and receiver aperture (or gain) terms appear in the numerator of the equation. This implies that you can compensate one for the other. In other words, you will receive the same amount of power by swapping the antennas, or by changing their values provided that the product is the same.

**Figure 2.17 — A large parabolic antenna for radio astronomy or deep space probe communications. [Photo courtesy NASA]**

Perhaps no better example of the concept of antenna aperture can be found in the parabolic (dish) antenna (**Figure 2.17**). The area of the dish is simply equated to the aperture of the antenna (assuming 100% *aperture* efficiency). Therefore the dBi gain of a parabolic antenna increases with increasing frequency while the aperture (in principle) remains constant.

Equation 2.3 points to another issue relating to wavelength. An often confused relationship is frequency (wavelength) and path loss. If all electromagnetic waves conform to the inverse-square law, why do lower frequencies seem to propagate "farther"? Understanding the relationship between gain and aperture solves this common source of confusion.

## Refinements of the Aperture Concept

The above discussion provides for an intuitive understanding of antenna aperture. However, this first approximation of this critical antenna parameter needs some refinement for a deeper understanding. Thus far we have only discussed *effective aperture* ($A_e$) but have also hinted at the effect efficiency has in reducing the effective aperture. As mentioned above, the gain and thus effective aperture of an antenna, decreases when an antenna becomes less efficient. Thus when an antenna has no loss and is perfectly matched to a load, the aperture is the *maximum effective aperture* or $A_{em}$. When power is lost to heat due to inefficiencies, the result is a "loss" aperture or $A_l$. Therefore,

$$A_e = A_{em} - A_l \quad \text{(Equation 2.9)}$$

Thus $A_l$ equates an equivalent area that captures power that is lost to heat. This is comparable to a solar cell's maximum possible aperture being reduced by inefficiencies.

For any "real" antenna:

$$\text{Efficiency} = \frac{A_e}{A_e + A_l} \quad \text{(Equation 2.10)}$$

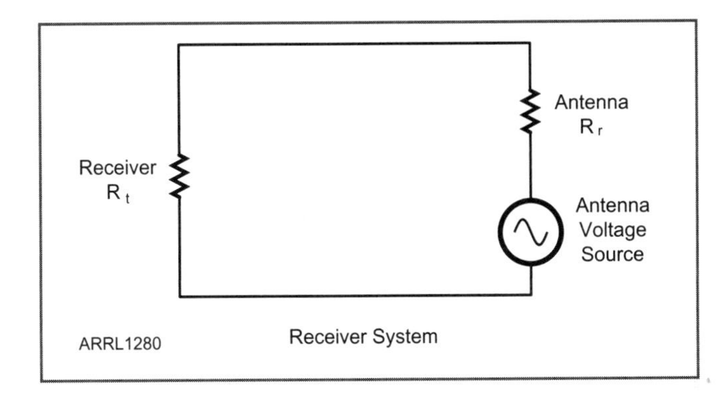


**Figure 2.18 — The equivalent circuit of an antenna and a receiver. In receive systems it is usually most advantageous to make the terminating impedance $R_t$ of the receiver equal to $R_r$ or a transformed $R_r$ value. This value is usually 50 or 75 Ω. Maximum power will be transferred when $R_t = R_r$.**

We have seen that power gain is proportional to area (aperture) in calculating antenna parameters. Therefore we can begin to formalize this relationship.

Let an incident wave produce a voltage ($V$) at the feed point of an antenna, as in **Figure 2.18**. For convenience we will use a dipole as an example. This voltage produces a current ($I$) in both $R_r$ and $R_t$. $R_r$ is defined as the *radiation resistance*, which in this case is also the feed point impedance value. In effect, $R_r$ is the source impedance and $R_t$ the receiver's input impedance (usually 50 or 75 Ω). $R_r$, a fundamental term, will be dealt with in detail later. In this discussion we will ignore any reactive parts of the source and load impedances, since they are routinely "tuned out" of the circuits. It is the real parts of these impedances that determine real power terms. Assuming no complex terms in the impedances (pure resistances) the power delivered to the load is thus:

$$P_{load} = I^2 R_t \quad \text{(Equation 2.11)}$$

and

$$I = \frac{V}{\sqrt{(R_r + R_t)^2}} \quad \text{(Equation 2.12)}$$

where $V$ is the induced antenna voltage. Thus

$$P_{load} = \frac{V^2 R_t}{(R_r + R_t)^2} \quad \text{(Equation 2.13)}$$

Assuming a conjugate match, $R_r = R_t$. Therefore,

$$P_{load} = \frac{V^2 R_t}{(2R_t)^2} = \frac{V^2 R_t}{4R_t^2} \quad \text{(Equation 2.14)}$$

or

$$P_{load} = \frac{V^2}{4R_t} \quad \text{(Equation 2.15)}$$

By our basic discussion of $A_e$ above, we know that

$$P_{load} = A_e \mathbf{S} \quad \text{(Equation 2.16)}$$

wher$4SR_t$ the Poynting vector, or power density per square meter, and $A_e$ is the aperture that captures power proportional to its area.

Therefore, with a conjugate match and no loss

$$A_{em} = \frac{V^2}{4SR_t} \quad \text{(Equation 2.17)}$$

This discussion points to a possible contradiction. Figure 2.18 shows that ½ of the received power is "lost" in the antenna's radiation resistance. If we assume a 100% efficient antenna, then the power is not "lost" to heat. Also, if we assume we only deliver 50% of the received power to the antenna, then the aperture and thus gain should only be ½ of what it "should" be. The answer is that an antenna perfectly

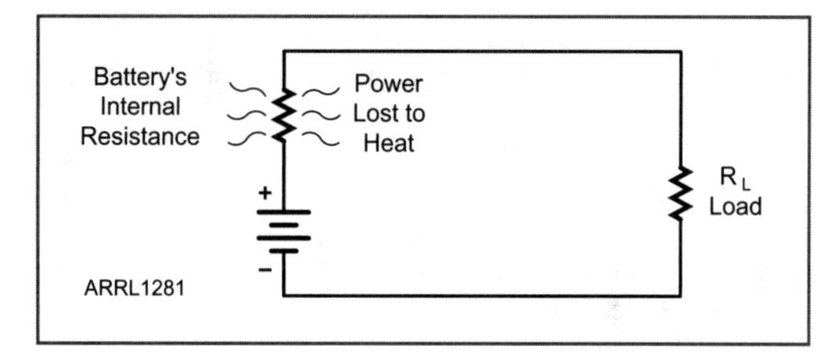


**Figure 2.19 — The equivalent circuit of a battery, its internal resistance and a load resistance.**

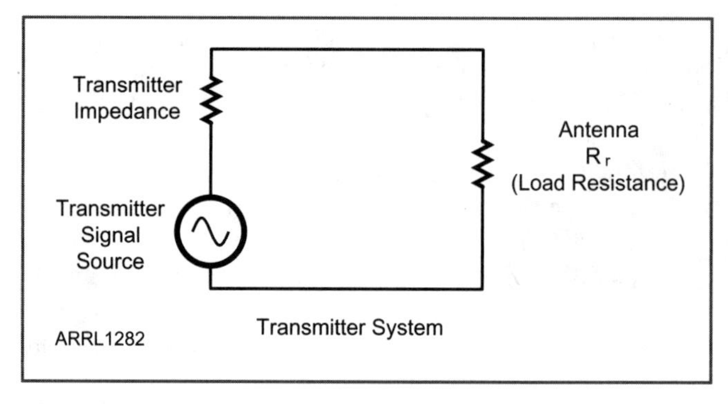


**Figure 2.20 — The equivalent circuit of an antenna and a transmitter. Unlike the receiver system above, the usual goal for matching in the transmit circuit is *efficiency*. The internal resistance of the transmitter is set to some value lower than the load resistance. A conjugate match will only provide 50% efficiency. A "perfect" transmitter would have an output impedance of 0 Ω.**

matched to its load reradiates ½ of the power by virtue of yet another aperture, the scattering aperture $A_s$.

In the case of the scattering aperture, the scattered power is simply *lost* back (scattered) as re-radiation into free space. The issue is comparable to a simple dc circuit consisting of a battery with an internal resistance and a load represented by $R_L$(load) (**Figure 2.19**). Maximum power will be delivered to the load when the source and load resistances are equal. In the battery circuit, ½ the power is dissipated as heat. In the Figure 2.18 antenna circuit, ½ of the power is re-radiated back into space.

The transmitting case is also similar to the dc battery case. The transmitter can be modeled as an RF voltage source with an internal resistance (source impedance). The antenna retains its internal resistance as defined by $R_r$. There is an important difference in the transmitter case, however. In the receiving case, the goal is usually to achieve a *conjugate match,* to make $R_t = R_r$. This will achieve *maximum power transfer* into the receiver resulting in a maximum *signal-to-noise ratio* (more on this subject in Chapter 10). In practice, the receiver's input impedance may be tuned a bit off the exact complex conjugate impedance, as minimum *noise figure* is seldom found at the exact conjugate match.

In a transmission system we are usually most interested in maximum *system efficiency.* For example, if the transmitter were 100% efficient and the source and load resistances were identical, the overall efficiency would be 50%. In such a conjugate match condition, real transmit systems would be very inefficient. For maximum efficiency the transmitter impedance is typically tuned to a *lower* impedance than the load, so a higher efficiency is established. Some transmitters can achieve very low source impedances at their maximum efficiency operating point, resulting in efficiencies of over 90%. However, the antenna "sees" a 50 Ω source impedance but all the power supplied by the line is radiated by the radiation resistance. $A_s$ remains present in the transmitting case, but only applies to power incident on the antenna *from* free space.

**Figure 2.20** shows the equivalent circuit of an antenna and a transmitter. Transmitters are usually specified as "50 Ω output." The real definition of this specification is that the transmitter is tuned for maximum efficiency assuming a 50 Ω load. Thus for a *given antenna source impedance*, the receiver input is tuned at or near the complex conjugate of the antenna impedance. In contrast, for a *given load impedance* the transmitter is tuned to some output impedance that will permit maximum efficiency of the system.

The mathematics that describes the transmit condition of maximum efficiency is rather complex and belongs more in a text on circuit theory rather than antenna physics. The terms involve the actual characteristics of the amplifier device proper (tube or transistor) and the LC tuning unit between that device(s), including its Q and the transmission line impedance. However, the above is an adequate simplification for conceptual understanding of these basic relationships.

$A_s$ is often used to great advantage in antenna design and is often misunderstood. In our basic discussion of the aperture concept we found the maximum effective aperture ($A_{em}$) of an isotropic and a dipole antenna. Now let's revisit this analysis.

A properly terminated antenna will deliver ½ its power to the load (usually a receiver) and re-radiates the other ½. The "total" antenna aperture that includes both the $A_e$ and $A_s$ is defined as an additional aperture, the *collecting aperture* ($A_c$). The collecting aperture is simply the sum of the thus far defined apertures:

$$A_c = A_e + A_l + A_s \quad \text{(Equation 2.18)}$$

Therefore, in a 100% efficient antenna, $A_c$ is about twice the area of $A_e$ since $A_c = A_e + A_s$ and $A_e = A_s$.

$A_s$ is an extremely useful property of antenna elements. In particular, parasitic elements use this property to maximum advantage. In a parasitic element, we set the load impedance to zero (a short circuit), as in a *director* or *reflector* element.

When the terminal impedance of a resonant or near-resonant dipole is zero (as in a parasitic element), $A_s$ is by Equation 2.13:

$$P_{rr} = \frac{V^2 R_r}{(R_r + 0)^2} \quad \text{(Equation 2.19)}$$

Since the terminal impedance is zero, then $A_e = 0$. Therefore all $P_r$ will be re-radiated, or $P_{rr}$.

$$P_{rr} = \frac{V^2}{R_r} \quad \text{(Equation 2.20)}$$

Thus the re-radiated power from a parasitic element is four times that of a conjugate matched antenna! We will explore this in more detail in Chapter 8.

$$A_s(maximum) = \frac{V^2}{SR_r} \quad \text{(Equation 2.21)}$$

where $V^2/R_r$ is the power term and $S$ is the power flux/square unit on the sphere, so the aperture is simply $P_{rr}/S$.

Again, compared to $A_s$ for a terminated antenna dipole, often referred to as a "driven" element, $A_e$ is four times as large for a "parasitic" element compared to a "driven" element shown again in Equation 2.22.

$$A_e = A_s = \frac{V^2}{4SR_r} \quad \text{(Equation 2.22)}$$

Thus, the scattering aperture ($A_s$) has a maximum area when the dipole feed point is shorted, or no power delivered to the load and all power re-radiated. On the other hand, maximum $A_{em}$ occurs when the dipole is matched into a load. The re-radiation from the parasitic element can be adjusted in its phase and amplitude response referenced to the *driven* dipole to create very desirable antenna patterns. The most common application of this principle is in creating a Yagi-Uda array discussed in Chapter 8.

Now consider the opposite condition: an open center to the dipole, where $R_t = \infty$. $A_s$ becomes very small because the antenna is no longer terminated. Mismatching and low $R_r$ prevent "capturing" incident wave power and also re-radiation of the received power. This double-effect renders the antenna to have negligible effect upon the incident wave. In effect, the ½ wave dipole is now two ¼ wave dipoles closely spaced end-to-end.

Now we are prepared to present a graph of how $A_e$, $A_s$, and $A_c$ change when the terminal resistance is changed. Note that $A_{em}$ appears only at one point, the maximum possible value for $A_e$.

**Figure 2.21** plots $A_e$, $A_s$, and $A_c$ against the normalized value of $A_{em}$. When $R_r = R_t$ then $A_e = A_{em}$ and maximum power transfer is accomplished.

Summary of the various types of antenna aperture:

**$A_e$ Effective Aperture**: the "effective" equivalent of an antenna's area that delivers power to a receiver or power coupled into free space by a transmitter.

**$A_l$ Loss Aperture**: the area of an antenna "lost" due to heat.

**$A_{em}$ Maximum Effective Aperture**: the maximum possible area of an antenna where $A_{em} = A_e + A_l$ when the antenna is matched to a load.

**$A_s$ Scattering Aperture**: the equivalent area of the antenna that re-radiates power back into space.

**$A_c$ Collecting Aperture**: the total equivalent area of an antenna that represents power delivered to the load, lost to heat, and re-radiated, where $A_c = A_e + A_l + A_s$.

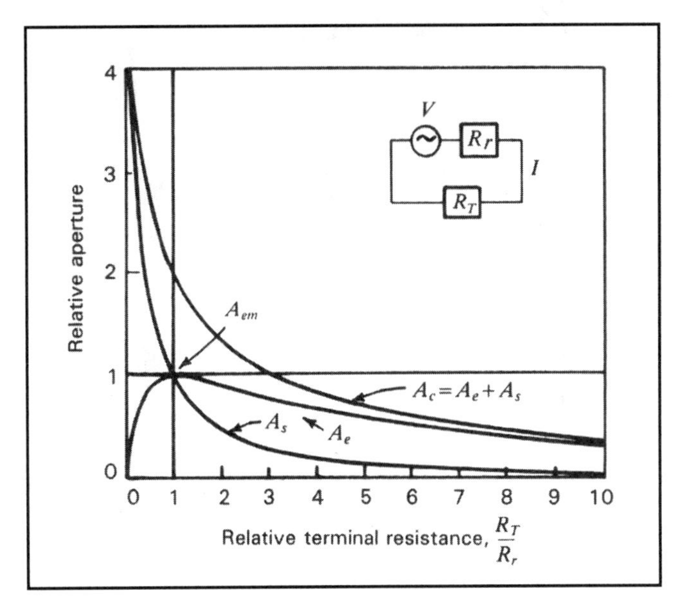


**Figure 2.21 — This figure plots the three key aperture parameters against terminal resistance and the relative aperture (to a conjugate matched ½ λ dipole). [From John D. Kraus, *Antennas*, 2nd ed, 1988, reprinted with permission of McGraw-Hill Education]**

**$A_p$ Physical Aperture**: There is also an additional aperture term, the *physical aperture.* This term defines the actual physical area of an antenna (two dimensional physical aperture). It is applied typically to "aperture" antennas, such as horns and parabolic "dish" antennas, where the physical aperture indeed resembles the collecting aperture. For example, if we consider a dish antenna 10 meters in diameter, then $A_p = \pi 5^2 = 75.5$ m². If the *aperture efficiency* were 100%, this would also be the $A_e$. If the efficiency were only 50%, $A_e$ would be 37.75 m². In contrast, the "area" of a wire dipole antenna is very small compared to its $A_c$ , therefore the physical aperture has little meaning for single-dimension antennas.

## Path Loss and Fixed Aperture Antennas

Many engineers prefer a simplified equation that uses dB rather than linear-type terms to calculate the received power from the transmitted power, antenna gains, free-space distance, and frequency. This is also the equation that appears in other ARRL publications.

$$P_r dBm = P_t(dBm) + G_{tant} + G_{rant} - 20log(km) - 20\log(MHz) - 32.44 \quad \text{(Equation 2.23)}$$

where $G_t(dBi)$ is the transmitter antenna gain, $G_r(dBi)$ is the receiver antenna gain. *km* is the free space distance in kilometers, and of course MHz is the frequency. The term dBm refers to the dB *power* referred to 1 mW, where 0 dBm = 1 mW. This term is discussed in more detail in Chapter 10.

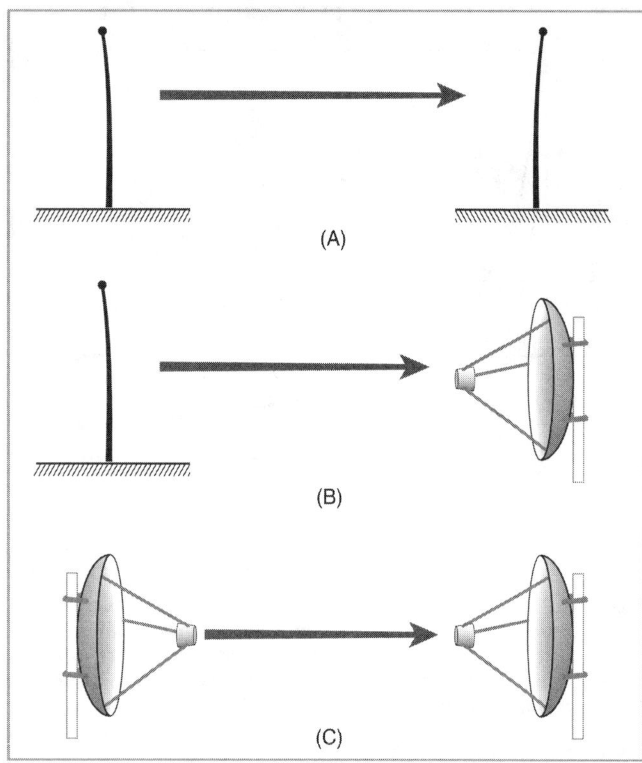


**Figure 2.22 — As a final example to the concept of antenna aperture in this chapter, we solve link budgets for three types of systems, using 1/4 λ whip antennas (constant gain vs. frequency) and fixed aperture parabolic reflector antennas. See text.**

Transmission line losses can be subtracted from the antenna gain figures. This is the common equation used to calculate link budgets. Much of this chapter is devoted to explaining *how* link budgets are defined. This equation is often used as a "short cut" without the user understanding the underlying physics.

As a final example to the concept of antenna aperture in this chapter, we solve link budgets for three types of systems, using 1/4 λ whip antennas (constant gain vs. frequency) and fixed aperture parabolic reflector antennas. See **Figure 2.22**. For simplicity we assume that our whip antennas are indeed isotropic with apertures of $\lambda^2/4\pi$.

**A.** First we consider two wavelength-aperture dependent "whip" antennas creating a transmit to receive link. The path loss is the solution given in Equation 2.6:

$$P_r = P_t\left(\frac{\lambda}{4\pi r}\right)^2$$

**B.** Now we consider an isotropic-to-fixed aperture link. Since $A_e$ of the parabola is fixed as simply the physical cross-sectional area (for a "perfect" parabola), the gain is defined by

$$G_{iso} = \frac{A_e}{\frac{\lambda^2}{4\pi}} = G_{iso} = \frac{A_e 4\pi}{\lambda^2} \qquad \text{(Equation 2.24)}$$

Where $G_{iso}$ is the linear gain relative to an isotropic antenna

Notice that the gain for a fixed aperture antenna *increases* with increasing frequency. However, the aperture of the parabola simple remains $A_e$. In this case the parabola is the receiving antenna therefore we use $A_{er}$.

Using the Friis equation (Equation 2.8) we recall

$$P_r = P_t\frac{A_{et}A_{er}}{r^2\lambda^2}$$

In this case,

$$A_{et} = \frac{\lambda^2}{4\pi}$$

but $A_{er}$ is not frequency independent, it remains simply $A_{er}$. So, we derive

$$P_r = P_t\frac{\frac{\lambda^2}{4\pi}A_{er}}{r^2\lambda^2} = P_t\frac{A_{er}}{4\pi r^2} \qquad \text{(Equation 2.25)}$$

Notice that the wavelength term cancels! Thus, if you use a 1/4 λ whip antenna (for the frequency in use, all 1/4 λ whips have the same gain at all frequencies) at one end of a link and a parabolic dish antenna at the other (with vertical polarization and feedhorns for the same frequency), there is no difference in the link budget when changing frequency.

**C.** Finally, for the two fixed parabolas, we simply use the Friis equation

$$P_r = P_t\frac{A_{et}A_{er}}{r^2\lambda^2} \qquad \text{(Equation 2.26)}$$

Here, with two fixed apertures (two parabolic antennas pointed at each other) the link budget *increases* as the frequency increases.

Conclusion: If we have a fixed distance (r), and use isotropic antennas and fixed aperture parabolas for a radio link, we can derive an important set of conclusions:

**A.** Using two isotropic (or any wavelength-depended aperture antenna types, i.e. 1/4 λ whip antennas) the link budget is proportional to $1/\lambda^2$ (the link loss *increases*) with higher frequencies.

**B.** Using one isotropic antenna and one fixed aperture antenna the link budget is *independent* of the operating wavelength (the link loss remains the same)

**C.** Using two fixed aperture antennas (wavelength independent) the link budget is *inversely proportional* to $\lambda^2$ and thus the losses *decrease* with increasing frequency.

# Radiation of Radio waves

In Chapter 2, some of the basic parameters of antenna physics were presented. In addition to the imaginary isotropic antenna, the "real" dipole antenna was discussed, with its radiation resistance, apertures, gain, and radiation pattern quantified. However, these parameters were not explained in detail. They were only presented as self-evident, empirical parameters — measured, but not explained in a convincing manner.

A deeper explanation requires more sophisticated tools. Therefore we will investigate the nature of Maxwell's equations and attempt to provide some intuitive knowledge of them.

## Modern Expression of Maxwell's Equations

While it is possible to express most of the simple parameters of antennas with relatively simple mathematics, it is almost impossible to present more subtle definitions without using Maxwell's equations. However, an intuitive insight is possible without a mastery of advanced mathematics. Although on first sight Maxwell's equations look somewhat formidable, the relationships that these equations represent are relatively simple to understand.

For those readers not familiar with calculus, Appendix A is provided. In addition, Chapter 2 has already provided the concept of an invisible sphere, with power flux projected through the sphere as a portion of the sphere's surface. In Chapter 2 we made no attempt to explain what this power is. We simply accepted that light or "RF power" was propagating through the surface of the sphere. We will now consider a similar concept, but rather than power we will consider the flux to be either an electric or magnetic field that, together, constitute the power of the wave.

Here we explain "Maxwell's" equations, which actually originate from Gauss, Faraday, and Ampere. These equations are applicable not only to static electric and magnetic fields, but also to electromagnetic radiated fields. First we will explain the totality of Maxwell's equations and then concentrate on the equations and their forms that are most useful for a discussion of antennas.

The modern expression of Maxwell's equations represents refinements made by Oliver Heaviside, who used vector calculus and some other notation changes. However, the essence of the relationships was initially developed by Maxwell.

## Maxwell's First Equation, Gauss's Law for Electricity

$$\oint \boldsymbol{E} \bullet d\boldsymbol{A} = \frac{q}{\varepsilon_0}$$

The closed integral (sum) of an electrical field

$$\oint \boldsymbol{E}$$

over the closed surface area ($dA$) is the sum of positive and negative electrical charges contained within the surface ($q$) and inversely proportional to the permittivity ($\varepsilon$) of the "material," or lack thereof, inside the surface.

In mathematics, the simplest form for "summing" is the addition sign (+). A shortcut can be used if the summed values take on a sequence conforming to some defined change ($\Sigma$), the Greek letter "sigma" for "summing." If we need to sum over some *function* (no discrete values), we use the integral calculus sign

$$\int f$$

This is a "smoothed sum" over some function $f$. For example, if we want to sum the area under a sine wave, we need a smoothed sum and thus a precise answer requires calculus.

An additional refinement

is simply the notation for a "smoothed sum" (in this case the electric field fluxes) over the surface of the same imaginary closed surface, usually a sphere as described in Chapter 2. The little circle is written over the integral sign indicates we are summing over a closed area. $q$ is an electrical charge in coulombs (can be either negative or positive; for example, an electron has the minimum electrical negative charge $q$). We begin with an electrical charge at the sphere's center (**Figure 3.1**). We measure the total flux of the electric field through the sphere by using the operator

$$\oint \boldsymbol{E} \bullet dA$$

This is very similar to the problem associated with an isotropic antenna, which radiates equally in all directions. A "point charge" is also isotropic in that the field flux will be identical at all points on the sphere (assuming of course we are observing it in free space and the charge is located at the center of the sphere). However, the total flux will always be the same over any size or shape of enclosure (here we are using a sphere as the example). Also, the total flux will always be the same no matter how we distribute the charge or charges within the enclosure. We can intuitively accept this by noting that a 100 W light bulb will radiate a total of 100 W through the sphere even if it is not located at the center. The flux will simply be greater on the surface nearer to the light bulb.

We use a dot product, which is just a special form of multiplication for vectors, in this case over the surface surrounding the total charge(s). The dot product simply accounts for the angle of an area that is not normal to the source of the electrical charge (**Figure 3.2**). This is much like our solar cell example from Chapter 2 being tilted somewhat away from the Sun, where:

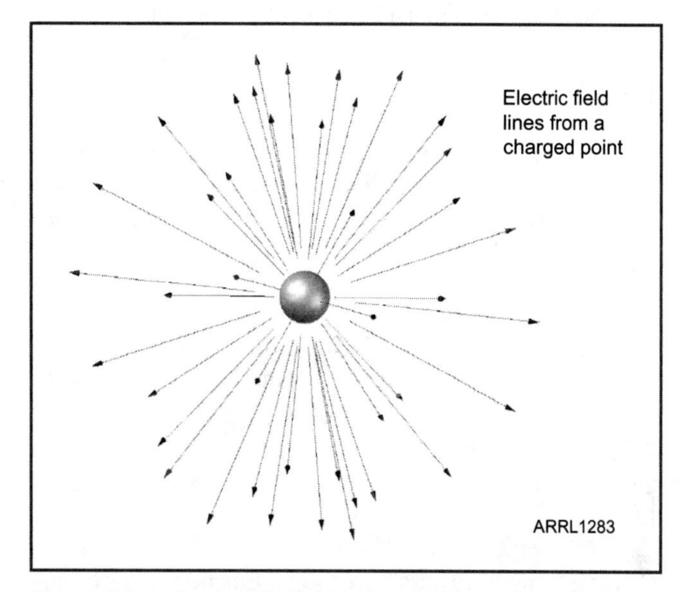


**Figure 3.1 — An electric field only requires a single point charge. The field radiates away from the charge in an isotropic pattern.**

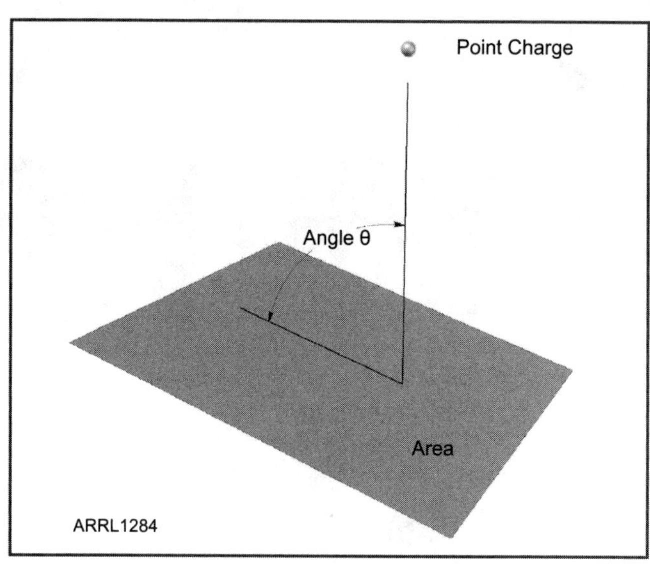


**Figure 3.2 — The dot product simply accounts for an irregular surface surrounding the charge(s).**

$$Aperture = area\ (m^2) \sin \theta \qquad \text{(Equation 3.1)}$$

Thus if the aperture is perpendicular to the charge (90 degrees), then the aperture is equal to the actual area of the surface. In this case, the dot product simplifies to a simple multiplication.

Finally, $q/\varepsilon_0$ represents the amount of charge (in coulombs divided by the permittivity of the medium), in this case free space where $\varepsilon = \varepsilon_0$ This is the essence of Maxwell's first equation: The total **E** field flux through a surface containing a total of $\pm q$ charge will be equal to the total charge divided by the permittivity of the material or free space inside the surface-defined volume.

If we place a negative and equal positive electric charge inside the sphere and integrate the total electric field flux, we will find the total flux will be zero, since the two charges cancel and will result in zero charge. Thus in this first equation, $q$ must account for the sum of all charges (negative and positive) within the closed surface.

## Maxwell's Second Equation, Gauss's Law for Magnetism

$$\oint \boldsymbol{B} \bullet dA = 0$$

The closed integral (sum) of a magnetic field

$$\oint \boldsymbol{B}$$

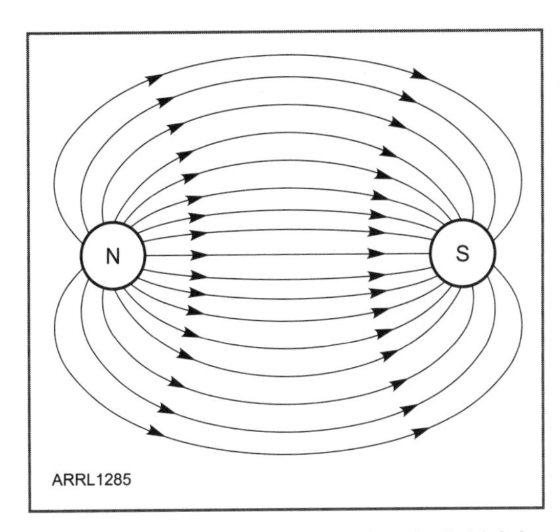


**Figure 3.3 — In contrast to an electric field, it is impossible for a magnetic field line to propagate infinitely into free space. It must always come back to its opposite pole.**

**Figure 3.4 — Modern power generators depend upon Faraday's Law of Induction. Turn a set of wire windings (closed loops of wire) within a magnetic field and an electrical current results.**

contained over a closed surface area (*dA*) is zero.

A critical difference between electric and magnetic fields is that an electric field can result from a *point charge.* If the universe were occupied by only one electron, the entire universe would be filled with a very small negative electric field. There is no such analogy with magnetic fields. There must always be an opposite charge of reverse value, exemplified by the north and south poles of a magnet. If we place such a *dipole* magnet inside our sphere, we find that over the entire surface of the sphere there will be zero net magnetic field simply because the two field fluxes sum to zero (**Figure 3.3**).

Unlike the straight field lines emanating from a point electric charge, the magnetic field lines form a concentric curvature structure. The only "straight" magnetic field line is found along the line directly connecting the opposite poles. So if we imagine a sphere surrounding our bar magnet, we can easily envision any magnetic field line that propagates "out" of the sphere must also come back "in," thus cancelling the net magnetic flux through its surface. The equation always equals zero. Consequently Maxwell's second equation is always equal to zero.

Now things start to get interesting!

## Maxwell's Third Equation, Faraday's Law of Induction

Maxwell's third equation:

$$\oint \boldsymbol{E} \bullet d\boldsymbol{l} = -\frac{d\boldsymbol{\Phi}_B}{dt}$$

Faraday's Law states that an electric field **E** is created when the magnetic flux $\Phi_B$ changes with time over some length (*dl*). In other words, a time-varying **B** field creates an **E** field. An example is pushing a magnet in and out of a coil of wire (inductor), creating an electric current (moving electrical charge) in the inductor. It also explains how an electric current is generated when rotating the inductor inside a magnetic field, thus creating a generator (**Figure 3.4**).

Maxwell acknowledged that it was the work of Faraday that led him to his revolutionary theory. Faraday was not an expert mathematician, so he was unable to express the subtleties of his insights in mathematical form. Maxwell was perhaps the first to realize the significance of Faraday's work. He understood it and was able to express it with complex mathematics, with some help from Heaviside.

## Ampere's Law

$$\oint \boldsymbol{B} \bullet d\boldsymbol{l} = \mu_0 I$$

Ampere's Law states that the magnetic field (in this case generated along a loop conductor of length *l*) is proportional to the current *I* in that loop and the permeability of the medium inside the loop. An example of this is the electromagnet, or solenoid, where a magnet is created by running a current through a conductor (the effect is increased by configuring the conductor into a solenoid). Notice that the third and fourth equations are reciprocal: an electric field is formed by a *moving* magnet and a magnetic field is formed by a *moving electrical charge*, or current. A stationary magnetic field produces no electric field and vice-versa.

It should now be apparent that there is some type of reciprocity between the electric and magnetic fields. We can generate an electric field by moving a magnet in and out of a solenoid, but we can also create a magnetic field by running an electric current through the same solenoid. An electromagnet such as the one shown in **Figure 3.5** is a practical example of Ampere's Law at work.

Although Maxwell's equations look formidable to anyone without knowledge of advanced calculus, their *meanings* are easily understood by anyone with a basic understanding

**Figure 3.5 — Ampere's Law at work in this giant electromagnet: a moving charge (current) flowing in a loop creates a magnet. Turn the current off and the magnetic field disappears. [Library of Congress, Prints & Photographs Division, FSA/OWI Collection, LC-DIG-fsa-8c34729]**

of electricity and magnetism. The development of these equations into forms that explain electromagnetic fields *does* require advanced mathematics. Indeed, most of the time a calculus student spends in advanced courses deals with learning how to develop and solve the bread and butter equations (solving for derivatives and anti-derivatives of functions). The concepts are far easier to understand intuitively. Therefore, when possible, I will skip most of the advanced mathematics and simply state the relevant results, but also attempt a simple explanation.

Maxwell had a problem with the fourth equation. In the case of a capacitor inserted into a wire loop, there is *zero loop current* in that a current does not flow around the complete length of the loop. However, a current is necessary to charge or discharge the capacitor (to displace charge); therefore Maxwell derived and added an additional term to Ampere's Law to account for this *displacement current.*

$$\varepsilon_0 \mu_0 \frac{d\Phi_E}{dt} \quad \text{(Equation 3.2)}$$

Thus. if we add this to Gauss's Law we derive Maxwell's fourth equation.

## Maxwell's Fourth Equation

$$\oint \mathbf{B} \bullet d\mathbf{l} = \mu_0 I + \varepsilon_0 \mu_0 \frac{d\Phi_E}{dt}$$

Thus the displacement current can be defined as an electric field flux $\Phi_E$ that changes with time (*dt* indicating the time derivative of $\Phi_E$). This is the reciprocal operation of Maxwell's third equation. The addition of the displacement current was the critical term for defining all the relationships within an electromagnetic *wave.* From the third and fourth equations we find that a *changing* electric field creates a *changing* magnetic field and vice-versa. This is precisely how electromagnetic waves *propagate* through free space, a continuous sequence of E-M-E-M... fields as described by the third and fourth equations.

Usually, we refine the word "change" to "accelerate" for a more precise description of how a field must behave in order to create a propagated wave. Usually we assume a radio wave to be sinusoidal, and indeed a sine wave is nearly always in a state of acceleration. (See Appendix A for a proper definition of acceleration.)

These four equations define exactly how fields are formed and how they behave. In this book we are concerned only with the *formation* and *propagation* of radio waves. In free space (vacuum) there is no matter and thus no possibility of a charged *particle* and no possibility of a magnet since both require some type of matter present (even an electron). Also there is no current present in free space because there is no conductor on which a charge (electrons) can travel. (Of course electrons can flow through a vacuum, but since electrons are matter, the vacuum is no longer a vacuum!) Therefore, the first two of Maxwell's equations, in free space, equate to zero. The fourth equation would also equate to zero, except for Maxwell's displacement current term, and this is the key equation to describe the *radiation of radio waves.*

Therefore, everything you need to know about an electromagnetic wave can be condensed into two equations:

$$\oint \mathbf{E} \bullet d\mathbf{l} = -\frac{d\Phi_B}{dt} \quad \text{(Equation 3.3)}$$

$$\oint \mathbf{B} \bullet d\mathbf{l} = \varepsilon_0 \mu_0 \frac{d\Phi_E}{dt} \quad \text{(Equation 3.4)}$$

On an antenna the situation is different. In order to create the fields in the first place, we must have an *accelerating* charge (current). As we will soon discover, a moving charge with constant velocity is equivalent to a dc current and will thus create only a static magnetic field around the conductor. However, an *accelerating charge* (current changing amplitude and/or direction) over time creates *waves that radiate.* An antenna must be made of matter and most are constructed from conducting material, such as metals. But how do we know these propagating fields constitute a wave?

## The Wave Equation

Imagine a string that is stretched along the X-axis (**Figure 3.6**). We "pluck" the string at one end thus create a wave that propagates along the string at velocity *v*. The wave exists in the X-Y plane, with the amplitude of the wave being a displacement in *Y* direction and the direction of travel along the string along the X-axis. More importantly we see that the amplitude of the wave is defined in both space (*X* vs *Y*) and time *Y* vs *T*) and the coefficient in the equation is velocity.

The wave equation had been known since the 18th century (see Chapter 1) when it was derived to describe the harmonic motion of a plucked string. Any phenomenon that takes the form of a wave must conform to the wave equation, a second-order differential equation of the form:

$$\frac{\partial^2 y}{\partial x^2} = \frac{1}{v^2}\frac{\partial^2 y}{\partial t^2} \qquad \text{(Equation 3.5)}$$

In order to actually solve this equation for a given set of physical circumstances, we would need to be able to solve a differential equation. However, we only need a brief explanation to understand what this equation means. Also the derivation of this equation requires some rather sophisticated calculus. I will omit this derivation and simply refer the reader to most advanced texts on electricity and magnetism (E&M) for the formal derivation.

The wave equation defines exactly what is happening in the Y-axis (where the instantaneous wave's amplitude exists). The first term

$$\frac{\partial^2 y}{\partial x^2}$$

explains what the wave is doing relative to the X-axis, which is the direction of propagation. The second term

$$\frac{1}{v^2}$$

is a function of the velocity of the wave on the string, and the third term

$$\frac{\partial^2 y}{\partial t^2}$$

informs what the wave is doing as a function of time. All other parameters of any wave can be derived from this simple set of relationships. This is the essence of the wave equation. Now let us explore a physical example for additional clarity and perform some important derivations.

The velocity of the wave along a string is dependent upon the tension on the string and the mass of the string/unit length. Anyone who has ever played a stringed instrument (**Figure 3.7**) knows this relationship: If you tighten a string,

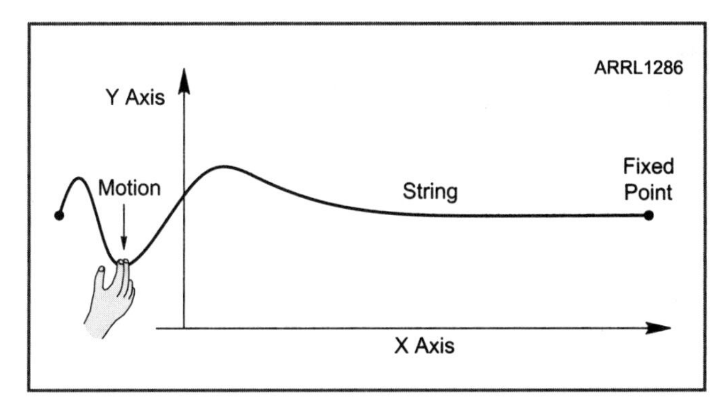


**Figure 3.6 — Imagine a string that is stretched along the X-axis. We "pluck" the string at one end, thus creating a wave that propagates along the string at velocity *v*. The wave exists in the X-Y plane, with the amplitude of the wave being a displacement in *Y* direction and the direction of travel along the string along the X-axis. More importantly, we see that the amplitude of the wave is defined in both space (*Y* vs. *X*) and time (*Y* vs. *T*) and the coefficient in the equation is velocity.**

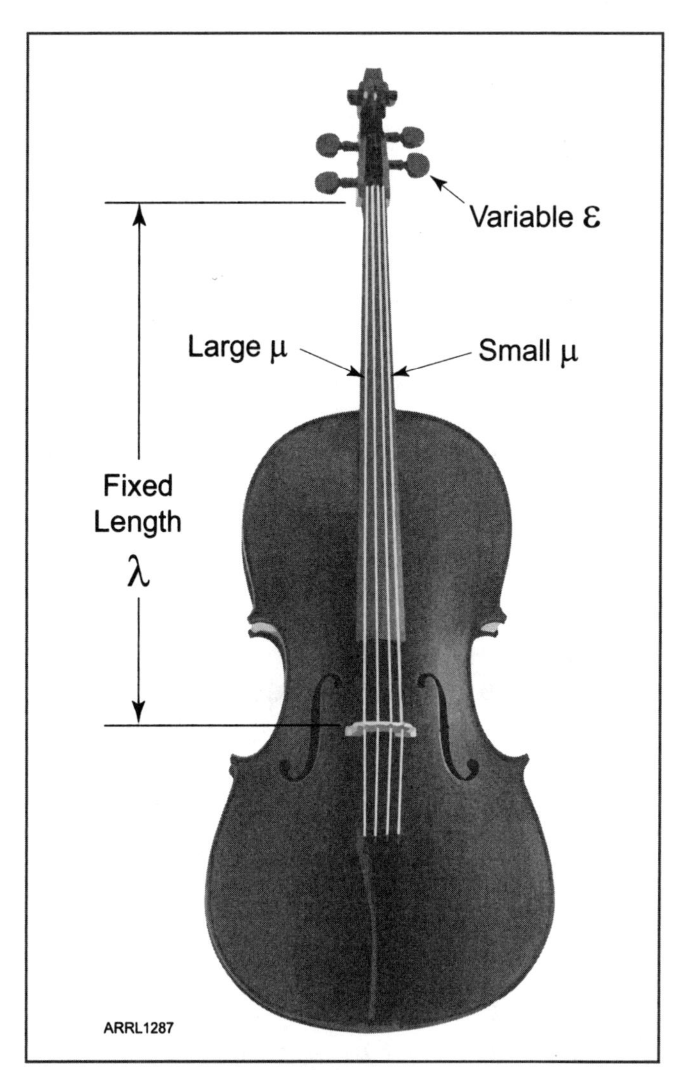


**Figure 3.7 — The velocity of the wave along a string is dependent upon the tension on the string and the mass of the string/unit length. Anyone who has ever played a stringed instrument knows this relationship.**

you get a higher frequency (note) where

$$f = \frac{v}{\lambda}$$

Many readers will recognize this equation as defining the relationship between a radio wave's wavelength and frequency where the velocity is usually taken to be constant (the speed of light, $c$). In the case of a violin, $v$ is also constant and we adjust the tone (frequency) by changing the $\lambda$ by shortening the vibrating string with our finger.

The other term that describes velocity is the mass of the string (more precisely the mass per length). The relationship between the "tightness" of the string, the mass per length of the string, and the velocity is

$$v^2 = \frac{\sigma}{\mu} \qquad \text{(Equation 3.6)}$$

where $\sigma$ is the "tightness" of the string and $\mu$ is the mass/length of the string.

Strings on musical instruments optimized for lower frequencies (notes) will be bigger (higher mass per length). If we convert our "tightness" term into "looseness" (tightness being the reciprocal of looseness) then the equation becomes

$$v^2 = \frac{1}{\mu\ (looseness)} \qquad \text{(Equation 3.7)}$$

Or,

$$v = \sqrt{\frac{1}{\mu\ (looseness)}} \qquad \text{(Equation 3.8)}$$

Also if we define "looseness" as $\varepsilon$, we get

$$v = \sqrt{\frac{1}{\mu\varepsilon}} \qquad \text{(Equation 3.9)}$$

Now that we have a velocity term for the mechanical version of the wave equation we can investigate the similarity (and differences) to the electromagnetic wave equation.

## Wave Equation for Electromagnetic Waves

Since we do not need the loop current term, we rewrite Maxwell's fourth equation without Ampere's current term.

$$\oint \mathbf{B} \bullet dl = \varepsilon_0 \mu_0 \frac{d\Phi_E}{dt} \qquad \text{(Equation 3.10)}$$

Again, the third equation is:

$$\oint \mathbf{E} \bullet dl = -\frac{d\Phi_B}{dt} \qquad \text{(Equation 3.11)}$$

The derivation of the wave equation from Maxwell's equations is quite complicated. For completeness, we present some of the important steps in the process. However, again, the important point is to understand that the third and fourth equations lead directly to a wave equation.

From the third equation we derive the following:

$$\oint \mathbf{E} \bullet dl = -\frac{d\Phi_B}{dt} = \frac{\partial \mathbf{E}}{\partial x} = -\frac{\partial \mathbf{B}}{\partial t} \qquad \text{(Equation 3.12)}$$

Here we start with Equation 3.11 and equate it to Equation 3.10 by using the reciprocal to the closed integral, or the differential. The integral (or anti-derivative) and the differential (or derivative) represent the two main divisions of calculus that solve problems by working in different directions. A rough comparison would be multiplication and division that solve for different circumstances in opposite mathematical "directions."

Also if we take the second derivative of

$$\frac{\partial \mathbf{E}}{\partial x}$$

and some advanced mathematics, we can derive an equation with two partial differential $E$ terms, one relating to the X-axis and the other to time:

$$\frac{\partial^2 \mathbf{E}}{\partial x^2} = \frac{\partial}{\partial x}\frac{\partial \mathbf{B}}{\partial t} = -\frac{\partial}{\partial t}\frac{\partial \mathbf{B}}{\partial x} = -\frac{\partial}{\partial t}\left(-\varepsilon_0 \mu_0 \frac{\partial \mathbf{E}}{\partial t}\right) \qquad \text{(Equation 3.13)}$$

If we take the partial time differential

$$\frac{\partial}{\partial t}$$

of another partial time differential

$$\frac{\partial \mathbf{E}}{\partial t}$$

we get a second partial time differential where

$$-\frac{\partial}{\partial t}\left(-\varepsilon_0 \mu_0 \frac{\partial \mathbf{E}}{\partial t}\right) = \varepsilon_0 \mu_0 \frac{\partial^2 \mathbf{E}}{\partial t^2} \qquad \text{(Equation 3.14)}$$

Therefore, from Equations 3.13 and 3.14 using simple algebra we derive

$$\frac{\partial^2 \mathbf{E}}{\partial x^2} = \varepsilon_0 \mu_0 \frac{\partial^2 \mathbf{E}}{\partial t^2} \qquad \text{(Equation 3.15)}$$

And finally from Equation 3.9 we get

$$\frac{\partial^2 \mathbf{E}}{\partial x^2} = \frac{1}{v^2}\frac{\partial^2 \mathbf{E}}{\partial t^2} \qquad \text{(Equation 3.16)}$$

This equation is *exactly* the form in Equation 3.5 and therefore satisfies the wave equation. Thus the **E** field (usually defined as a sine function) is a wave. Also, by Maxwell's fourth equation we know that an accelerating **E** field creates a **B** field and so on. Thus *electromagnetic radiation must be a wave.* Again, we skipped much of the complex mathematics to arrive at this conclusion for simplicity.

In Chapter 1 we described how Maxwell derived the speed of light in the form:

$$c = \sqrt{\frac{1}{\mu_0 \varepsilon_0}} \qquad \text{(Equation 3.17)}$$

This form of the velocity term has now been shown to be a direct analogy to the wave action on a string. However, rather than describing μ and ε as a mass/length and as "looseness" of the string's physical condition, we now define space itself in terms of different physical conditions: permeability and permittivity. As Einstein later postulated, the presence of a field "bends" space, so an oscillating field "oscillates" space itself, in space-time. There is no matter in free space (including the imaginary "ether" once thought to exist); therefore space itself is the medium for electromagnetic waves.

As Maxwell rightly credited Faraday for the breakthrough that led to Maxwell's theory of electromagnetism, Einstein rightly credited Maxwell with the foundation of his relativity theory. Einstein, in turn, discovered the *photoelectric* effect, which is the mechanism for the solar cells we discussed in Chapter 2. Also, he showed that light has *momentum*, and therefore light is also a particle! The particle of light (or any other wavelength of electromagnetic radiation) is called a *photon.* When a photon has enough momentum (energy) it will dislodge an electron, creating the photoelectric effect. This effect is much like a billiard ball striking another billiard ball, but at the sub-atomic level. This leads to the peculiar fact that electromagnetic waves exhibit a *wave-particle duality.* In Chapter 10 we will explore this duality in some more detail. However, we know it is a wave because it conforms to the wave equation, and it is also a particle because solar cells would be impossible without the photon.

Thus Einstein played a pivotal role in formulating the basis for the two great subdivisions of modern physics: relativity and quantum mechanics. You may be pleased that for antennas operating in the RF portion of the spectrum, Maxwell is all that is required for even an advanced intuitive and practical understanding. Maxwell can be thought of the "bridge" between Newtonian classical physics and modern physics pioneered by Einstein (and of course many others).

## Relationships Between the E and B Fields

The wave equation is expressed in *differential* form. The differential was briefly shown to be the reciprocal of the *integral* form. For the wave equation and some additional explanations, it is easier to use the differential form of the equations. We now know that there must be an **E** and also a **B** field to constitute an electromagnetic wave. The relative amplitudes of the two fields will be derived later and this ratio uses relatively simple mathematics. However, we need differential vector calculus to describe another important relationship: the relative *polarity* between the fields.

Again, we are concerned only with electromagnetic radiation in this text, so we can focus upon the equivalent differential forms of only the third and fourth equations. Again after some rather difficult calculus we can re-write the third and fourth equations as:

Maxwell's third equation

$$\nabla \times \boldsymbol{E} = -\frac{\partial \boldsymbol{B}}{\partial t} \qquad \text{(Equation 3.18)}$$

and Maxwell's fourth equation

$$\nabla \times \boldsymbol{H} = \frac{\partial \boldsymbol{D}}{\partial t} + \boldsymbol{J} \qquad \text{(Equation 3.19)}$$

As before, we can eliminate the Ampere current term, here represented by $\boldsymbol{J}$. So the differential form of Maxwell's fourth equation simplifies to

$$\nabla \times \boldsymbol{H} = \frac{\partial \boldsymbol{D}}{\partial t} \qquad \text{(Equation 3.20)}$$

This form of differential equation uses the ($\nabla\times$) operator also known as the *curl.* A simple intuitive demonstration of the curl is called the "right-hand rule." If you shape your right hand as shown in **Figure 3.8**, you can see clearly the curl result. The thumb points in the direction of the accelerating charge (function of time), for example from Equation 3.18

$$\left(-\frac{\partial \boldsymbol{B}}{\partial t}\right)$$

and $\nabla \times \boldsymbol{E}$ (the curl of **E**) is represented by the four fingers encircling (curling) around the thumb line. These two equations show that the opposite fields tend to wrap themselves

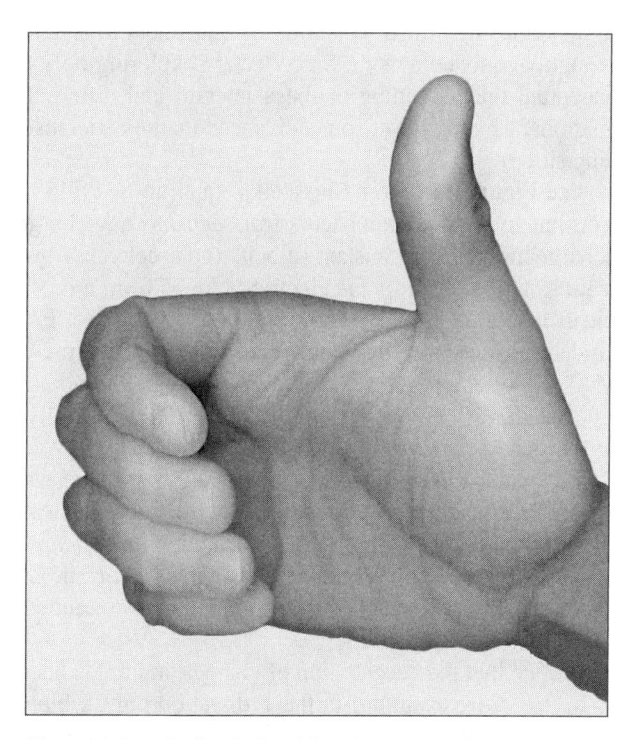

**Figure 3.8 — A simple intuitive demonstration of the curl is called the "right-hand rule." If you shape your right hand as shown, you can see clearly the curl result.**

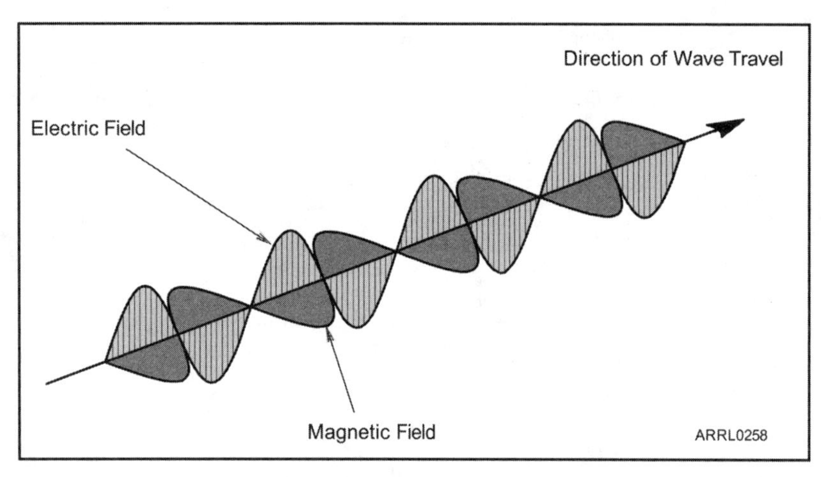


**Figure 3.9 — Representation of the magnetic and electric field strengths of an electromagnetic wave. In the diagram, the electric field is oriented vertically and the magnetic field horizontally.**

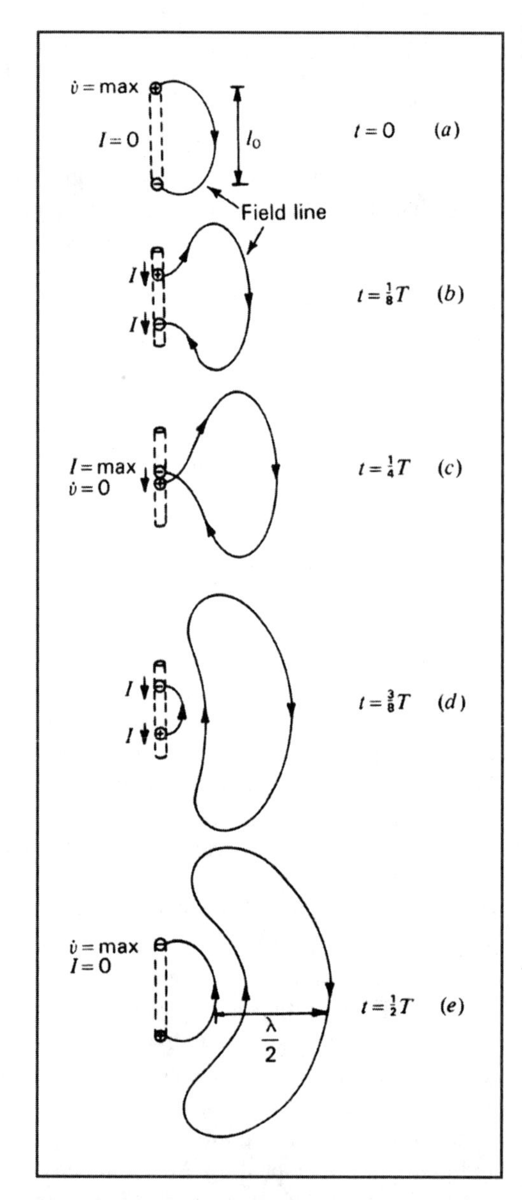


**Figure 3.10 — The electric field is "detached" from the antenna and wave propagation commences due to a changing (accelerating) charge. [From John D. Kraus, *Antennas*, 2nd ed, 1988, reprinted with permission of McGraw-Hill Education]**

around a concentrated accelerating field. When a field becomes a plane, as in a "plane wave," the opposite field cannot wrap itself around, and thus it also becomes a plane wave.

There is another important result to the curl. Notice the *direction* the thumb points relative to the *direction* the fingers point. This implies, correctly, that the **E** and **B** fields are always orthogonal, having a 90 degree difference — ie "vertical" and "horizontal." Again, the equations use very sophisticated mathematics, but express a very simple concept.

For mathematical correctness, these equations are presented using some different terms. For our purposes **H** is roughly equivalent to **B**, **D** is roughly equivalent to **E**, and **J** is roughly equivalent to $\mu I$. Since we are exploring only the conceptual understanding of these integral and differential equations, we may upset some advanced mathematicians by being sloppy!

See **Figure 3.9**. From Maxwell's equations we find that electromagnetic radiation occurs from *accelerating* charges. A charge moving at a constant velocity (no acceleration) will create a static magnetic and electric field, as from a dc current in a solenoid: constant current, constant charge. From classical physics we know that acceleration is the time derivative of velocity or, $A = dV/dt$ (Appendix A). For radiation to occur, the velocity of the charge must change (usually at the frequency of the RF signal). Charge that changes velocity (accelerating) at a time-dependent value of a sine wave will produce a field directly analogous to the same sine wave. Since current ($I$) is moving charge, a changing *rate of moving charge* is the cause of radiation. Thus *antenna current* is of fundamental importance for determining the key parameters of how an antenna behaves.

Notice that the acceleration of charges and fields occurs (as in the string example) in the $Y$ direction only, which is the change in the rate of *amplitude.* In contrast, an electromagnetic wave's velocity (also in the string example) remains *constant* (the speed of light) in the $X$ direction, or the direction of propagation. As a second step to the intuitive right-hand rule, image that the fist throws a punch. The direction of the punch (the direction of propagation) is orthogonal to the direction of both the thumb and the fingers.

The fields that propagate outward from the charge's location become detached at their source, since polarity is changing at the rate of twice the RF frequency (**Figure 3.10**). The result is that we create "waves" of electric and magnetic fields, not a constant static field. The electric field can be compared to a soap bubble first appearing on the surface of soapy water from a bubbling air source underneath. The bubble expands until it finally detaches from the water, propagating away from its source, "up," only to be followed by the next bubble already being formed. Eventually a series of bubbles is formed. Of course bubbles tend to become small spheres, while the shape of the field "bubbles" will conform to the directivity of the antenna. As they propagate away from the antenna, they converge into "plane waves" as the wavelength becomes much greater than the distance propagated as along the expanding sphere.

## Impedance of Free Space

Maxwell's equations provide the relationships (direction and ratio) between the magnetic and electric fields associated with an electromagnetic wave. The ratio of the two fields' amplitudes remains constant and the two fields are *always* orthogonal, at least in the far field. This is the condition for *plane waves*. As we will show later, there are exceptions to this general rule.

Since the ratio of these two fields is constant in a free-space electromagnetic wave, those familiar with Ohm's Law (Z = E/I) will correctly postulate that some type of resistance is present, but in free space?

In order for propagation of electromagnetic waves to occur in any medium (including free space), there must exist a finite intrinsic impedance. If the impedance is zero, or if it is infinite, the magnetic and electric fields would have to be either infinite or zero. If one or the other cannot exist, there can be no electromagnetic radiation. In order to have electromagnetic radiation in free space, free space must have some finite real impedance, greater than zero but less than infinity. Indeed, the universe would be a *very* different place without a finite impedance of empty space!

We have already defined two physical characteristics of free space, permeability and permittivity. That should give us our first clue that there is more to free space than just these two simple terms, but we begin with these terms, and use some algebra and equations from fundamental classical physics.

Ignoring the coefficients for simplicity, we show how we derive ohms (a pure resistance Ω) from the permittivity and permeability of free space:

$$\varepsilon_0 = \frac{F}{m} \text{ and } \mu_0 = \frac{H}{m} \qquad \text{(Equation 3.21)}$$

and thus

$$\sqrt{\frac{\mu_0}{\varepsilon_0}} = \sqrt{\frac{H}{F}} = \sqrt{\frac{J/I^2}{J/V^2}} = \sqrt{\frac{V^2}{I^2}} = \frac{V}{I} = \Omega$$

where $F$ is Farads, $H$ is Henrys, and $J$ is Joules where

$$J = Nm \qquad \text{(Equation 3.22)}$$

where $N$ is Newtons (a physical force) and $m$ is meters. Thus we can actually relate linear distance and physical force to the electromagnetic terms that describe free space.

From Chapter 1 we found that

$$\frac{1}{\sqrt{\varepsilon_0 \mu_0}}$$

reduces to a velocity (of light) in free space, now we find that

$$\sqrt{\frac{\mu_0}{\varepsilon_0}}$$

reduces to an impedance (of free space). Thus it is clear that the two terms are intrinsically related where:

$$\sqrt{\frac{\mu_0}{\varepsilon_0}} = \frac{\mathbf{E}}{\mathbf{H}} \qquad \text{(Equation 3.23)}$$

Then

$$\frac{\mathbf{E}}{\mathbf{H}} = Z_0 \qquad \text{(Equation 3.24)}$$

This is the ratio of the values of the electric and magnetic fields, again comparable to the ratio of voltage and current equating to resistance.

We can formalize the relationship between the speed of light in free space and the impedance of free space by:

$$\sqrt{\frac{\mu_0}{\varepsilon_0}} = \frac{1}{c\varepsilon_0} = Z_0 \qquad \text{(Equation 3.25)}$$

Because

$$\frac{1}{\sqrt{\varepsilon_0 \mu_0}} = c \qquad \text{(Equation 3.26)}$$

consequently

$$c = \frac{Z_0}{\mu_0} \qquad \text{(Equation 3.27)}$$

and if

$$\frac{1}{\sqrt{\varepsilon_0 \mu_0}} = c \qquad \text{(Equation 3.28)}$$

then

$$\sqrt{\frac{\mu_0}{\varepsilon_0}} \qquad \text{(Equation 3.29)}$$

must be $Z_0$.

Plugging the coefficients of these terms back in, we can calculate the actual impedance of free space:

$$Z_0 = \sqrt{\frac{\mu_0}{\varepsilon_0}} \approx 377\,\Omega \qquad \text{(Equation 3.30)}$$

The speed of light is related to the impedance of free space by the simple term $1/\mu_0$. Since 377 ≈ 120 π, $Z_0$ often appears as either term.

Notice the similarity to Ohm's Law, in that **H** roughly coincides with current and **E** roughly coincides with voltage.

Also, for any space occupied by matter rather than a vacuum, the values of μ and ε are somewhat larger. Thus the speed of light inside the material and a material's intrinsic impedance are calculated by the same simple equations. This will become an important concept in later chapters.

For most practical purposes, air's terms are sufficiently close to a pure vacuum to be ignored. The exceptions are almost always at frequencies above about 1 GHz, where the size of raindrops approach the operating wavelength, or at frequencies where air molecules (particularly oxygen) render the air as a partial absorption medium. The atmosphere can also affect the propagation of nearly all frequencies, but the antenna itself is seldom directly affected by its operation in air.

Above, we derived the impedance of free space from a derivation of Ohm's Law. We now have an electromagnetic specification for free space! Such a specification is necessary to quantify how we couple RF power into free space and how we capture RF power from free space. In effect, free space becomes part of the entire radio system "circuit." As such, this number is critical to all aspects of antenna design and performance.

At this point we simply accept that "space" indeed has physical characteristics, even when there is no matter present. To explore this non-intuitive reality, classical physics fails and the curious are forced into the non-intuitive and often bizarre world of modern physics. People often feel a need to understand fields and radiation. We cannot experience first-hand electric, magnetic, or even gravitational fields. We can only observe the effects they have upon matter. However, science allows us to *quantify* the actual physical characteristics of these non-sensual phenomena. The proof of our physical understanding is using this understanding to duplicate experiments and then actually utilize this physics by engineering radio equipment, including antennas.

## Fields, Power, and Energy

The *field strength* of an electromagnetic wave can be measured either by the electric field (volts/meter) or the magnetic field (amps/meter). Usually RF field strength is stated only in volts/meter (the electric field) since it is easier (technological limitations) to directly measure than the magnetic field. Notice that these terms are measured in single-dimensional space (meter). If we multiply the electric and magnetic fields, we are left with *watts/meter*$^2$, a two-dimensional (area) function reminiscent of the surface of a sphere and also defining a *power* term.

$$V/_m \; I/_m = \frac{W}{m^2} \qquad \text{(Equation 3.31)}$$

Above we see that the relationship between field strength and power is really a special calculation of the *power law*, but also involves a dimensional transformation. The *direct* calculation of power density from a field strength requires knowledge of the exact relationship between the field strength and power. For this we need a combination of the power law and Ohm's Law. The form we base this calculation upon is

$$\boldsymbol{S}_r = \frac{\boldsymbol{E}_r^{\,2}}{Z_0} \qquad \text{(Equation 3.32)}$$

In this case $Z_0$ is the impedance of free space ($120\pi\Omega$), **E** is the electric field strength, (volts/meter) and $\boldsymbol{S}_r$ is the power density (*watts/meter*$^2$) both at the point of measurement at distance *r* from the source, on the sphere where the measurement takes place. Again, this is simply a combination of the power law and Ohm's Law.

$$\boldsymbol{E}_r = \sqrt{\boldsymbol{S}_r Z_0} \qquad \text{(Equation 3.33)}$$

or,

$$\boldsymbol{E}_r = \sqrt{120\pi \boldsymbol{S}_r} \qquad \text{(Equation 3.34)}$$

For example, we can calculate the effective radiated power from an isotropic (EIRP) antenna in the direction of the measured point. If we can measure the field strength at a distance *r* ($\boldsymbol{E}_r$) we find the power density by $\boldsymbol{S}_r = \boldsymbol{E}_r^{\,2}/Z_0$, and then multiply $\boldsymbol{S}_r$ by the total area $4\pi r^2$. Thus the total transmitter power (EIRP) is:

$$P_{EIRP} = 4\pi r^2 S_r \qquad \text{(Equation 3.35)}$$

Thus we can calculate through these simple equations the relationships among power density, EIRP, field strength, and/or the impedance of free space.

We have to know that the transmitter is using an isotropic antenna to make the isotropic assumption. In any event, if we know the distance from the transmitter and the field strength at that distance we can directly calculate the EIRP at the point (also defining the direction from the transmitter) where we took the measurement despite the directivity of the antenna. Furthermore, if we also know the power of the transmitter, we can also calculate the gain of the transmit antenna, in dBi. So this is just an extension of the intuitive reasoning from Chapter 2 to now include fields into our arsenal of terms. Equation 3.31 applies to both peak and average power. "Average power" is used for "actual power" calculations. Peak power is simply taken from peak current and peak voltage in circuits and for power calculations from field strength (peak **E** and/or peak **B** fields). Average power is taken from the rms values of voltage and/or current values and in calculations for power from field strengths, the rms values of the field strength, *rms* = .707I

x 707V = ½ *peakW*, or

$$\frac{V^2 \text{rms}}{Z_0} = \frac{1}{2}\, peakW$$

Although the energy term is not often used in conjunction with antenna problems, we can complete our explanations by defining the term. *Energy* is defined as the time integral of power. For example, you are billed for electric energy, not electric power. Electric energy is defined in watt-hours, or more conveniently kW-hours. Electromagnetic energy can only exist as a radiating set of fields propagating at the speed of light. If we have a 100 W isotropic light bulb burning for 1000 hours in free space, then the total radiated electromagnetic energy will be 100 kWhr. This energy will be contained within a sphere with a radius of 1000 light-hours, at least for that particular instant in time. So electromagnetic *energy* is not only a time integral, but also a space integral. Where the **E** and **H** fields are measured in *m,* power is measured in $m^2$ and energy in $m^3$, in this case the *volume* of an expanding sphere

$$\frac{4}{3}\pi r^3$$

Such measurements are sometimes useful in astronomical research, such as finding residual energy from the *Big Bang* (Chapter 10).

## Radiation Resistance

We will encounter several terms in antenna physics whose very title can imply an incorrect intuitive understanding. The term *radiation resistance* is often a source of confusion. It is easy upon first encountering this term to interpret it as a *resistance to* radiation. In other words, the higher the resistance the more difficult it becomes to radiate. This is an understandable yet mistaken interpretation of the definition of radiation resistance. Unfortunately the confusion does not stop with the title. We introduced this term in Chapter 2. Now we will expand upon it, and specific applications will be presented in later chapters to further improve understanding.

If we place an RF voltage source on a conductor (antenna) and measure its impedance, we will see the real part of the impedance (a resistance), and, likely, a measureable reactance. This is the case even for a theoretical lossless conductor, perhaps constructed of a super-conducting material. How can a zero resistance conductor register some real value of impedance? The answer, of course, is that power is being "lost" to free space by radiation. The resistive value of an antenna at a specific frequency that is associated with radiated power is called the radiation resistance ($R_r$) of the antenna.

The formal definition of radiation resistance is very often misunderstood, even by professional radio engineers. $R_r$ is a critical term for calculating antenna efficiency which, of course, directly affects effective aperture, gain and overall antenna efficiency.

Antenna efficiency is defined as:

$$eff = \frac{R_r}{R_r + R_l} \quad \text{(Equation 3.36)}$$

where ($R_l$) is loss resistance, or resistance resulting in RF power conversion into heat. This is a simple, straightforward equation, but the determination of $R_r$ can be complicated and therefore often mistaken. It is obvious that you must use the correct antenna $R_r$ value to get a proper calculation of efficiency. But how do you determine the correct $R_r$ value? In Equation 3.36, how can you determine the difference between $R_r$ and $R_l$?

In the most basic model, $R_r$, being a resistive value, may be calculated through the power law where

$$R_r = \frac{P_r}{I^2} \quad \text{(Equation 3.37)}$$

where $P_r$ is the radiated power and $I$ is the current at a point along the antenna. However, this only provides the $R_r$ *at that* point on the antenna, not the antenna as a whole, which is necessary for a proper calculation of efficiency. The first step toward a proper definition of $R_r$ is to measure $I$ at a current maximum point along the antenna. Second, to calculate $P_r$ it is necessary to integrate (sum) all the power being radiated through the sphere. The process is relatively easy if the antenna and wavelength are small. In such cases the far field Poynting vector may be integrated inside an anechoic chamber. For antennas and wavelengths too large for practical chamber sizes, power integration is exceedingly difficult, especially when the antenna is mounted close to the ground or other material that causes power loss.

On the other hand, $I$ is a straightforward measurement at the point of current maximum. Thus this equation has three variables, two of which are very difficult to directly measure. Consequently it is difficult to set up a closed form equation that correctly defines $R_r$ using Equation 3.37. Furthermore, this equation does not explain other factors that shape the value of $R_r$. Therefore it cannot explain *why* $R_r$ can change, even if $P_r$ remains constant but $I$ changes. To resolve this problem, Kraus gives a more refined definition for the general definition (Equation 3.38) where an antenna's all-important 3 dB beamwidth (gain) is included in the definition. The idea is that the total power radiated will usually

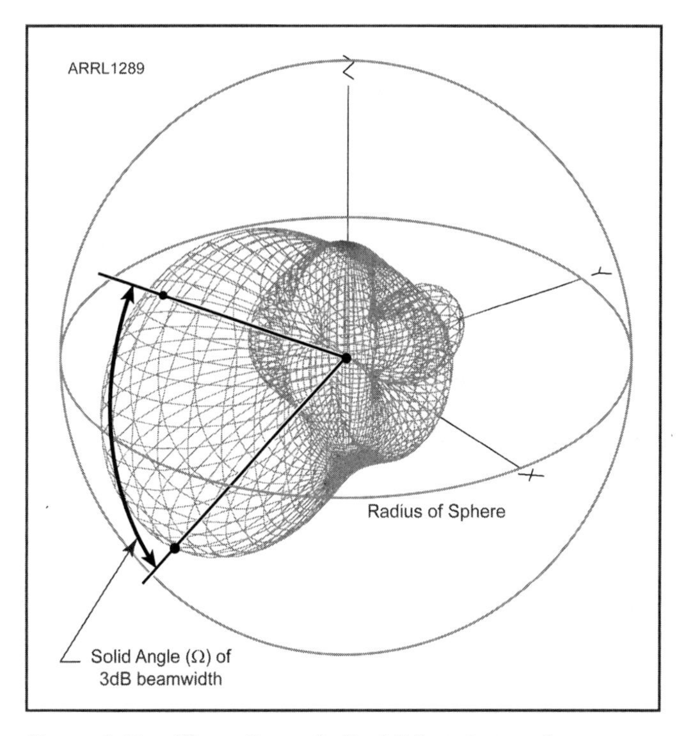


**Figure 3.11 — The pattern of a Yagi-Uda antenna showing the maximum lobe. The area that is within 3 dB of the maximum gain point defines the solid angle (steradian) Ω for that antenna.**

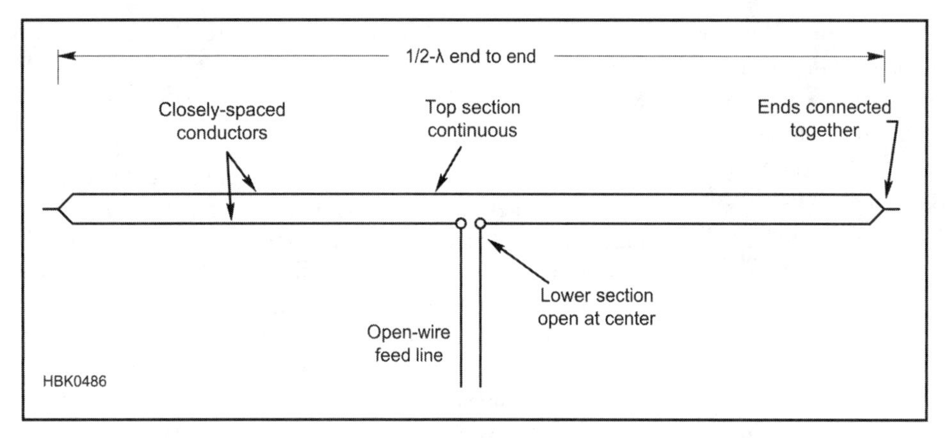


**Figure 3.12 — The feed point impedance of a center-fed ½ λ folded dipole is 4× the radiation resistance because the current at the feed point is ½ the total current at the equivalent point of the feed point. $R_r$ is identical to a single wire dipole, or about 73 Ω.**

entire sphere (average power) will be ¹⁄₁₀ the peak radiated power on the sphere.

$I^2$ is the RF current squared appearing on the antenna element at a *current maximum*, as detailed below. Thus, in the general Equation 3.38 above, $R_r$ depends upon the steradian (**Ω**) beamwidth of the maximum radiated power, which also directly relates to gain and thus aperture as well. Therefore $R_r$ is a function of antenna current, free-space impedance, power radiated and antenna gain. We will later develop this general equation for $R_r$ for specific antenna types. In many cases, $R_r$ is the critical term for designing and building efficient antennas. Kraus specifies the general definition of radiation resistance (Equation 3.38) because it represents a far more utilitarian definition than Equation 3.37. Equation 3.37 cannot inform if the terms are being influenced by antenna gain and/or loss, whereas Equation 3.38 isolates these two terms.

If we know the total power radiated from a *lossless* antenna and its current maximum, we can directly calculate $R_r$ by Equation 3.37. However, care must be taken in the computation of this current value. For example, if we feed a folded dipole (**Figure 3.12**) at a current maximum, we are only measuring ½ the current flowing at that "point" (compared to the wavelength the "folded element" is essentially at the same points along the antenna as the fed element). This will give a false $R_r$ value of about 292 Ω rather than the correct $R_r$ for a ½ λ dipole in free space of about 73 Ω. Therefore, again, $R_r$ *is not necessarily equal to a power law calculation taken at the antenna's feed point.* For any arbitrary point on the antenna, the radiation resistance at that point is $R_{rp}$.

closely approximate the solid angle that is defined by the 3 dB beamwidth. In effect, the assumption is that the power remaining outside the 3 dB beamwidth will "fill in" the solid angle, as shown in Figure 2.8 in Chapter 2 instead of the "actual" pattern shown in **Figure 3.11**.

$$R_r = \frac{S(\theta,\phi)_{max} r^2 \Omega_a}{I^2} \quad \text{(Equation 3.38)}$$

Equations 3.38 and 3.37 are approximately equal. The *(max)* term in $S(\theta,\phi)_{max}$ designates a specific area on the sphere: the steradian value ($\Omega_a$) of the 3 dB beamwidth of the maximum gain lobe. In other words, $S(\theta,\phi)_{max}$ is the area on the sphere that contains the main lobe of the antenna pattern. When the antenna pattern is almost isotropic (nulls are less than 3 dB difference than the peak), then **Ω** is the same as an isotropic or the 4π solid angle defining the entire sphere's surface. If the gain pattern of the antenna is a distinct lobe (like the pattern from a Yagi-Uda antenna), **Ω** would be defined as the solid angle containing the maximum gain point and surrounding area within 3 dB of the maximum gain point.

On the other hand, Equation 3.38 is precisely equal to Equation 3.37 when the steradian contains precisely all radiated power as in Figure 2.8 (in contrast to Figure 3.11). For example, let us assume that the main lobe has a maximum gain of 10 dB, which is also a linear power gain of 10. However, this maximum gain location typically defines a *point* on the sphere. Around this point we must define an area that is defined by the steradian value. Let us assume that this area corresponds with the gain number of the steradian value, or ¹⁄₁₀ the area of the sphere, or

$$\Omega_a = \frac{4\pi}{10}$$

This model assumes that *all* the radiated power is contained in that steradian constraint and that the area on the sphere defined by that steradian and that the power density is uniform over that area with zero power appearing over the balance of the sphere. Therefore the power integration over the

## Relationship Between Radiation Resistance and Feed Point Impedance in Linear Antenna Structures

The real portion of the feed point impedance equals the radiation resistance plus losses of the antenna *only* for single-conductor antennas fed at a current maximum along the antenna. It is a *very* common mistake to automatically equate the feed point impedance (real value) to the antenna radiation resistance measured at a current maximum or worse, elsewhere along the antenna.

$R_r$ will coincide with the real part of the feed point impedance at the center of an antenna that is less than or equal to ½ electrical wavelength long, or an antenna that is an odd number times ½ wavelength. For all other antenna types, great care should be used when calculating or measuring radiation resistance, since simply using power input and feed point

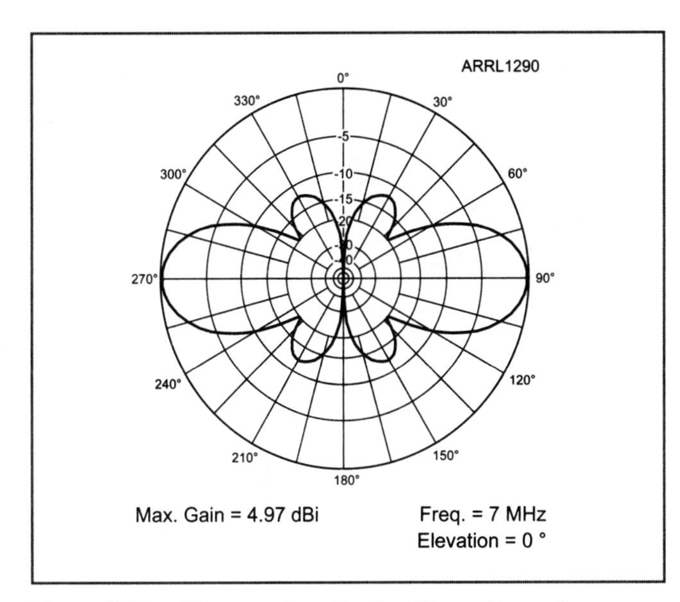


**Figure 3.13 — Here we show the familiar pattern of an extended double Zepp antenna in free space fed at the center. The total length of the antenna is 10/8 wavelength and is center fed.**

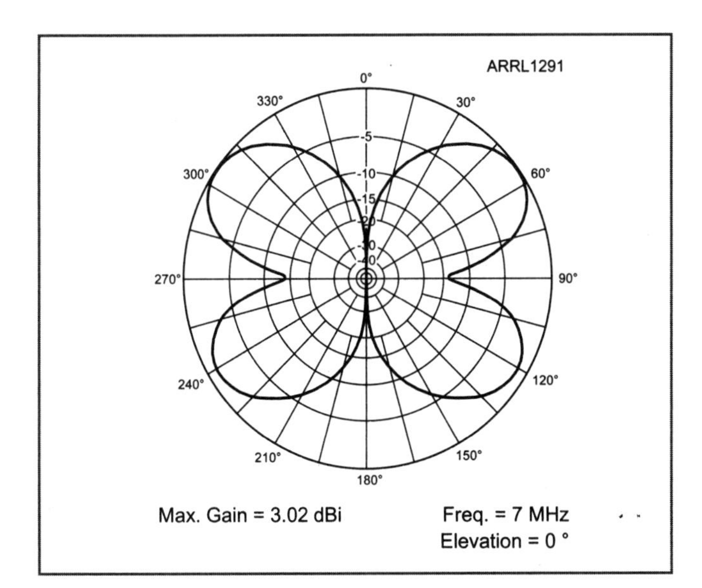


**Figure 3.14 — This is the same antenna as above. By moving the feed point off-center (¼ the distance from one end rather than ½ the distance from one end) we discover a different feed point impedance, a different pattern, and $R_r$ has also changed. Notice the very different pattern due to a different distribution of current along the antenna.**

current does not necessarily yield $R_r$. Indeed, the difference between the feed point impedance $R_{rp}$ and $R_r$ can be very large. For example, a single wire ½ λ dipole in free space has the same value of $R_r$ no matter where the feed is connected. The *only* point where the two values coincide is at the center of the dipole, which is also the current maximum. $R_{rp} = R_r$ only at a current maximum on the antenna. $R_{rp}$ is usually associated with a feed point that is not at a current maximum.

The resonant center-fed ½ λ dipole has an input impedance of about 73 Ω, which is also the $R_r$ of the antenna. Now if we move the feed point impedance to 25% the distance from one end (off center), the feed point impedance at $R_{rp}$ is now about 137 Ω, but the antenna $R_r$ remains 73 Ω.

Another complexity: In antennas longer than ½ wavelength the current distribution (and thus the radiation resistance) can be changed by changing the feed point location. So, when calculating and/or measuring the location(s) of current maximum(s) along an antenna element, be careful that key terms that define $R_r$ are not changed by changing the feed point. Thus if the current is to be measured at a current maximum by moving the feed point, the current maximum may be moved as a result of moving the feed point! See **Figures 3.13** and **3.14**.

There are few applications where $R_r$ becomes so critical as it does in ground-mounted vertical antennas and antennas that are very small compared to the wavelength of operation. This will be treated with great detail in later chapters. For any experimenter working with antennas generally and these types of antennas in particular, it is essential to master the proper definition or radiation resistance.

Now let us return to Equation 3.38. Assume that the 3 dB three-dimensional beamwidth (equivalent area) is

$$\frac{4\pi\Omega}{8}$$

In other words, the area is ⅛ the total surface area of the sphere. Since we know our steradian value, we also know we have a power gain of about 8, or about 9 dBi. We carefully measure our feed point current as 1 A.

$$S(\theta,\phi)_{max} r^2 \Omega_a = \left(\frac{\pi}{2}\Omega\right)$$

results in 200 W EIRP with a gain of 8. Thus we can calculate

$$R_r = \frac{200\Omega\left(\frac{1}{8}\right)}{1^2} = 25\ \Omega \qquad \text{(Equation 3.39)}$$

Thus $R_r$ is inversely proportional to the antenna gain and therefore also approximately proportional to the 3 dB beamwidth. Notice that we have isolated and precisely defined these terms by including antenna gain (or aperture) in the equation. Measured deviations from these predicted values indicate loss and/or mistakes in the measurement. As such, it is a far more useful general definition than Equation 3.37. Regarding ground-mounted vertical antennas, a yet more special form is presented later.

As a final thought, it is interesting to note that if we extrapolate the characteristics of a ½ λ dipole with infinitely thin wires in free space (gain and $R_r$) to a hypothetical isotropic antenna, we would find the $R_r$ = 119.9 Ω, or exactly $Z_0/\pi$, where $Z_0$ is the impedance of free space. To derive this number for the isotropic, we simply adjust the gain from 2.15 dBi to 0 dBi and the consequence of this adjustment to the steradian value. Once again we see an example of the beautiful symmetry of antenna physics, particularly when we use proper terms that lead directly to closed-form equations.

## Antenna Pattern Formation

The accelerating charges along the length of an antenna element represent the source of the radiated wave. But what do the resulting patterns look like? Thus far we have shown several antenna patterns. In Chapter 1 we presented some specific antennas (isotropic and a ½ λ dipole) and their respective patterns. We also just showed the pattern of the 10/8 λ linear antenna fed at two different points. The question is: What is the mechanism for producing these patterns?

We begin by introducing an imaginary antenna *segment* called a *short dipole.* A short dipole can be thought to have some amount of accelerating charge or instantaneously changing current traveling along in a very short length of the antenna. Indeed, with the help of integral calculus we can make the length essentially zero to create an *infinitesimal dipole* (see Appendix A). The short dipole, although very short, is also assumed to be straight. In other words, the charges on the dipole can only flow in two directions, along the length of the straight segment. We know that an accelerating charge produces a radiated electric field parallel to direction of the acceleration and a radiated magnetic field in an orthogonal direction, and the direction of radiation is orthogonal to both fields. Indeed, we can calculate the pattern of such a short dipole.

If we place many short dipoles in series, we can model a "real" antenna at an arbitrary length.

## PADL

The *phase and amplitude* of these accelerating charges from each of these short dipoles (antenna segments) contribute to the *total* pattern of the entire antenna in the far field. There are also two other critical parameters that define the resulting far field: the *direction* of the accelerating charges and the 3D *location* of these accelerating charges. Thus we need to know four variables to determine all the parameters of an antenna's far field: *Phase*, *Amplitude*, *Direction*, and *Location* — *PADL.* The relationships of the electric field, magnetic field, and direction of propagation become deterministic depending upon the PADL variables.

Returning to the infinitesimal dipole antenna, the phase is simply the point along the RF wave defining that phase point (0-2π radians) along the RF sine wave. The amplitude is the instantaneous RF current amplitude. The instantaneous direction of the current is always the same (defined by a point on the antenna element) changing twice per RF cycle and thus reversing the polarity twice per cycle. There are only two directions the current can accelerate on a linear element: left or right (as viewed from the side). The location of the infinitesimal dipole is a point in free space. All calculations for any antenna in free space simply sum the PADL terms (using integral calculus) to form a far field pattern. But first we need to determine the pattern of an infinitesimal dipole so we will know what to sum in more complex antenna structures.

**Table 3.1**
**Directivity and Gain of a Dipole**

| *Antenna* | *Directivity* | *dBi gain** |
|---|---|---|
| Infinitesimal dipole | 1.50 | 1.73 dBi |
| ½ λ dipole | 1.64 | 2.15 dBi |

*assuming 100% efficiency

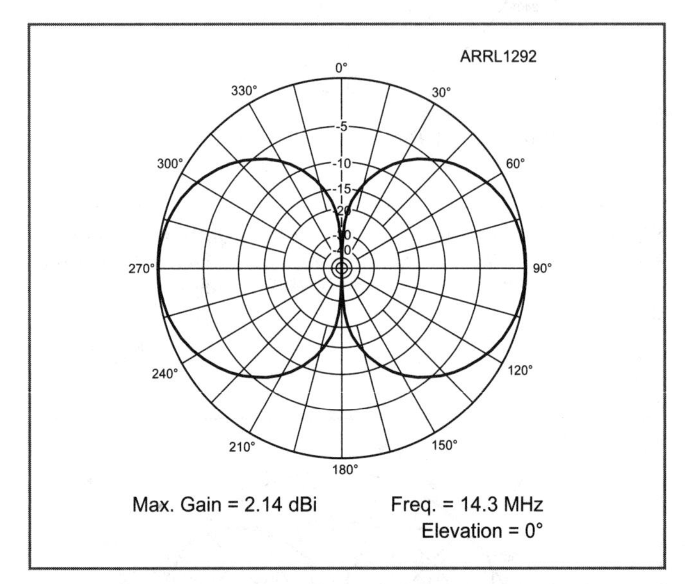


**Figure 3.15 — Free-space doughnut slice of a ½ λ dipole in free space showing the 2.15 dBi maximum gain. The dipole is vertically oriented at the circle's center.**

Ignoring the integral calculus to derive this important term, we find that the linear (power) pattern of the infinitesimal dipole follows a curve defined by a $sin^2$ function. The pattern is a doughnut pattern very similar to the ½ λ dipole, but has less directivity. The directivity of an isotropic antenna is 1. Therefore the dBi directivity of an antenna is simply:

$$dBi_a = 10 log D_a \quad \text{(Equation 3.40)}$$

where again,

$$Gain = Directivity \times Efficiency \ (expressed\ as\ fraction) \quad \text{(Equation 3.41)}$$

See **Table 3.1** and **Figures 3.15** to **3.17**.

So, as we build a full-size ½ λ dipole from many short dipole segments, we find we increase the directivity by about 0.42 dB. The amplitude of the current along a ½ λ dipole is sinusoidal, and the amplitude of the current in the short dipole is assumed to be constant along its length (a small portion of a sine wave). Therefore, the amplitudes of each of the long series of short dipoles will have to be adjusted to the sinusoidal value at the same point on the ½ λ dipole

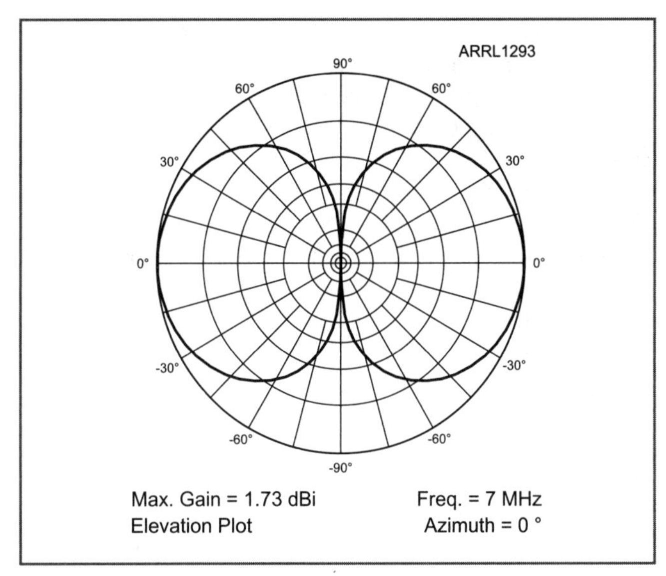


**Figure 3.16 — Free-space doughnut slice of a very short dipole at 7 MHz. The dipole is vertically oriented at the center of the circle. Also the direction of antenna current is flowing in a vertical direction (oscillating up and down). Therefore, the direction of each segment of current is always perpendicular (orthogonal) to the maximum directivity (conforming to Maxwell's equations). As the length is reduced to an infinitesimal value, the directivity remains essentially the same. The linear curve (not shown by this dB scale) of this pattern follows a *sin²* function.**

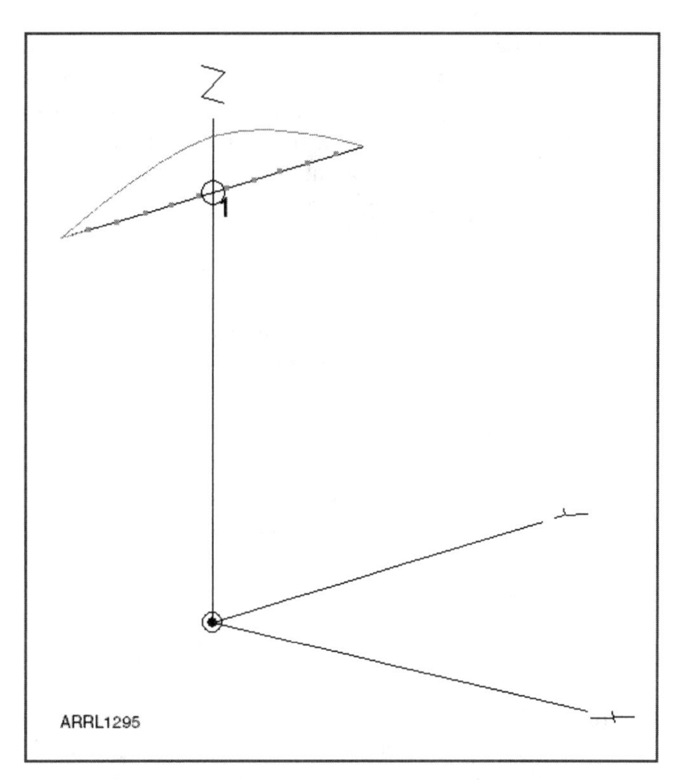


**Figure 3.18 — Here is a typical *EZNEC* free-space model of a ½ λ dipole. Notice the sinusoidal current distribution shown along the antenna (a very nice feature) as well as the dots along the antenna showing segmentation using, in this case, 11 short dipoles in series. *EZNEC* uses the 11 PADL values to calculate the resulting far field.**

**Figure 3.17 — This figure plots the broadside directivity of a center fed linear antenna as a function of the overall length of the antenna. The directivity remains essentially constant from an infinitesimal length (<< ½ λ) to a length of approximately ⅛ λ. The broadside directivity peaks at a length of 1¼ λ. This length center fed antenna is called an extended double Zepp. As the antenna is made longer, its broadside gain quickly reduces and the pattern becomes a "four leaf clover" shape as in Figure 3.13.**

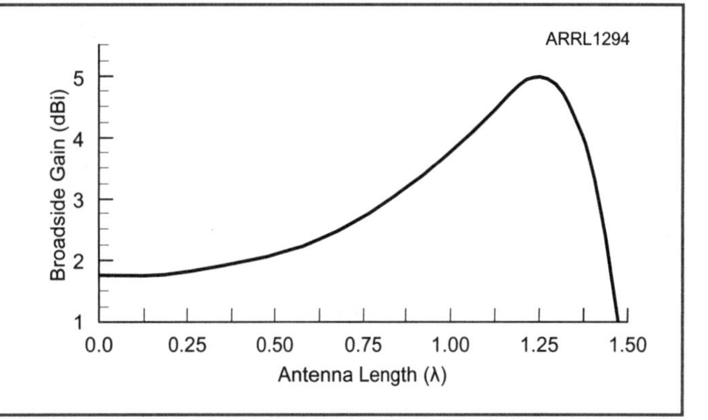


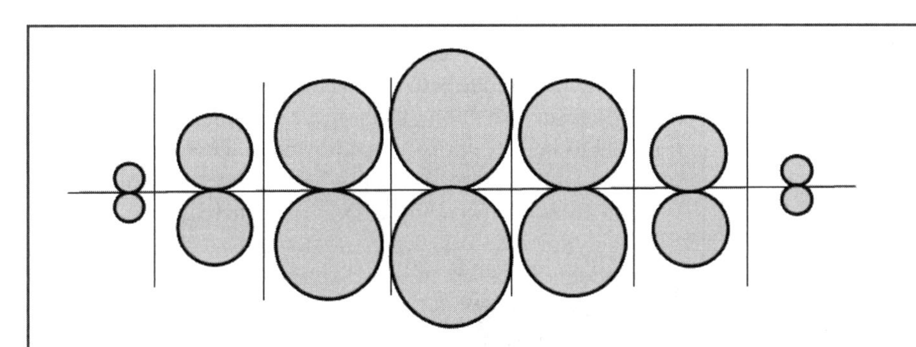

**Figure 3.19 — This is a conceptual diagram of a seven segment ½ λ dipole antenna with the size of the segment doughnut patterns shown to approximate a sinusoid distribution of current along the antenna.**

in order for the definite integral to provide the correct result. For example, the current amplitude value will be maximum at the segment at the center of the ½ λ dipole. As we move away from the center, the segments' current amplitudes will decrease as a sin function, and at the ends of the dipole, the segments will have close to a zero value.

The concept of the short dipole segmentation of larger antennas is very convenient for direct calculation of patterns by hand (using calculus) or by advanced mathematical modeling tools such as *MATLAB*. Antenna modeling tools such as *EZNEC* also use segmentation for equivalent calculations and as shown in **Figure 3.18**.

Although finite-length segments used in antenna modeling, by definition, offer only an approximation, surprisingly few segments offer results very close to results obtained through an exact solution from integral calculus. Thus we begin with a formal definition using calculus, but then discover that adequate accuracy can be derived from much simpler summing techniques.

The total antenna pattern is simply the sum of these smaller patterns at a far distance from the antenna (shown in **Figure 3.19**). The doughnuts also contain phase information that can lead to either field reinforcement or cancellation in the far field. Therefore, the complete antenna pattern is simply a smoothed sum of the PADL parameters. Of course, the resulting fields from each of the segmented currents are assumed to propagate at the speed of light. All antenna patterns are thus the result of the current distribution in amplitude and phase at all points on the antenna and the position of those points in 3D space. In the case of a ½ λ dipole, none of the doughnuts are far enough apart to provide complete cancellation. As the length is made longer than 10/8 wavelength, the phase shift along the antenna is enough to rapidly increase nulling in the broadside direction.

## Current and Voltage Distribution

The previous section emphasizes the primary importance of current amplitude and phase along an antenna element for producing the antenna's directivity. The *mechanism* that results in the current distribution is key to antenna performance.

It is important to think of antenna "voltage" and "current" in terms of RMS values. The *instantaneous* values are constantly changing in accordance with forward and reflected waves traveling up and down the elements. For this discussion we will only consider the RMS values.

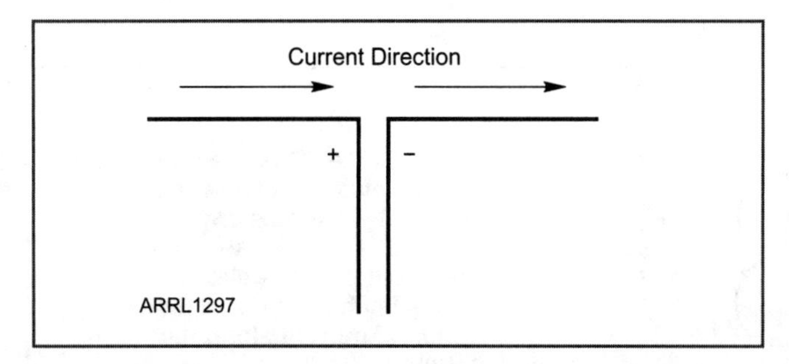


**Figure 3.20 — This figure shows the instantaneous direction of current flow. The direction changes twice for each complete cycle of the RF signal.**

Again, if we consider a simple center-fed ½ λ dipole, current will always flow in the same direction on both sides of the dipole. We assume that the transmission line is applying equal but out-of-phase current to the antenna's center. Since the wires point from the feed point in opposite directions, the instantaneous currents run in the same direction (**Figure 3.20**).

When the accelerating charges reach the ends of the dipole, they are reflected and start back toward the center. Since there is an equal amount of charge flowing "forward" and "backward," there is no RMS current at the end of the antenna and thus no radiation from the end of the antenna. This is the mechanism that results in the sinusoid distribution of current along the dipole. It also explains why the center of a ½ λ dipole is the point of maximum current and thus contributes the largest amount of radiated power compared to other antenna segments.

Measurement of RF current is relatively easy. We can simply place a sample loop around the element and measure the induced voltage (Faraday's Law). If we move the sampling loop up and down the element we discover the current distribution shown in Figure 3.21, which is sinusoidal. In a practical linear antenna, the current and voltage values are not quire sinusoidal due to a finite diameter of the linear elements. However, for our purposes, we will assume a sinusoidal distribution.

Voltage is represented as "force" per charge (coulomb). As a charge is accelerated down a linear antenna element, it eventually comes to the end of the antenna element. Since the charge cannot continue flowing, the density increases and this creates a high voltage point. An analogy is the displacement current, which is simply charging and discharging a capacitor. If the "end" of the antenna is thought to be a capacitor (which it is!) then accelerating charges stop, before being reflected at the antenna end, and create a oscillating charge in the capacitor where $V_c = q/C$, where $V_c$ is the voltage a the antenna element's end, $q$ is the total charge (large) and $C$ is the capacitance value (small) at the antenna element's end. By this equation, assuming the end of a thin dipole represents a small capacitance and the amount of charge can be very large, it becomes apparent that very high voltages will exist even for modest transmit power.

When a charge exists, but there is no current, there *must* be a capacitance involved. The instantaneous voltage at the end of the dipole increases while current is flowing toward that end, and reduces to near zero when the current flows in the opposite direction. Thus, if we consider the ends of the dipole to be capacitors, the voltages on these two capacitors charge and discharge in an out-of-phase relationship. As we move away from an end, RMS current increases and thus the voltage decreases. The current flowing on a dipole can be considered to be Maxwell's displacement current as it never completes a "loop" as in Gauss's original equation.

Not surprisingly the center feed point resistance (in this

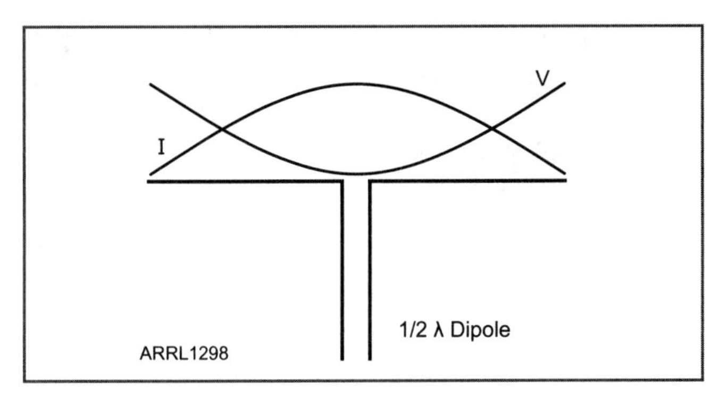


**Figure 3.21 — The relative current and voltage values along a ½ λ dipole.**

case also very close to $R_r$) is simply

$$R_r = \frac{V_f}{I_f}$$

where *f* indicates the feed point.

A general assumption is that the maximum voltage appearing on an antenna's element appears at its end point. This assumption is helpful in developing an intuitive understanding of voltage and thus current distribution along a linear antenna element. The sinusoid distribution of antenna *I* and *V* are usually 180 degrees out-of-phase along a linear antenna element (**Figure 3.21**). Thus you can estimate voltage and current peaks and nulls by working back from and antenna's end point (if there is one!) following a sinusoidal distribution at the assumed RF wavelength.

## Antenna Polarization

Thus far we have assumed that our antenna is linear, an antenna element in a straight line. Such an antenna is said to have *linear polarization.* In a linear polarized antenna, the current can only flow along that line, either one way or the other. The resulting radiated fields' directions, according to Maxwell's equations are exactly defined: The electric field (**E**) is *parallel* to the direction of accelerating charge (parallel to the wire). The magnetic field (**H**) is perpendicular to the line of the wire. See **Figure 3.22**. In "outer" free space a linear polarized wave's *direction* is arbitrary because there is no reference for a line in space. Near or on the Earth, the Earth's surface defines a plane. With terrestrial reference, on Earth if the E field is vertical it is said to have *vertical polarization.* If the **E** field is horizontal, it is said to have *horizontal polarization.*

The other type of polarization is *circular polarization (CP).* If we take two dipoles, one horizontal and one vertical, and feed them 90 degrees out-of-phase shown in **Figure 3.23**, the amplitude of the *instantaneous* radiated fields will follow the sinusoid acceleration of the current on the elements. Since

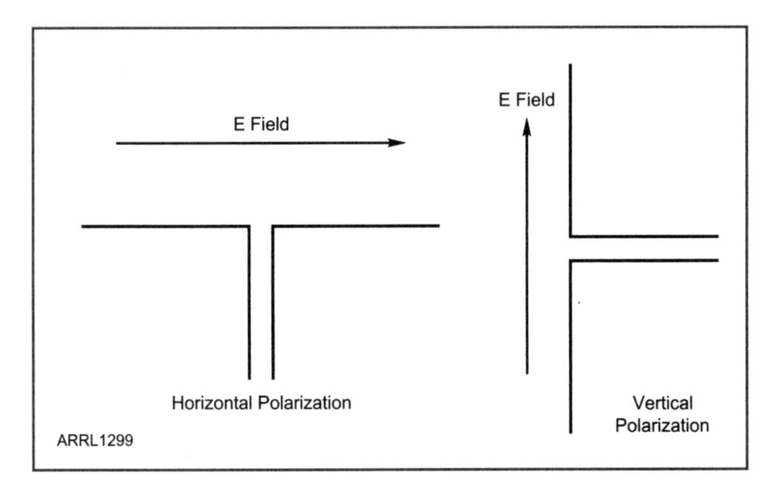


**Figure 3.22 — Presuming that the reader is sitting or standing upright on the Earth, we can see the difference between horizontal and vertical polarization.**

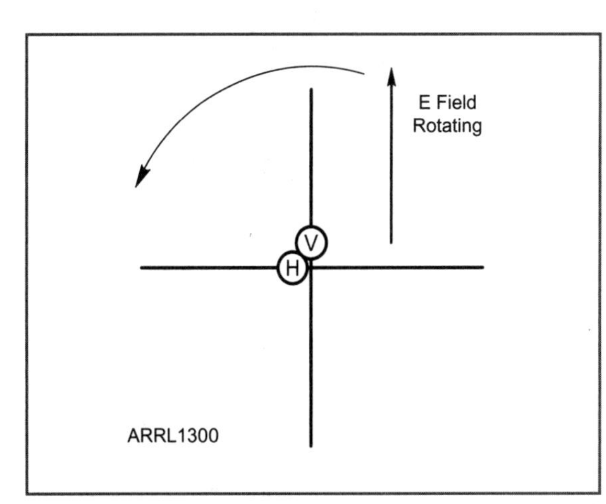


**Figure 3.23 — Two crossed dipoles fed in quadrature (90 degrees out-of-phase) will produce a circular polarized wave.**

## The Poincaré Sphere

Wave polarization is often defined as a point on a *Poincaré Sphere*. This spherical presentation was introduced in 1892 by French mathematician Henri Poincaré as a conceptual tool for defining any particular form of polarization. At the time there was considerable work being performed on the nature of light and the implications Maxwell's equations were having upon these studies. Since RF waves are also electromagnetic waves, the Poincaré Sphere was a perfect means of defining polarization of RF waves as well. *Any* possible form of polarization can be defined as a simple point on the sphere described as a "latitude and longitude".

See **Figure 3.24**. The "north pole" signifies a perfect right-hand circular polarization while the south pole, the left-hand polarization. The "equator" designates linear polarization from "horizontal" where horizontal polarization is at (0°, 0°), vertical polarization at (0°, 180°), negative latitude indicates a sloped linear polarization (/) and positive latitude (\) from 0-90 degrees on both sides. Between the poles and the equator are all possible elliptical polarizations, which naturally form linear changes from pure linear to pure circular as a function of latitude and "tilt" of the ellipses corresponding to their longitude.

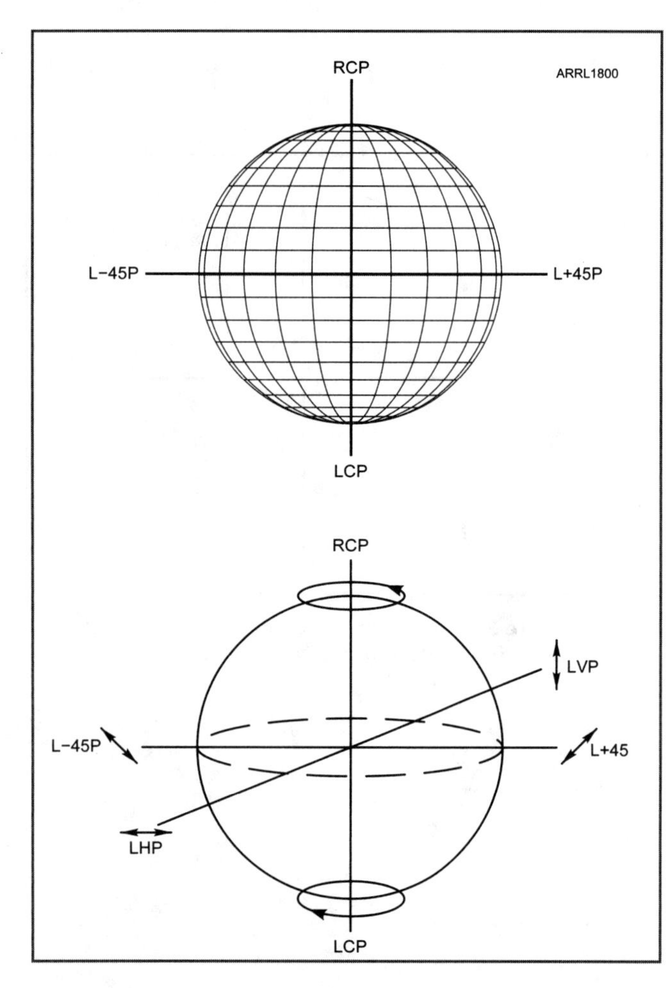


**Figure 3.24 — The Poincaré Sphere.**

## Radiating and Non-Radiating Fields

The first two of Maxwell's equations define the first critical principles of the Electric and Magnetic fields. The first and simplest form of a field is the static **E** field defined by Maxwell's first equation:

$$\oint \boldsymbol{E} \bullet d\boldsymbol{A} = \frac{q}{\varepsilon_0}$$

Notice that the electric field integrated over a closed surface area is proportional to the electrical charge inside that surface, with the dielectric constant (in this case free space) as a coefficient. This analysis of this simplest form of static fields is referred to as *electrostatics.* As discussed before, an **E** field may exist from a static charge, but a **B** field can only be produced by a moving charge, even if at an atomic level, such as a permanent magnet. In effect, if you perform detailed analysis of the static electrical qualities of a charged comb, you are engaged in electrostatics.

According to Maxwell's third equation, to produce a **B** field, we need charges that are moving *dq/dt* (electrical current), where

$$\oint \boldsymbol{B} \bullet dl = \mu_0 I$$

The study of this situation is *magnetostatics.* If the net charge inside a closed surface is non-zero, then the first equation holds, and an **E** field is also produced. If the currents sourcing **E** and **B** fields are constant but result from a non-accelerating current (dc), static fields are produced. Figure 3.3 shows the creation of a static **B** from a constant current. In effect, if you perform detailed analysis of a fixed bar magnet or a solenoid with a dc current creating the equivalent of a bar magnet, you are engaged in magnetostatics.

The current may also be alternating current (ac) including RF current. In such a case we are now involved with *electrodynamics* which involve all four of Maxwell's equations. Notice the similarity of terms (*static and dynamic)* which are also used to define the first two sub-divisions of mechanical engineering. When **E** and **B** fields are produced by an RF current, radiation can only occur when a $R_r$ exists where the current is flowing. In effect, there must be sufficient length, area or volume for the current to travel (as a function of wavelength) for the resulting fields to detach themselves from the current source(s) and thus become radiated fields. If $R_r$ is very low, as in a typical lumped inductor or capacitor, then the energy in the fields in and around these components will be "trapped" by their current sources. On a 1-dimentional conductor (such as a wire or tube) the critical figure of merit for this "length" is called the *effective height or effective length.* Effective height $h_e$ is often used since its traditional application has historically been in the analysis of ground mounted vertical antenna. Therefore, a lengthy description of $h_e$*'s* relationship with $R_r$ is presented in Chapter 6. However, for this general discussion, we only need to understand the above concept that the more "space" (1D, 2D or 3D) an RF current has to spread itself, the more effective will be the

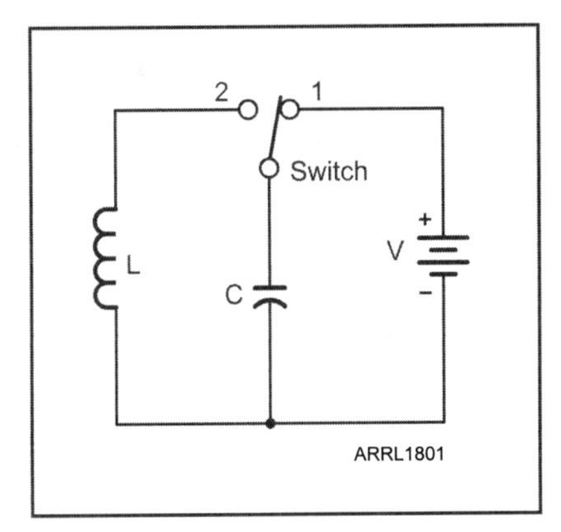


Figure 3.25 — In the initial condition, we charge up a capacitor with a battery. The capacitor stores its energy as an E field. When we throw the switch to form a parallel LC tank, the capacitor discharges itself as a current through the inductor. For this moment, the E field disappears, and a B field builds up on the inductor. Then the B field collapses with a reverse current again charging the capacitor, but with the field lines reversed. This process continues until resistive ($R_l$) and/or radiation ($R_r$) losses deplete the charge. See text.

*radiation* of waves. The "detachment" of the **E** field from its current source is shown in Figure 3.10. Notice a finite amount of length is necessary for the field to detach itself from the sinusoidal varying flow of charges (an RF current).

See **Figure 3.25**. In the initial condition, we charge up a capacitor with a battery. The capacitor stores its energy as an **E** field. When we throw the switch to form a parallel LC tank, the capacitor discharges itself as a current through the inductor. For this moment, the **E** field disappears and a **B** field builds up on the inductor. Then the **B** field collapses with a reverse current again charging the capacitor, but with the field lines reversed. This process continues until resistive ($R_l$) and/or radiation ($R_r$) losses deplete the charge. Notice that this circuit has become a passive LC oscillator with a resonant frequency of

$$2\pi f = \frac{1}{\sqrt{LC}} \quad \text{(Equation 3.42)}$$

If there are no losses present, this oscillation will continue unabated. An **E** field trapped inside a capacitor has potential energy and the impedance of the oscillating field is completely reactive (capacitive reactance). The oscillating current on the inductor creates an oscillating **B** field which contains kinetic energy and the impedance of the field is reactive (inductive reactance).

Notice that the electric and magnetic fields remain completely separated (in reality there is always some electric field on an inductor and some magnetic field on a capacitor). In contrast, in a radiated wave composed of both fields, the **E** and **B** fields are intrinsically related through Maxwell's equations. In a *radiated wave* $\mathbf{W} = \mathbf{E} \times \mathbf{B}$ where **W** has no reactive value, the power in an electromagnetic wave is real power.

## Pendulum Analogy

A physical analogy to the oscillation of fields in an LC circuit is the pendulum (**Figure 3.26**), where

$$2\pi f = \sqrt{\frac{g}{L}} \quad \text{(Equation 3.43)}$$

where L is the length of the pendulum in same units as g, the pendulum mass is a point mass at the end of the pendulum, g is the gravitational acceleration 32 $ft^{-2}$ or 9.84 $m^{-2}$. This is without any air or frictional resistance. Notice the mathematic similarity of the LC circuit and the pendulum resonant frequencies, where the length of the pendulum corresponds to the inductance and the inverse of gravitational acceleration corresponds to capacitance. The resonant frequency of the pendulum is *independent* of the mass.

## Field Regions

The types of fields existing around a radiating antenna change gradually as you move the measurement point closer to or farther from the antenna. Close to the antenna there is

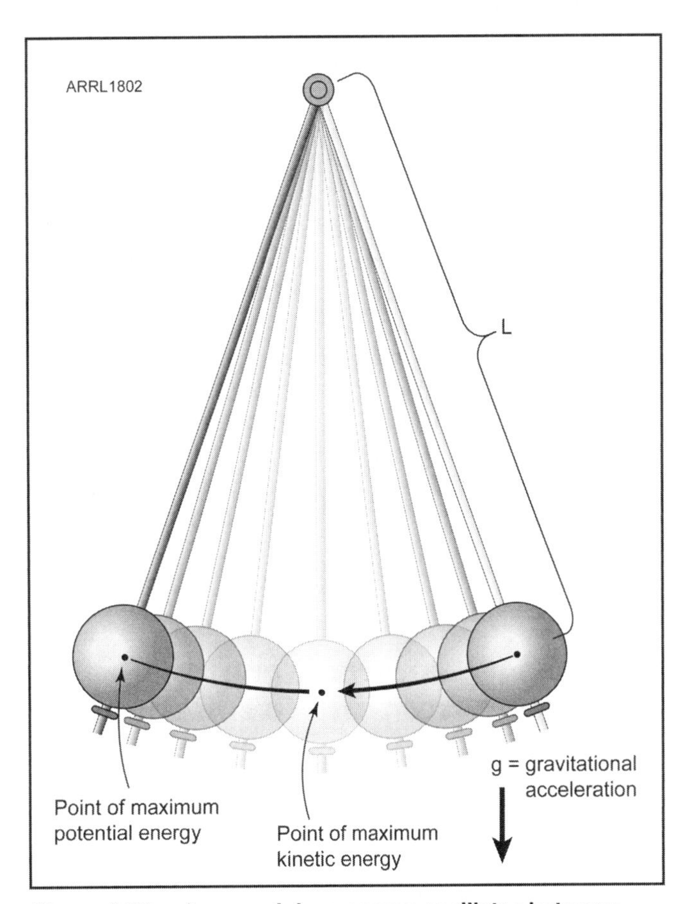


Figure 3.26 — In a pendulum, energy oscillates between potential (at the mass's highest point and zero velocity) and kinetic (at the mass's lowest point and highest velocity). This is analogous to the LC tank oscillator where energy oscillates between potential energy (capacitor filly charged, maximum E) and kinetic (current maximum through inductor maximum B).

usually a complex mixture of "induction" fields and radiated fields. The radiated fields can be considered to be originating at an infinitesimal scale.

Maxwell's equations show that all changing fields **E** and **B** will produce the other field. In an induction field the fields are 90 degrees out-of-phase, or reactive. In a *wave*, the fields are either in phase or 180 degrees out-of-phase, i.e. a reflected guided TEM wave. The simple power factor equation: $PF = \cos\theta$ where $\theta$ is the phase difference between the voltage and current also holds for the phase relationship between oscillating fields where PF of 1 is "real" power and a power factor of 0 is imaginary power.

The geometry of the "antenna" along with the wavelength determine the free space $R_r$ of the antenna, and thus how "easily" radiation fields are set up. Induction fields *store* energy in the near field, while *radiation* fields *dissipate* energy into free space through radiation. This explains how an antenna with no losses will radiate all the energy applied to it, but also be storing energy. When radiation fields are allowed to form, Maxwell's third and fourth equations describe the formation of an electromagnetic wave complete with its Poynting vector.

Thus the "near" field can be defined as actually two separate "regions"; one containing the induction fields and the other radiated fields that have not yet formed the final "settled" far field. Only as they radiate away from the antenna do the radiation fields sum and eventually form the far field directivity and gain of the antenna. The summing is usually calculated by integrating the field intensity of either the magnetic or electric field over the sphere surrounding the antenna (as in Chapter 2). Only one field integration is necessary to calculate power much like you can calculate power dissipated in a resistor by using $I^2R$ or $V^2/R$ where the current term is a rough equivalent to the magnetic field calculation and the voltage term analogous to the electric field term and R being analogous to the impedance of free space.

The electrical and magnetic induction fields lose their field strengths at a $1/d^2$ rate, and the radiated fields both attenuate at a $1/d$ rate. This is why as you measure farther from the antenna the induction fields lose their strength much faster than the radiated fields, until only the radiated fields remain.

Thus, we have defined three "regions" surrounding the antenna. See **Figure 3.27**

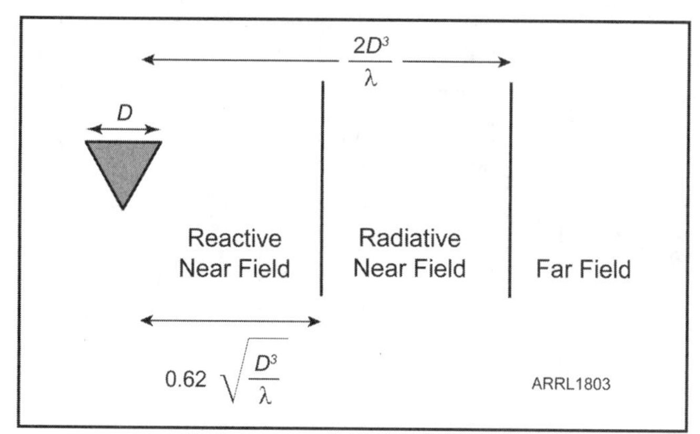


**Figure 3.27 — The Near Fields and the Far field.**

## Summing Electromagnetic Fields and Power from and to Antennas

A common source of confusion when considering radiating and receiving radio waves is the relationship between how power (**P**) and the fields associated with the wave (here we will only deal with the electric field (**E**)) behave. Here we explain how waves re-enforce and cancel and offer several examples for clarification.

The characteristics of electromagnetic waves are precisely defined by a set of linear equations commonly known as Maxwell's Equations. In physics, the principle of superposition applies to linear equations, therefore also applies to problems involving electromagnetic fields. The magnetic (**B**) and electric (**E**) fields which comprise an electromagnetic wave are mathematically defined as vectors. When superposition applies, simple vector summation and subtraction is valid for problems involving wave re-enforcement and cancellation. Examples of wave re-enforcement and cancellation are familiar to amateurs who study the formation of antenna patterns from multi-element arrays as well as ground reflections. Therefore, the examples presented here will include both of these situations.

From the above discussion we can now state the fundamental mechanism for the summing and nulling of radio waves: *At a given point in space, the total value of the E field is the sum of all E fields from all sources at that instant in time.* For our purposes, we may add the caveat that the sum of these E fields come from the same source, i.e. the same transmitter (but in some examples, the transmit power will feed more than one antenna element).

Confusion arises from the fact that if you sum two identical **E** fields, the **E** value doubles, but that means the *power flux* quadruples! So, the typical question is: How can I quadruple the power (6 dB) at a distant point by using a 2-element driven broadside array if it will only have a gain of 3 dB over a single element? 3 dB gain is *doubling the power!*

First, we define the power that constitutes a radio wave as:

$$\frac{W}{m^2} = \frac{E^2}{Z_0} = \frac{V^2}{Z_0 m^2} \qquad \text{(Equation 3.44)}$$

where (**E**) is defined by the term *V/m* or, volts/meter and is referred to as the *field strength* and $Z_0$ is the impedance of free space, or about 377 Ω. Notice that this is simply the well-known power law, but as it applies to a free-space radio wave. P, or the power flux is watts/square meter.

1. The simple case of a 1/2 wave transmit dipole and an identical 1/2 wave dipole receive antenna.

This first example will introduce the terms used in all examples and provide a reference for the subject power and gain terms. All examples will assume 100 watts output from the transmitter. The distance between the transmit and receive antennas is *d* such that 100 watts × the gain of the dipole will result in an E field of one volt/meter at our test receiver locations. We will also normalize the wavelength λ to 1 meter (300

MHz) to simplify the equations. Since we are only interested in the discussion of how power and fields are related in these examples, we only need the above terms. Some additional terms will be derived from the above in later examples.

In **Figure 3.28**, the gain of the receive dipole is 1.64 (2.15 dBi). Therefore, the power delivered to the load impedance (the receiver) will be the power flux of the radio wave at the receiver location (Equation 1) multiplied by the aperture of the receive antenna (a function of gain). So,

$$P_{received} = (SA_e) = \left(\frac{1V^2}{377\Omega m^2}\right)\left(\frac{1.64\lambda^2}{4\pi}\right) = 346\ microwatts$$

(Equation 3.45)

where

$$\left(\frac{1V^2}{377\Omega m^2}\right)$$

defines the power flux in watts/square meter, also defined as ($S$) (Equation 3.32) and

$$\left(\frac{1.64\lambda^2}{4\pi}\right)$$

is the aperture ($A_e$) of a 1/2 wave dipole. The aperture of a 1/2 wave dipole is the numerical power gain relative to an isotropic antenna (dBi), in this case 1.64, multiplied by the aperture of an isotropic antenna

$$\left(\frac{\lambda^2}{4\pi}\right)$$

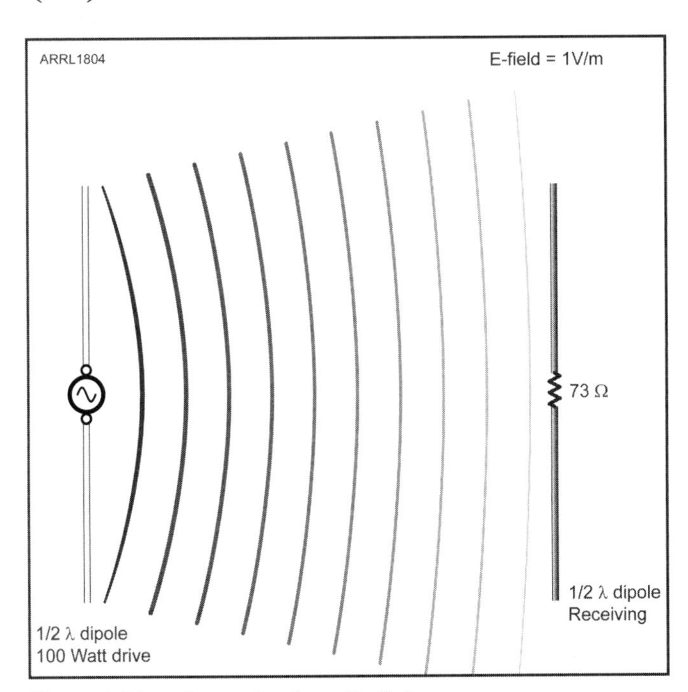


**Figure 3.28 — A very basic radio link.**

Thus the $m^2$ terms cancel, and we are left with $V^2/R$ which of course is the received power. This basic example shows the basic mechanisms of the radio link

Having defined these terms, we can now address some more complex examples.

2. Power re-enforcement at the elevation angle where the incident and ground-reflected waves perfectly re-enforce.

In this example we use an *EZNEC* plot of the broadside gain of our 1/2 wave dipole (**Figure 3.29**). Since the ground is an infinite perfect conductor, the frequency does not matter, so 7 MHz will show identical results to 300 MHz. In this case, the power gain from the incident field will simply be the broadside gain of the dipole alone (2.15 dBi). However, the reflected field *doubles* the value of **E** in the far field thus quadrupling the EIRP at this particular angle (in this case 30 degrees elevation). Thus, we have:

$$E_{total} = E_{incident} + E_{reflected} \quad \text{(Equation 3.46)}$$

Since we have perfect reflection, the total field is simply $2E_{incident}$. Again, since power is a function of $E^2$ the total power at the elevation angle where the fields add is *four* times the line-of-site (as in a free space wave) power, *not* a simple doubling of power, 6 dB rather than 3 dB.

Thus, we would expect the total gain to be

$$G_{total} = 2.15\ dBi + 6\ dB = 8.15\ dBi \quad \text{(Equation 3.47)}$$

We actually see 8.39 dBi. The small difference is due to the effects of a mutual impedance between the dipole and ground. In this case the dipole is 1/2 wave above the ground.

This example shows why ground reflected re-enforcement can generate very impressive EIRP values at specific

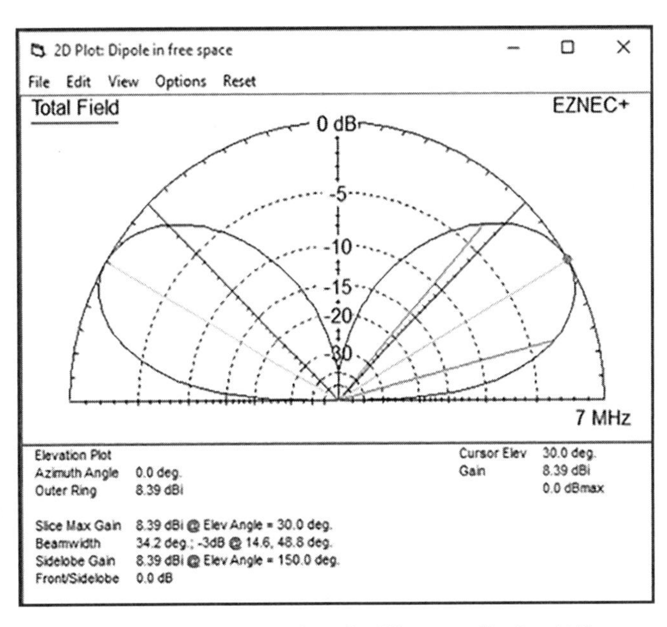


**Figure 3.29 — An *EZNEC* plot of a 1/2 wave dipole at1/2 wavelength height above a perfectly conducting ground plane.**

take-off angles. Over "real" ground the reflection coefficient becomes less than 1, so the resulting gain is lower than this idealized case. The actual reflection coefficient then becomes a function of the conductivity of the dielectric (soil) and the frequency.

The plot in **Figure 3.30** plot shows the effect of a perfectly reflected wave (ground reflection is a common example) upon an incident wave. The Y axis is the relative received power (linear), with 1 being the normalized power value for the radiating antenna in free space. The X axis plots the phase difference between the reflected and the incident wave plotted from 0-2π radians (0-360 degrees). 0 radians represents perfect re-enforcement, while π radians (180 degrees) represents perfect cancellation. The curve plots the function

$$Y = (1 + cosX)^2 \quad \text{(Equation 3.48)}$$

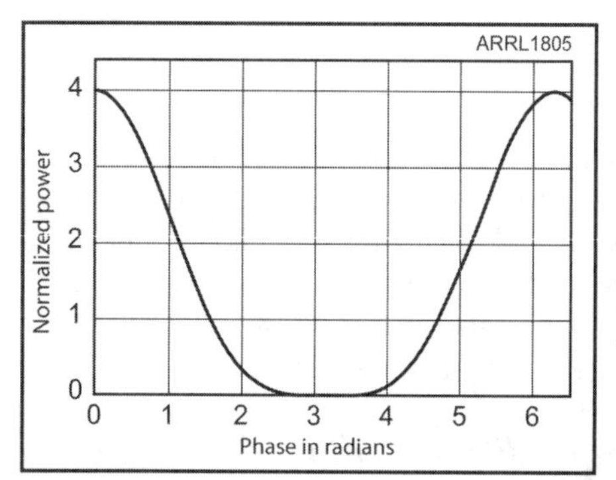


**Figure 3.30 — This plot shows the effect of a perfectly reflected wave (ground reflection is a common example) upon an incident wave.**

where 1 represents the normalized field strength of the incident wave and *cosX* represents the relative reflected field strength value as a function of the phase difference. The squaring term simply provides the relative power value. The distance of the incident and reflected waves are assumed to be the same, and the reflection efficiency is perfect. Both these terms can be easily applied to the equation with a simple coefficient in front of the cosine term.

3. Power received from a 2-element array of 1/2 wave dipoles fed in-phase with equal currents and spaced 1/2 wavelength.

**Figure 3.31** shows two 1/2 wave dipoles configured as a 2-element broadside array. The gain of such an antenna has a theoretical maximum gain of about 3 dBd, or twice the gain of a single dipole.

With the single dipole transmit antenna in Figure 3.28, *all* 100 watts is fed to the single antenna. This configuration was assumed to provide 1 V/meter field strength at the distance d. However, in the 2-element array, only 50 watts is fed to each antenna. Therefore, each antenna's contribution to the field at distance d will be 0.707 V/meter. These fields will simply add, providing 1.414 V/meter. Squaring the relative field strength, we get a relative power of 2 (compared to Figure 3.28). Therefore, the array has a power gain of 2 over a single dipole antenna or 3 dBd. A common mistake is *not* remembering the necessary power splitter and thus 100 watts is assumed in *both* antennas, and thus an array gain of 6 dBd is falsely calculated.

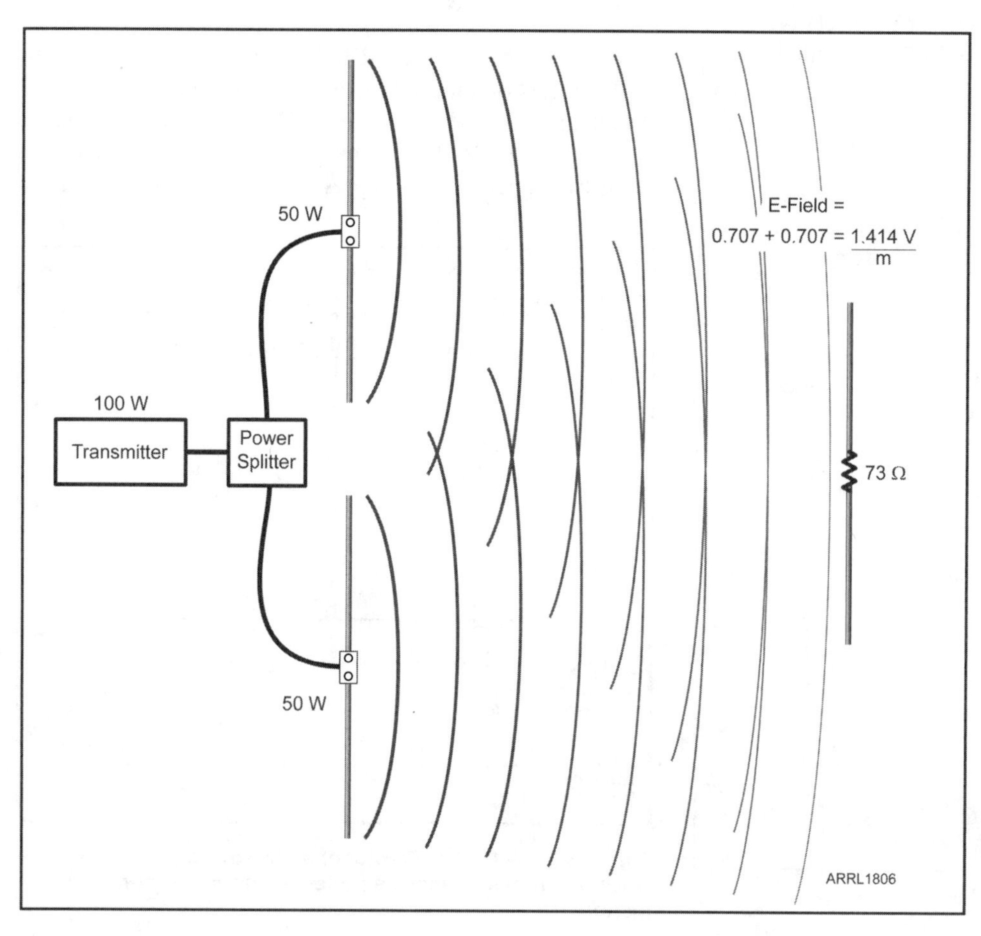


**Figure 3.31 — Two 1/2 wave dipoles configured as a 2-element broadside array. The gain of such an antenna has a theoretical maximum gain of about 3 dBd, or twice the gain of a single dipole.**

4. Power received by a 2-element broadside dipole array from a single dipole element transmit array. This example is the reciprocal of Figure 3.31.

In this example, the same 1 V/meter field strength is incident on both receive dipoles. However, when considering a receive antenna, it is necessary to realize that a single antenna is a *power detector*. The power it receives is calculated by Equation 2.1. The concept is quite simple. RF power at a given point is defined a watts/square meter, or power/area. Therefore, the power received is directly proportion to the antenna's aperture (or area).

To calculate the actual power received, we simply use Equation 3.45

$$P_{received} = \left(\frac{1V^2}{377\Omega m^2}\right)\left(\frac{1.64\lambda^2}{4\pi}\right) = 346\ microwatts$$

(Equation 3.49)

And multiply by 2 (since there are now two power detectors = 692 microwatts, or 3 dBd gain, similar to Figure 3.31.

Using two separate antennas the total power received is the sum of the two dipoles.

5. Power received by a 2-element co-linear receive antenna and a single dipole transmit antenna.

**Figure 3.32** shows a 2-element collinear antenna which is actually a full wave dipole. Indeed, the voltage induced in this *antenna will* be twice that of a 1/2 wave dipole. Therefore, we *might* expect the 2-element collinear to have 6 dB higher gain than a 1/2 wave dipole. However, in this case, we need some more terms to calculate the actual approximate gain of 2 dBd.

See **Figure 3.33**. A critical term to be defined is the effective height, or in this case more appropriately called effective length of an antenna. A passing wave, with identical polarization and the antenna presenting its maximum aperture toward the wavefront, will induce a voltage in the antenna proportional to

$$V = h_e \boldsymbol{E}$$ (Equation 3.50)

where V is the voltage induced in the antenna and $h_e$ is the effective height of the antenna, and again $\boldsymbol{E}$ is the field strength at the antenna. Effective height (or length) is

$$\frac{I_{ave}}{I_{max}} \frac{l_{physical}}{\lambda}$$

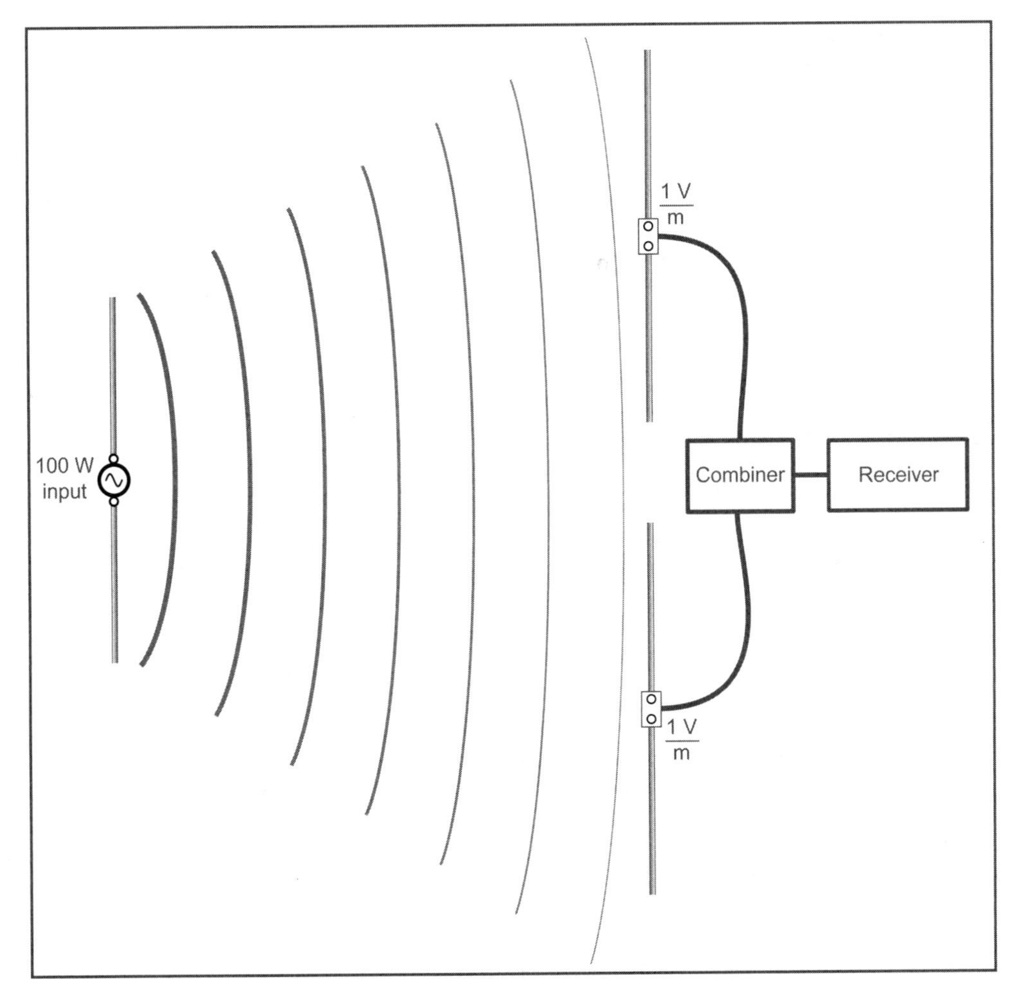


**Figure 3.32 — A single dipole transmitter and a 2-element broadside dipole receiver.**

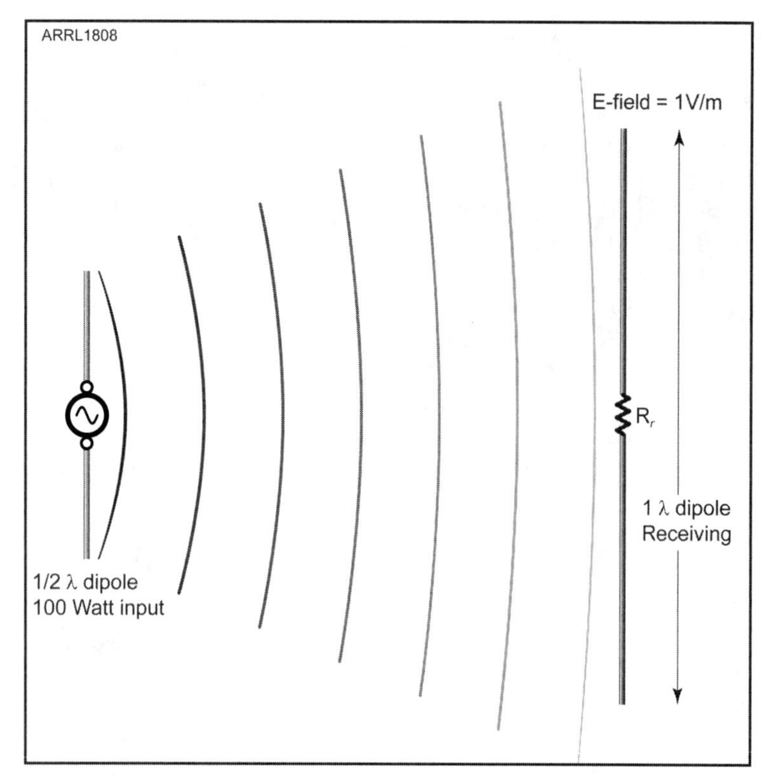


**Figure 3.33 — A full wave receive dipole antenna (2-element collinear) array.**

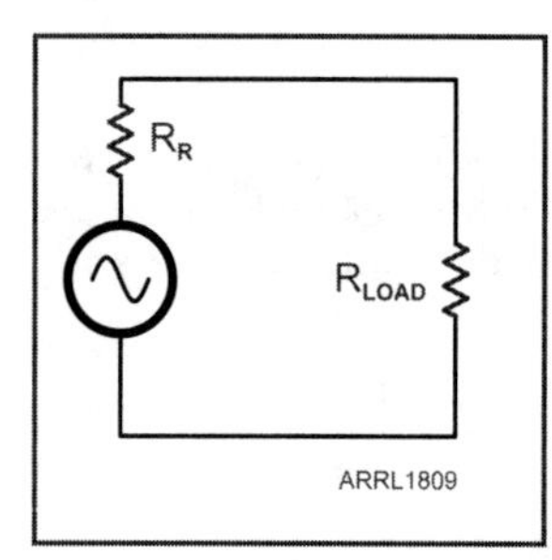


**Figure 3.34 — An equivalent receive antenna connected (usually through a transmission line) to a load, which is usually a receiver.**

or the average current along the antenna length/maximum antenna current, then multiplied by the physical length of the antenna in λ. A formal definition and calculation can be found in Kraus' text.

The current distribution along a dipole and a 2-element collinear is nearly sinusoidal. In the case of a 1/2 wave dipole, the current value forms a 1/2 wave sinusoid distribution. The average value of a sine wave (normalized to a peak value of 1) is 0.64. Therefore, the average value of the current on both antennas is 0.64 (assuming we normalize the maximum current to be 1 amp. However, $h_e$ is 0.32 for the 1/2 wave dipole and 0.64 for the full wave dipole since we must multiply the 1/2 wave dipole average current by 0.5, and we have normalized λ to 1 meter for simplicity.

Therefore, by equation 3.50, the resulting voltage on the full wave dipole will be twice the value of the 1/2 wave dipole assuming the field strength is the same for both antennas, again, for simplicity we have normalized it to 1 V/meter. This would imply that the power received by the full wave 2-element collinear would be *four* times that of a 1/2 wave dipole. This is not the case. The resolution of this involves some additional algebra than presented above.

**Figure 3.34** shows an equivalent receive antenna connected (usually through a transmission line) to a load, which is usually a receiver. The antenna itself can be modeled as a voltage source along with a source impedance defined as the radiation resistance $R_r$. The receiver is $R_{load}$. When the two are equal (a conjugate match) maximum power is transferred to the receiver which is the usual desired condition. However, the power provided by the antenna to the receiver is a function of both the antenna voltage *and* these resistances, and this is the resolution to the apparent paradox. But we must know the radiation resistance of the antenna to calculate the power received, and thus also the antenna gain.

Radiation resistance can most easily be computed for this type of problem by the equation

$$R_r = \frac{h_e^2 Z_0}{4A_e} \quad \text{(Equation 3.51)}$$

And thus, we can also provide another definition for $A_e$. This equation will be discussed in detail in Chapter 6.

$$A_e = \frac{h_e^2 Z_0}{4R_r} = \frac{G\lambda^2}{4\pi} \quad \text{(Equation 3.52)}$$

where $R_r$ is the radiation resistance, $h_e$ the effective height, $Z_0$ the impedance of free space (377 Ω), G is the numeric gain of the antenna and $A_e$ the effective aperture of the antenna. We can find the gain, and thus the effective aperture of both the 1/2 λ dipole and the 1 λ collinear/dipole from *EZNEC*. The dipole numeric gain is 1.64 and the collinear is 2.39 (both referenced to an isotropic antenna)

For a 1/2 wave dipole this works out to about 73 Ω. By re-arranging terms from Equations 2, 7 and 9 we can write

$$P_{received} = SA_e = \left(\frac{E^2}{Z_0}\right)\left(\frac{h_e^2 Z_0}{4R_r}\right) = \frac{V^2}{4R_r} = 504\ microwatts \quad \text{(Equation 3.53)}$$

for the 2-element collinear. Notice that 504 microwatts is definitely *not* four times 346 microwatts from the dipole example. Rather it conforms nicely, as we would expect, to the gain ratio between the two antennas about 2 dB.

To conclude, for a single antenna element that generates twice the voltage from a field strength due to twice its effective length, the radiation resistance limits the power gain. Also, up to relatively long linear (straight wire) antennas, as the antenna length is made physically longer *both* the effective length *and* the radiation resistance increase thus limiting the power gain. The simple term

$$\frac{V^2}{4R_r}$$

explains the paradox.

# Transmission Lines

In Chapter 2 we stated that an antenna "…may be defined as the structure associated with the region of transition between a guided wave and a free-space wave, or vice-versa." Following that definition, most of the discussion has been on the antenna and its relationship to free space. We alluded to the existence of the "guided wave" and "transmission line," but did not deal with this important topic in detail. Although not strictly part of an antenna proper, the transmission line represents the indispensable link whose properties share many characteristics with antennas, but also many that do not.

We begin with (a) some general principles, then (b) link line characteristics with principles developed in earlier chapters, (c) derive the critical specifications of coaxial and open wire lines, (d) provide a brief introduction to transmission line modes, (e) provide a detailed explanation of line losses, reflections, and how they relate to VSWR, (f) finish the chapter with a detailed analysis of open wire lines, including a new program in EZNEC to extensively model open wire lines. Many charts and conclusions are included for those who may be interested in general results, but not needing to actually model specific designs.

## The Guided Wave

Most often the antenna is not co-located with the radio and thus a physical connection must be provided. Whereas an antenna's function is to form the transition between guided and radiated waves, it is the function of the transmission line to deliver the guided wave to/from the antenna to/from the radio. There are two basic objectives within the line: minimize power loss to heat $R_L$ and minimize power loss to radiation Rr. Therefore, the ideal line will exhibit zero for both values. Also, a line that does not radiate will also not behave as a receive antenna, due to the reciprocity theorem. This becomes an important characteristic to minimize the introduction of undesirable noise into a receiver while simultaneously losing power received by the antenna through re-radiation. An excellent antenna design can be ruined by loss and pattern distortion from a poorly designed or installed line.

Therefore, the design goal of a transmission line is to configure the PADL parameters (Phase, Amplitude, Direction, and Location — introduced in Chapter 3) so that no radiation appears in the far field from the line. If two conductors are spaced << $\lambda$ and the RF current is out-of-phase and equal in amplitude, then minimum radiation will occur. In effect, the entire B and E fields will be canceled according to Maxwell's first two equations at some finite distance from the line. This is the essence of the PADL distribution in a transmission line. The actual location (L) is irrelevant since at any location in free space the conditions of phase/amplitude cancellation will be the same.

Therefore, no matter which direction we "view" the

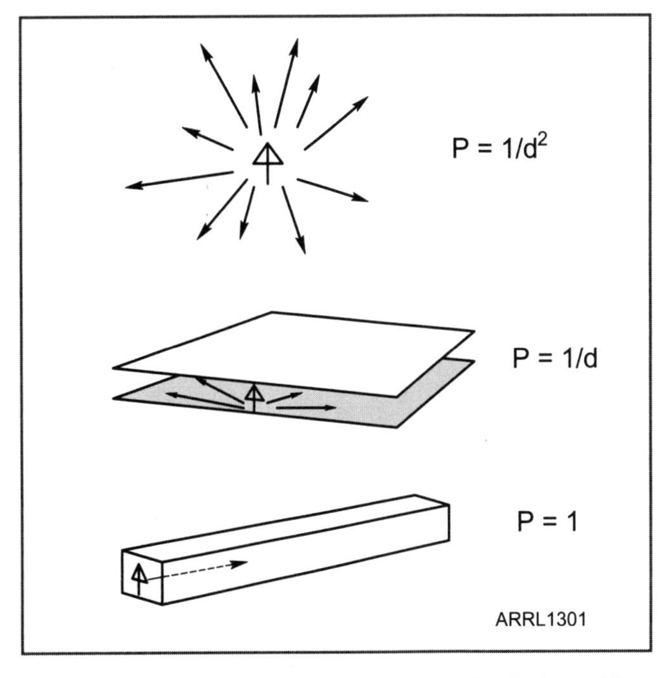


**Figure 4.1 — This figure shows how power is dissipated in 3D free space, (A) where the inverse-square law holds 1/*d*$^2$ (where d is the distance from the source). If the antenna source is confined within two closely-spaced conductive planes (B) the power is distributed at a 1/*d* rate since the dispersion is now confined to 2D. When confined to a 1D "wave guide" or transmission line (C), there is no loss of power with distance, of course assuming a perfect line.**

transmission line (from the surface of the sphere), we observe almost completely canceled fields and thus no radiation. If we measure the input impedance of a perfect unterminated line, there will be no real part of the impedance since Rr is zero. The reactive values along an unterminated (or shorted) line change between inductive and capacitive as we gradually change the physical length of the line or the wavelength. Therefore, we can also use a length of transmission line as a capacitor, inductor, parallel resonant circuit, or series resonant circuit.

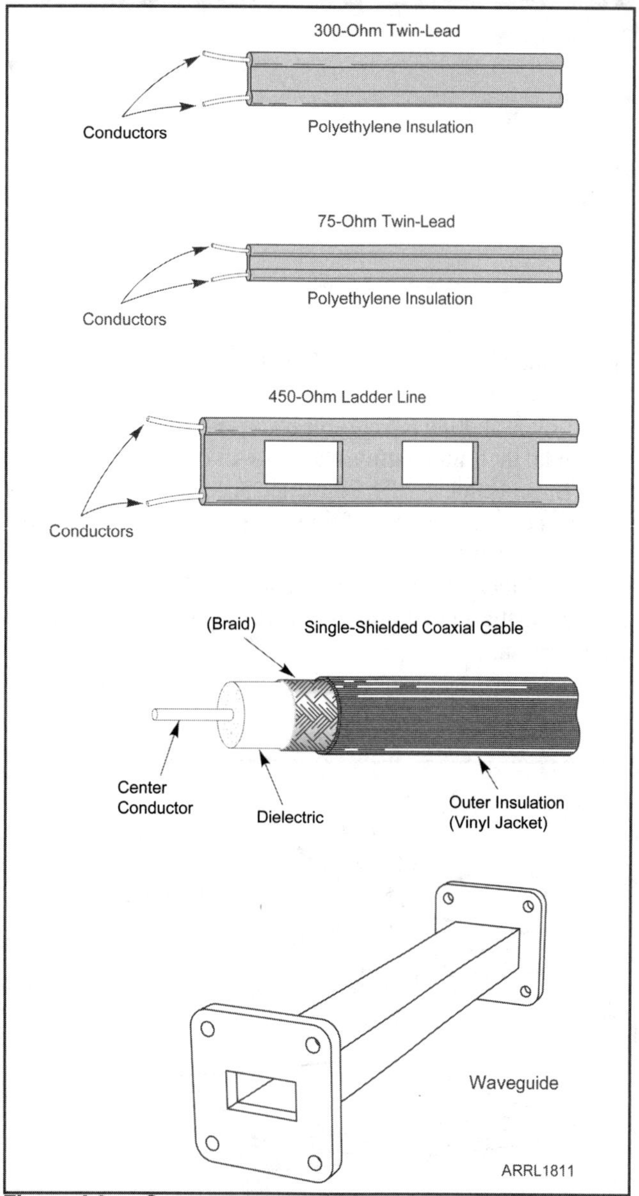

Figure 4.2 — Common types of transmission lines.

An antenna typically radiates or receives RF power from three-dimensional free space, but the transmission line concentrates the power into a one-dimensional structure and thus, in principle, does not permit the RF power flux to attenuate due to dispersion (**Figure 4.1**). This figure shows how power from an antenna in free space dissipates as (3-dimensional dispersion) as discussed in Chapter 2. However, if we limit the radiation between two conductive plates (with proper spacing relative to wavelength) the loss decreases to (2-dimensional dispersion) and finally if we concentrate the wave into a single dimensional guide, we create a transmission line with zero loss of the wave since there is no dispersion (spreading out) of the wave. Therefore, a transmission line can be considered a 1-dimensional waveguide with no dispersion and no loss. Again, this is assuming a lossless line.

## Preventing Radiation from Transmission Lines

In the previous chapters we developed the concept of radiation resistance as the term that defines the coupling of an RF wave into space. If the goal of a transmission line (see the common types shown in **Figure 4.2**) is to not radiate, then the Rr of the line should be 0 Ω. We are now familiar with the characteristics of a 1/2 λ center fed dipole antenna. The Rr is about 73 Ω when the wires on either side of the feed point are aligned in opposite directions, or θ = 180° as in **Figure 4.3** and Table 4.1. As we rotate the wires from 180 degrees to smaller angles, Rr begins to drop. At 0 degrees, Rr falls to nearly 0 Ω which is exactly what we want in a transmission line: no radiation, because Rr = 0 Ω. This is a simple derivation of how we can create a very effective

**Table 4.1**
**$R_r$ as a Function of Angle**

For the cases of a ½ wave dipole, an open-ended ¼ wave transmission line and any other intervening angle θ, the feed point impedance = $R_r$. Also see Figure 4.2.

| *Angle θ (degrees)* | *$R_r$ (Ω)* |
|---|---|
| 180 | 73 |
| 135 | 62 |
| 90 | 40 |
| 67.5 | 26 |
| 45 | 13 |
| 0 | 0 |

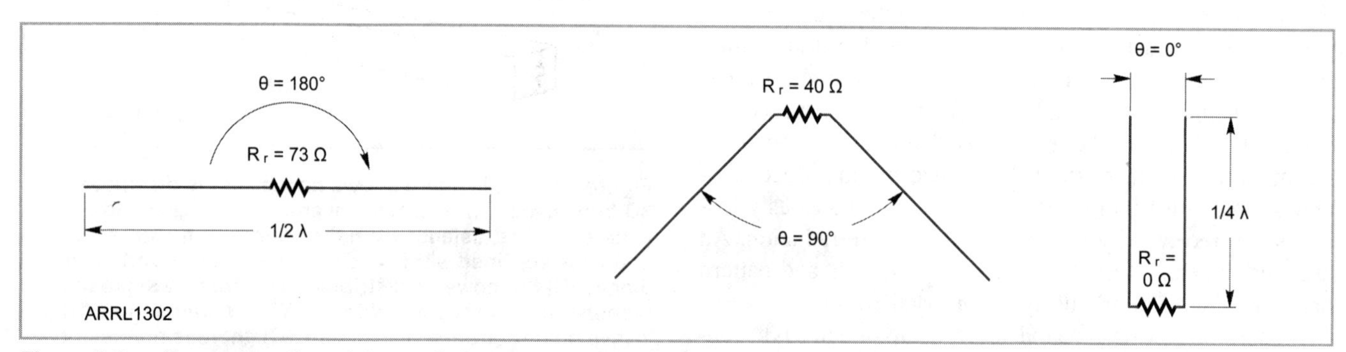


**Figure 4.3 — $R_r$ as a function of the angle formed by two ¼ λ wires. Also see Table 4.1.**

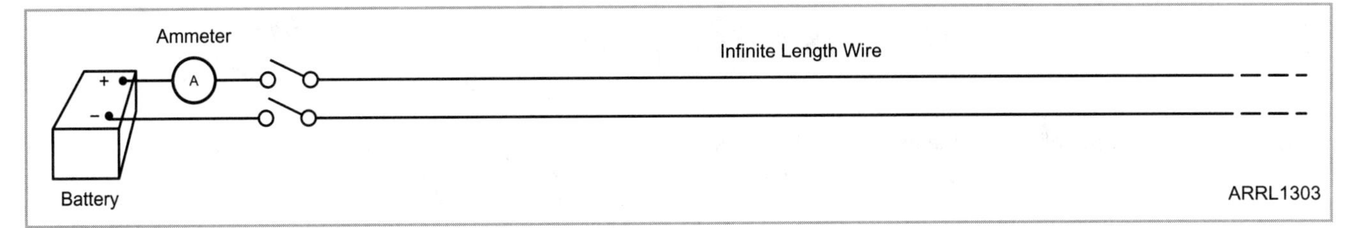


Figure 4.4 — An infinite length two-wire line.

radiator from two conductors, and then by reconfiguring the same conductors, we create a transmission line with no radiation. (Of course, all practical lines will have some small value of Rr and thus exhibit some finite amount of radiation.) The transmission line in this case is referred to as "open wire line." Later in this chapter we will formally analyze radiation loss, especially for open wire lines.

## Characteristic Impedance

Let us assume we have an infinite length two-wire line with zero resistance, constant spacing between the wires, and the wire diameters do not change for this very long line (**Figure 4.4**). The two wires each have an inductance/unit length, and the two wires also form a capacitance/unit length. The line is also assumed to be lossless, in that the resistance of the wire is zero and the resistance of the dielectric is infinite. More often the dielectric is specified as a conductance rather than a resistance and also a dielectric constant (treated in detail in Chapter 6) which is a function of the permittivity of the material or free space. Intuitively we can imagine that two parallel wires will have a capacitance between them much the way two parallel conducting plates will form a capacitor. They also are inductors. Since the line is infinite, we designate the capacitance and inductance in terms of C and L per unit length. Therefore, the electrical equivalent of the line is shown in **Figure 4.5**. If we connect a dc voltage source to one end of the line, we might expect to measure a displacement current (Maxwell's fourth equation), as one capacitor after another is charged through the endless series of Ls and parallel Cs. Indeed we will measure a *constant* dc current value. When we apply a source voltage *V* and it results in a current *I*, we can conclude that there is some corresponding impedance to this line.

$$Z_0 = \frac{V}{I} \quad \text{(Equation 4.1)}$$

From Equation 3.30 we derived the equation for free space:

$$Z_0 = \sqrt{\frac{\mu_0}{\varepsilon_0}} = \frac{V}{I} = \sqrt{\frac{L}{C}} = 120\pi\ \Omega \quad \text{(Equation 4.2)}$$

This term not only determines the characteristics of a radiated wave, but also of a guided wave. If the *E* and *B* fields are constrained between two conductors separated by free space, then intuition suggests that the impedance of the line will be a function of 120πΩ and the characteristics of the two conductors.

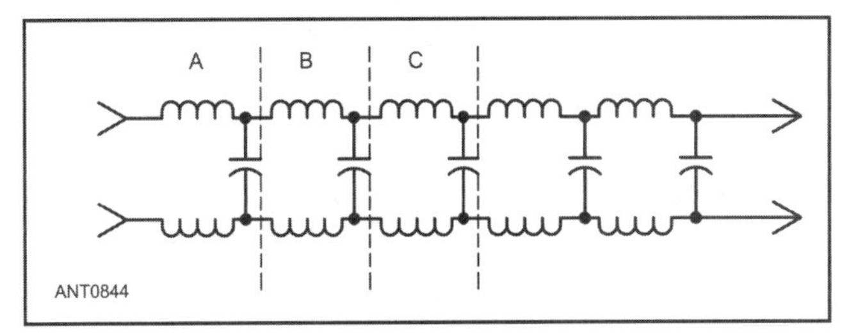


Figure 4.5 — The electrical equivalent of an ideal (lossless) transmission line in terms of ordinary circuit elements (lumped constants). The values of inductance and capacitance depend on the line construction.

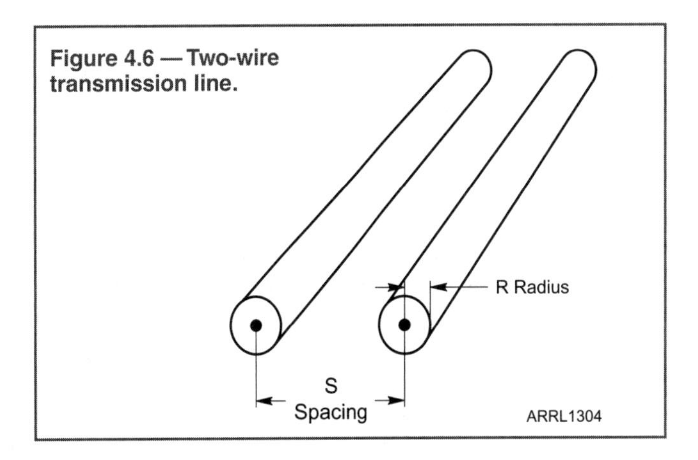


Figure 4.6 — Two-wire transmission line.

There is a rather non-intuitive equation from Equations 4.2 and 3.21

$$\sqrt{\frac{\mu_0}{\varepsilon_0}} = \sqrt{\frac{L}{C}} \quad \text{(Equation 4.3)}$$

Equation 4.3 is derived directly from Maxwell's equations and an important result is that the L and C distribution along a uniform line is *always* reciprocal. Reciprocity, in this case, means that for a given uniform line, the load and source may be "swapped" at opposite ends with no effect upon the resulting parameters, much like a resistor can be reversed in a circuit with no change in parameters.

Note: there are two definitions for the term $Z_0$ — the impedance of free space and the characteristic impedance of a transmission line. From Equation 4.4 through the rest of Chapter 4, $Z_0$ will represent the characteristic impedance of a transmission line.

For transmission lines, the form for the equation:

$$\sqrt{\frac{L}{C}} = Z_0 \quad \text{(Equation 4.4)}$$

is particularly attractive since the inductance and capacitance per length are both well defined by the geometry of the line. We will use this equation to find $Z_0$ for the most common transmission lines: two-wire open line and coaxial cable.

In Figure 4.3, if we switch off the source voltage (to 0 V), the capacitance in the line will discharge at the same rate, resulting in the same dc value for $Z_0$. Behavior is also the same for a continuously varying wave such as an RF sine wave, thus transmission line characteristics are valid for all types of applied voltages. This proves the line's reciprocity.

## Characteristic Impedance of Two-Wire Transmission Line

Again, through some rather complex mathematics, we can derive the inductance/length and capacitance/length for two uniform parallel wires with constant spacing. The resulting equations are:

$$\frac{C}{l} = \frac{\pi\varepsilon_0}{ln \, S/r} \quad \text{(Equation 4.5)}$$

where $S$ is the spacing between the two wires and $r$ is the radius of the wires (we assume both wires to have the same radius). And

$$\frac{L}{l} = \frac{\mu_0}{\pi} ln \, S/r \quad \text{(Equation 4.6)}$$

Substituting these equations into Equation 4.4 we get

$$\sqrt{\frac{L}{C}} = \sqrt{\frac{\frac{\mu_0}{\pi} ln \, S/r}{\frac{\pi\varepsilon_0}{ln \, S/r}}} = \frac{1}{\pi}\sqrt{\frac{\mu_0}{\varepsilon_0}} ln \, S/r = Z_0 \quad \text{(Equation 4.7)}$$

Notice that this equation for the impedance of an air core twin line cable contains the impedance for free space,

$$\sqrt{\frac{\mu_0}{\varepsilon_0}}$$

and the appropriate coefficient

$$\frac{1}{\pi} ln \, S/r$$

that accounts for the geometry of the twin line.

The length terms cancel and

$$Z_0 = \sqrt{\frac{\mu_0}{\varepsilon_0}} \frac{1}{\pi} ln \, S/r \; \Omega \quad \text{(Equation 4.8)}$$

where

$$\sqrt{\frac{\mu_0}{\varepsilon_0}} = 377 \; \Omega \quad \text{(Equation 4.9)}$$

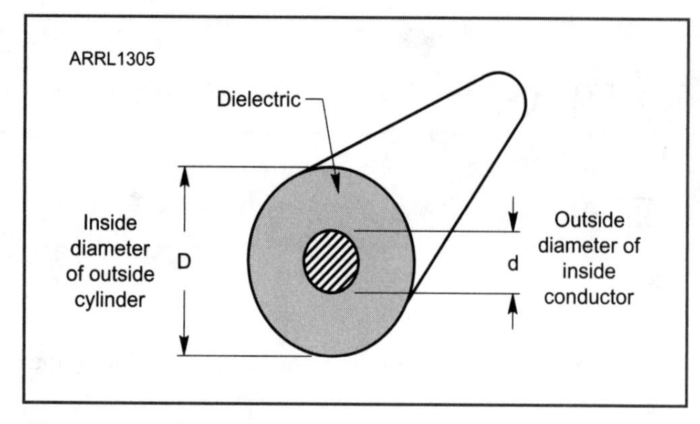


**Figure 4.7 — Coaxial transmission line.**

and

$$Z_0 = 120 ln \, S/r \; \Omega \quad \text{(Equation 4.10)}$$

or

$$Z_0 = 276 log_{10} \, S/r \; \Omega \quad \text{(Equation 4.11)}$$

where 377 Ω is the impedance of free space and 2.3 is the conversion factor from *ln* to $log_{10}$.

The $log_{10}$ term is often the more familiar form. Sometimes there is some confusion when using this equation. Rather than using the radius of the wires, occasionally the diameter *d* is used where

$$S/r = 2S/d \quad \text{(Equation 4.12)}$$

Therefore

$$Z_0 = 276 log_{10} \, 2S/d \; \Omega \quad \text{(Equation 4.13)}$$

These are the equations for the characteristic impedance of a two-wire open line.

## Characteristic Impedance of Coaxial Cable

Using a similar process to the open wire example, we can calculate the capacitance and inductance per unit length of a coaxial cable. See **Figure 4.7**.

The capacitance/length of a coaxial cable is

$$\frac{C}{l} = \frac{2\pi\varepsilon_0}{ln \, D/d} \quad \text{(Equation 4.14)}$$

And the inductance/length is

$$\frac{L}{l} = \frac{\mu_0}{2\pi} ln \, D/d \quad \text{(Equation 4.15)}$$

Again we substitute these L and C terms into Equation 4.4 to derive:

$$\sqrt{\frac{L}{C}} = \sqrt{\frac{\frac{\mu_0}{2\pi} \ln D/d}{\frac{2\pi\varepsilon_0}{\ln D/d}}} = \frac{1}{2\pi}\sqrt{\frac{\mu_0}{\varepsilon_0}} \ln D/d = Z_0 \qquad \text{(Equation 4.16)}$$

This can be simplified to the approximation

$$Z_0 = \frac{1}{2\pi}(377)2.3 \ln D/d\ \Omega \qquad \text{(Equation 4.17)}$$

Thus we also have

$$Z_0 = \frac{138}{\sqrt{k_e}} \log_{10} D/d\ \Omega \qquad \text{(Equation 4.18)}$$

$\varepsilon_r$ represents the *relative permittivity* of the medium separating the coaxial conductors, which is also the same as the *dielectric constant* of the material where $k_e = \varepsilon_r / \varepsilon_0$.

$k_e = 1$ for free space. For any other dielectric, $k_e$ will be some value more than 1 since $\varepsilon_r$ is *always* greater than $\varepsilon_0$. Typical values of plastic insulators used in coaxial cable have $k_e$ values near 2. For most practical purposes we can assume $\mu$ to be the free space value of $\mu_0$, thus typically we only need be concerned with $k_e$. In a two-plate capacitor the capacitance is directly proportional to $k_e$. For example, if we have a two plate capacitor of 100 pF with an air dielectric ($k_e = 1$) and insert a sheet of plastic with ($k_e = 2$), then the capacitance is now 200 pF.

Also, for open wire lines that use a dielectric other than free space, we use the same term

$$\frac{1}{\sqrt{k_e}}$$

to adjust the equation for $Z_0$.

The electromagnetic fields along a properly operating coaxial line are constrained between the conductors by virtue of its geometry. This is of great advantage in practice over open wire line whose fields propagate well beyond the immediate vicinity of the line. Only in the far field do the fields of open wire line cancel. Open wire line requires installation some "stand-off" distance from other objects to avoid distortion of the fields in the line's close proximity. Furthermore, by the reciprocity theorem, coaxial cable also has much better rejection of undesired RF sources, making it superior in spurious-free receiver applications. Mainly for these reasons coaxial cable has become the most widely used type of transmission line.

## Velocity Factor

When $\varepsilon_r$ increases by replacing the free space dielectric with a material, usually a plastic dielectric insulator, the impedance decreases. There is another effect upon the characteristics of the line: the velocity of the wave along the line. We again recall that:

$$\frac{1}{\sqrt{\varepsilon_0 \mu_0}} = c \qquad \text{(Equation 4.19)}$$

We assume that $\mu$ will be very close to $\mu_0$ in free space, therefore, the velocity along the line will be:

$$V_{line} = \frac{c}{\sqrt{k_e}} \qquad \text{(Equation 4.20)}$$

where $c$ is the free space speed of light. The *velocity factor* of a line is thus

$$VF = \frac{V_{line}}{c} \qquad \text{(Equation 4.21)}$$

For example, if we use a dielectric with a $k_e$ of 4, the velocity factor will be 0.5, or ½ the free-space speed of light. A common value for $k_e$ in coaxial cables is 2.3, and thus the result is a VF of about 0.66.

The velocity factor of the line is related to the characteristic impedance of the line only by

$$\frac{1}{\sqrt{k_e}}$$

and not the geometry of the line *per se*.

## Fields, Currents, and Voltages on Transmission Lines Derived from Maxwell's Equations

Maxwell's equations and their derivative equations prove that electromagnetic waves are indeed waves. They also define how EM waves behave in 3D free space, 2D space and along a 1D transmission line. When an EM wave encounters conductors, current and voltages will be produced (exactly as a receive antenna works). These equations also inform that the opposite is true: currents and voltages will also produce **E** and **B** fields (exactly how a transmitting antenna works). Thus, Maxwell's equations are reciprocal and invariant. This relationship between the fields (the **E** and **B** set) and the I and V set is simply ignored in a free space wave since there are no conductors, only the EM wave (**E** and **B** fields). In a transmission lines and antennas *both* sets are present: a *guided* EM wave and the associated voltages and currents in the conductors. Indeed, they cannot exist without each other when conductors are present. There are also dielectric waveguides, where the RF wave is constrained by the discontinuity between free space and a dielectric tube or pipe.

As we have shown in some cases, valid descriptions of antenna performance often "borrow" from electric circuit models. The use of circuit models becomes more prevalent with transmission line descriptions, since we are now dealing with a "circuit" which is also a *wave guide.* (Technically all types of transmission lines are "waveguides" however, the term "waveguide" is typically used to define a hollow tube transmission line as in Figure 4.2). Modifications of both the "circuit" and "waveguide" terms and methods of solutions are needed for accuracy and/or mathematical convenience. Since a transmission line forms the "link" between the radio circuit and the antenna, an attempt will be made to intertwine a description of both the set of current and voltages with the set of the **E** and **H** fields, and the Poynting vector **S**.

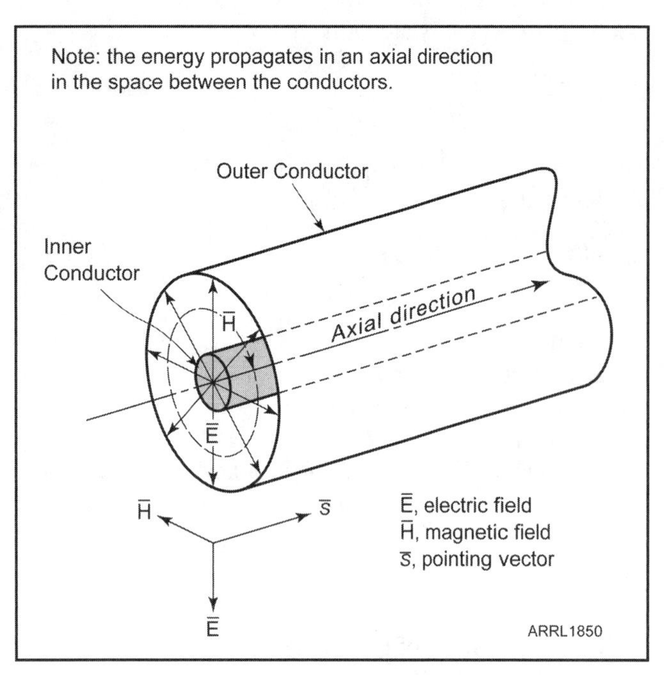


**Figure 4.8 — This figure shows the typical TEM relationship between the E and H vectors with S pointing down the cable.**

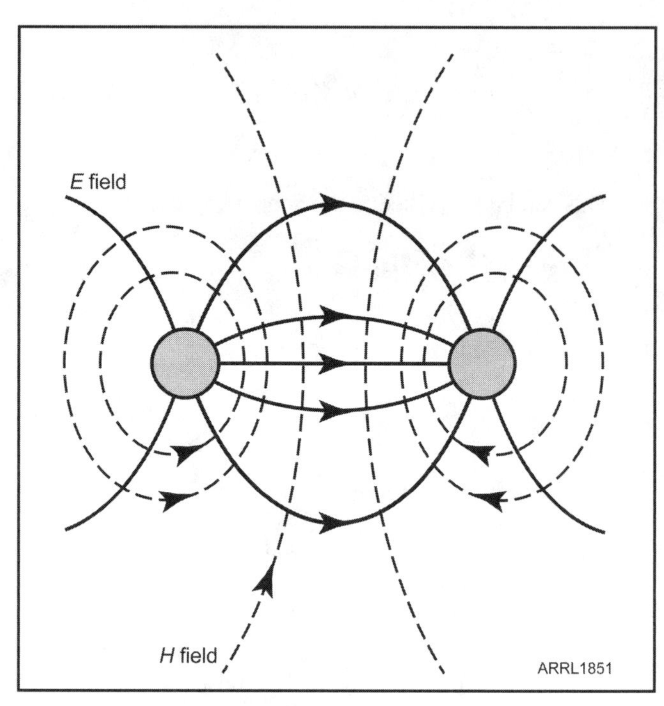


**Figure 4.9 — This shows the relationship between the E and H fields on an open wire line with S pointing into the page.**

In solving Maxwell's equations, with transmission lines it is more convenient to use **H** (magnetic flux density) rather than **B** (magnetic field). In free space, **H** = **B**. Within materials, such as transmission line dielectrics, **B** is a more convenient term. For open wire lines, the two terms are interchangeable.

In **Figure 4.8** we see the instantaneous position of the **E** and **H** fields resulting in the direction of propagation **S** (Poynting vector) to the right along the line for a free space wave in a coaxial cable. Inside a transmission line the directions of the **E** and **B** fields remain the same compared with **S**, although since they are now guided waves, the fields may be bent depending on the physical shape of the transmission line being used. However, for TEM waves (the usual mode for coaxial and open wire lines), both the **E** and **B** field lines must define orthogonal planes both of which are also orthogonal to **S** (as is the case with free-space radiated waves). When the RF signal changes polarity (twice during a single cycle of the RF sine wave), both fields change by 180 degrees and **S** remains pointing to the right. However, if only the **E** or only the **H** vector changes direction by 180 degrees, **S** also changes direction. This fact becomes important when discussing reflected waves, as will be soon discussed. The orientation of fields on a transmission line are typically more complex than a free space TEM wave as when a comparison is drawn between Figure 4.8 and **Figure 4.9** vs. free space vectors. However, it is apparent that in all these figures the orthogonal arrangement for the TEM mode is maintained.

Above a certain frequency, coaxial lines begin to act as waveguides, since their shields begin to act as conductive guides and thus the TEM field configuration in **Figure 4.10** begins to reconfigure as a TE wave. The cut-off frequency is, in practice, not "distinct." Coaxial cables are not normally used in the TE mode as waveguides specifically designed for such modes are preferred.

$$f_c \approx \frac{2c}{\pi\sqrt{e_d}(D+d)} \qquad \text{(Equation 4.22)}$$

where $f_c$ is the cut off frequency in Hz in a coaxial cable (approximate maximum frequency for TEM mode), c is the speed of light, $e_d$ is the dielectric constant of the dielectric, $D$ is the outer conductor inside diameter and $d$ is the inner conductor diameter. $C$, $D$ and $d$ use the same units. For coaxial cables, this change of mode produces a dilemma. As the frequency increases it becomes necessary to reduce diameter of the coaxial cable for it to maintain the TEM wave structure. However, as we shall see, line losses also increase in smaller cables. Furthermore, power capability decreases in smaller cables. Consequently, as the frequency increases (typically into the microwave region) waveguide is required as coaxial cable becomes impractical or useless.

## Resistive Losses in Transmission Lines

Thus far we have assumed that the conductors in our examples are perfect conductors (where resistance is zero) and the dielectric material separating them are perfect insulators (where the conductance is zero). This of course, is never the case in "real" transmission lines. These two causes of loss have an effect upon the characteristic impedance of the line where:

$$Z_0 = \sqrt{\frac{R + j\omega L}{G + j\omega C}} \qquad \text{(Equation 4.23)}$$

where $R$ is the RF resistance in the conductors at ω where ω

$= 2\pi f$, and $G$ is the conductance in the dielectric at the same $\omega$. Notice two results of this equation: The resistive loss in a transmission line is dependent upon frequency (usually the higher the frequency, the greater the loss due to skin effect) and when the line is perfect, both $R$ and $G$ become zero, and

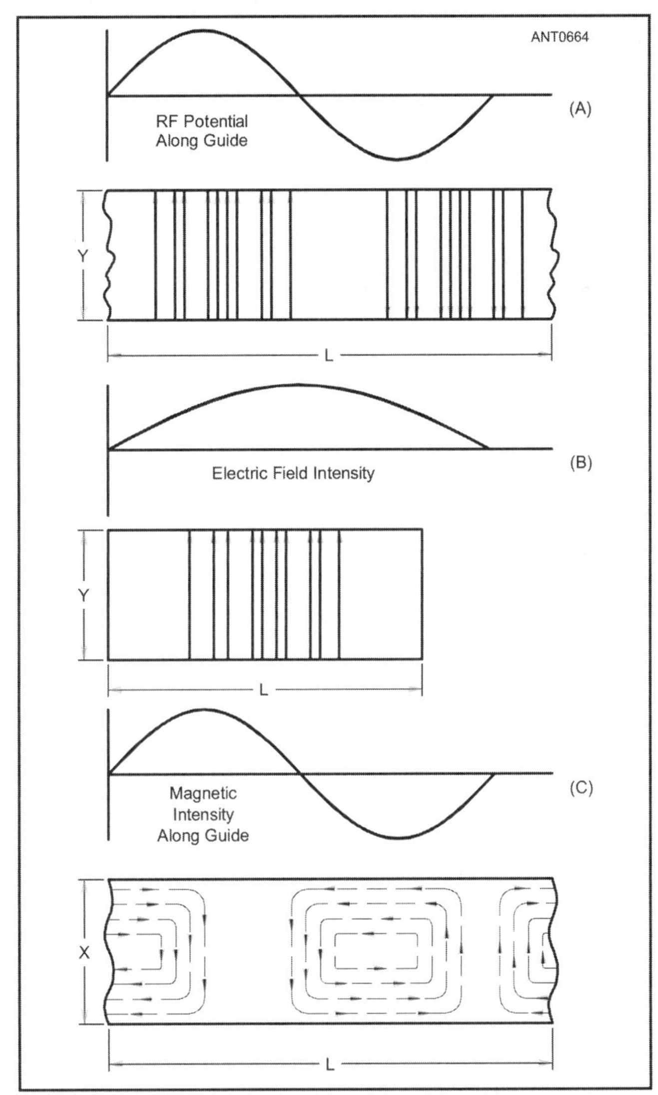


**Figure 4.10 — This diagram shows the field distribution inside a waveguide exemplifying a TE mode. "TE" is transverse electric, meaning that the electric field is orthogonal to the direction of propagation (either to the right or left). However, as can be seen, the magnetic field (part C) is partially oriented in the same direction as the direction of propagation, an impossible condition in the far field in free space. Notice also the electric field distribution (part B) only shows 1/2 wavelength while the other two show a full wavelength.**

thus Equation 4.23 reduces to Equation 4.4 (the $j\omega$ terms cancel in Equation 4.23).

The letter "$j$" in the equation represents "complex notation." The use of $j$ is common in engineering in contrast to the "true" pure mathematical "i". These two notations both represent $\sqrt{-1}$. The use of this "imaginary" number is used to greatly simplify equations across a wide variety of applications and is a very important term in mathematics. For example, the often-complicated mathematics used to derive the complex conjugate of an impedance distills to the very simple relationship:

$Z_0 = R - jX$ is the complex conjugate of $Z_0 = R + jX$. In this case if we are presented with an impedance of $R + jX$ meaning the impedance has a real part $R$ and inductive reactance $+jX$, the complex conjugate (matching) impedance will be $R - jX$ again, $R$ being the real part and $-jX$ being the same value reactance but capacitive. Therefore, $-j$ implies capacitive reactance and $+j$ inductive reactance. More complicated calculations involve algebra (sometimes linear algebra) rather than trigonometric identities which can become quite nasty when not using complex notation. It is remarkable that this use of the "imaginary" provides a very useful shortcut in mathematics, much like reaching into another dimension to find better tools to solve problems back here in our reality. A basic familiarity with $\pi$, $j$ and $e$ do *not* require an understanding of mathematics beyond high school.

Developing a knowledge of these numbers, what they mean, and how to use them will dramatically increase both an appreciation of mathematics and provide tools for a wide range of applications beside electronics, i.e. compound interest calculations become trivial on a calculator (using $e$, the root of natural logarithms). It is now common use to write complex impedances with complex notation. For example, *EZNEC* uses complex notation ($j$) for stating feed point impedance of antennas (see **Figure 4.11**).

The feed point of this antenna from *EZNEC* is 34.99 – $j$32.36 Ω. This equates to 34.99 Ω resistive, and –$j$32.36 Ω capacitive reactance.

There is a third mechanism that results in line loss that we introduced earlier in this chapter. Power is lost from a line that does not perfectly cancel the opposing fields traveling on it through radiation ($R_r$). This is of particular concern in open wire lines but is also a problem in coaxial lines which we will explore in detail later. Thus, we can define the three key sources of loss from a transmission line: $R_l$ from the conductors $G$, from the dielectric loss (actually $G = 1/R_{dielectric}$) and $R_r$ (radiation resistance) appearing when the line is also partially acting as an antenna.

```
Frequency = 7 MHz

Source 1        Voltage = 47.66 V at -42.77 deg.
                Current = 1 A at 0.0 deg.
                Impedance = 34.99 - J 32.36 ohms
                Power = 34.99 watts
                SWR (50 ohm system) =  2.291  (75 ohm system) =  2.629
```

**Figure 4.11—The feed point of this antenna from *EZNEC* is 34.99 –$j$32.36 Ω. This equates to 34.99 Ω resistive, and –$j$32.36 Ω capacitive reactance.**

## Causes of Reflected Waves on Transmission Lines

There are few topics in transmission line theory which have generated more debate, misunderstanding and misinformation than the causes and effects of the reflected wave. For nearly a century countless papers and texts have been written about this important topic, but precious few are without errors or misleading information. Therefore, we will look closely at reflected waves and find that most of the confusion is based on the fact that this phenomenon is both non-intuitive and complicated!

**Note:** These complicated interrelationships necessarily result in complicated explanations, and I make no claim to have simplified the explanation here much better than previous accurate treatments. Later in this section a diagram description is presented and hopefully this rather complicated "verbal" description together with the diagrams will help clarify these complicated relationships. I carefully read the similar description in Walter Maxwell, W2DU's classic text *Reflections* which gave me some clues for simplifying the explanation, but I was unable to simplify it much better than W2DU. His text is now out of print, but discussions at a comparable technical level deserve contemporary treatment.

First, we will explore how reflected waves are generated. Indeed, the very existence of reflected waves is a bit non-intuitive. How does a guided wave travel down a line suddenly reverse direction 180 degrees when it reaches the end of a mismatched line?

Let us assume that the line pictured in Figure 4.4 is *not* infinite. The ratio of the *applied* RF V/I in the initial wave traveling down a transmission line is determined by the $Z_0$ of the line which is simply defined by Ohm's Law. In other words, the V/I ratio of the guided wave traveling down a transmission line is *always* defined by the line impedance. Thus, the *initial condition* of the circuit is simply defined by $Z_0$. This is the condition of the line until the initial wave encounters a *change* in impedance. Along the line there is a current and a voltage $Z_0 = V/I$. The *actual* values of $V$ and $I$ also determine the power of the wave accepted by the line and the energy distributed along the line. There is no way for a wave to "know" there is a discontinuity down the line somewhere, even if it's a short circuit. It simply travels along the $Z_0$ until it encounters something different. The above statements are true for all waves (forward and reflected) traveling down a transmission line with constant $Z_0$.

**Matched line termination:** If at the end of the line there is a terminating resistance equal to $Z_0$, then the incident voltage and current upon this resistance will match the terminating resistance and all power on the line will be dissipated in the load. The ratio of $V/I$ propagating down the line will be the same as the ratio seen in the load. If the load resistance is equal to the $Z_0$ of the line, then all the power propagating down the line will be dissipated by the load; in effect, the load simply "looks" like a continuation of the line. From basic circuit theory, all power is absorbed by the load when the load's impedance (usually an antenna) is the complex conjugate of the source (the line). The use of "complex conjugate" is a bit misleading in this case, since there is no reactive part of the load or the source, assuming the line itself in non-reactive. However, "complex conjugate" is often used in such cases and is necessary when reactance does exist. The following additional examples will only use "real" load impedance values to simplify the examples. From a practical viewpoint, a finite reactive component is always present in the feed point impedance of an antenna and must be considered if it is a significant value.

**Open line termination:** Let us now assume that the line shown in Figure 4.4 is of finite length and left *unterminated.* Since the wave propagating down the line contains energy, that energy cannot simply "disappear". Since there is no resistive load (resistor) or radiating load (antenna), the only path for the energy is to be reflected back up the line, but how does this happen?

The forward current stops, but a reflected current is created since there is no load to dissipate it. Suddenly there is twice the charge at the end of the line (charge from the incident wave, and charge from the now-reflected wave). In effect a "traffic jam" of charge has been created, similar to suddenly closing off a road, with cars trying to back up and more cars slamming on their brakes. In this case the drivers are all very polite (since they all want to avoid each other) and the forward cars allow the backing up cars to pass, but the number of cars *at that point* has doubled.

Since there is an equal number of charges moving forward and backwards *in the same lane (same conductor)*, there are two equal currents but moving in opposite directions so there is no *summed* current at that point along the line. Because the actual number of charges has doubled, the

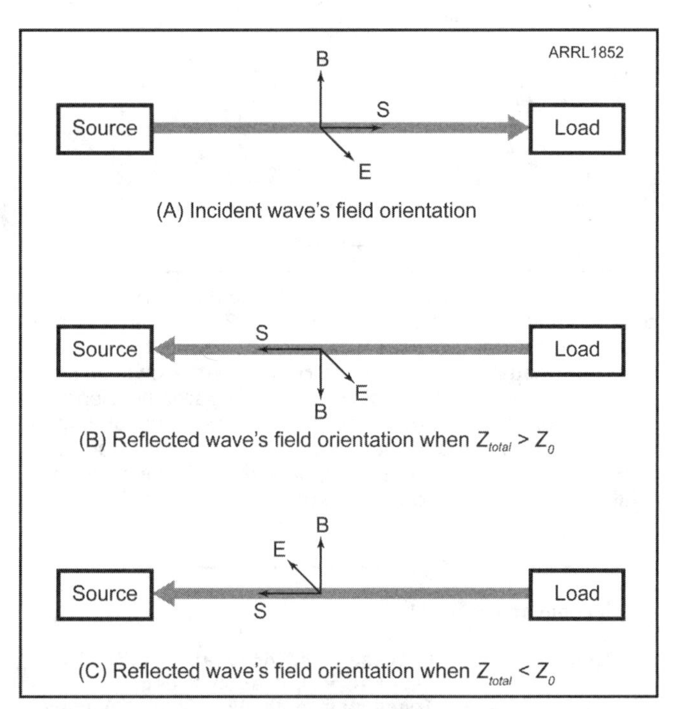


**Figure 4.12 — In (A) you can see the related field vectors for a forward wave, while (B) shows the field relationships for a reflected wave when $Z_{load} < Z_0$ and (C) shows the field relationships when $Z_{load} > Z_0$.**

voltage has also doubled, and **E** has doubled. The current has fallen to zero, therefore the *summed* **H** has disappeared. Now two key interrelationships occur simultaneously all defined by Maxwell's equations, which again, express these cause and effect relationships. First, since the current has reversed a new **H** field appears, but the **H** vector has changed direction by 180 degrees, since the current has also changed 180 degrees. However, the **E** vector still points in the same direction. Thus, **S** also reverses 180 degrees (see **Figure 4.12**), and a new wave along with its accompanying set of voltages and currents starts its journey back to the source. Second, the changing amplitude of the **E** vector results in the creation of the new **B** field (again by Maxwell's equations). If we assume a perfect line, the reverse wave's **E** field and line voltage value are the same as the incident wave's, and they sum, and the **H** field and the line current are also the same, but they cancel. Therefore, energy is conserved, and Maxwell's equations are obeyed, all equations are satisfied by converging into a beautiful closed form solution. Of course, the situation just described is valid only for 1/2 of a single RF sinusoidal waveform. When the incident RF wave changes polarity, the fields and line voltage and current also change polarity. Beside a reversing of direction for the reflected wave, the current and voltage associated with the reflected wave are 180 degrees out-of-phase in contrast to the in-phase forward wave where voltage and current are in-phase.

**Shorted line termination:** The second extreme case uses a shorted line, as opposed to an open-ended line. Again, let's assume the finite line shown in Figure 4.4 is shorted at the end rather than left open. In this case, as in the open-ended line, the incident wave travels along the line with $Z_0$ and is not yet affected by the short. At the short, the road again dead ends, but the cars don't stop. They simply turn around through the short (U-turn across the median strip, the dielectric) and head back the other way in the source-bound lanes. The current at this point on the line is doubled, half going toward the short (forward wave) and half heading back (reflected wave) toward the source. In this case the voltage and **E** field disappear at the short since there can be no voltage across a short circuit. However, there are two **E** fields, but they are identical and 180 degrees out-of-phase and thus cancel. Since the current polarity of the reflected wave is now moving on opposite sides of the transmission line (Figure 4.8) the direction of the E vector lines have reversed, but the **H** vector lines are in-phase with the original incident wave. As in the open wire line example, since only one field's vector lines changed 180 degrees, the **S** vector reverses direction (back toward the source). Again, review Figure 4.12.

The forward and reverse current however, are *in-phase*, but again in the reflected wave voltage and current are *always* 180 degrees out-of-phase. Imagine the cars traveling toward the short and an equal number of cars driving away in their proper directional lanes. When encountering the short, the cars change direction 180 degrees (and thus reverse phase 180 degrees) but also enter the reverse-directional lanes, rather than backing up in the same lane (as in the open line termination). In this case the cars have reversed direction but *in the correct lanes.* Again, through Maxwell's equations we can calculate how the reverse wave and its accompanying set of current and voltage are set up.

The analogy of cars on a highway analogy for both the open and shorted lines can be further explained by imagining the road is a divided highway. If the highway ends there is no way for cars to cross the median and get to the reverse lanes (open termination), the cars must back up. If the lane ends with a turn-around (short) the easiest way is to drive across the median strip (short circuit) and politely join the traffic moving in the same direction. Now simply extend this analogy for a two-conductor transmission line (coaxial or open wire): the two lanes are analogous to the two conductors, one is a dead end, the other a turn-around path across the median strip.

The electromagnetic mechanism for the open wire condition is precisely the way current reverses itself at the end of antenna elements as explained in Chapter 3 resulting in very high voltages at the element's end and no current. However, there is a key difference that limits voltages to much lower values on transmission lines when compared to the ends of antenna elements. Voltages (on a transmission line) at the end of the open-ended transmission line are more limited since opposite charges appear adjacent to each other (two-lead capacitor) and transmission line capacitance is much greater than the self-capacitance (single lead capacitor) at the end of a single wire. This can be explained by the simple equation:

$$V = \frac{q}{C}$$

where $q$ is the charge on the wire and $C$ is capacitance in farads. This condition is comparable to a comb holding a charge after combing your hair. The charge may be quite small, but the self-capacitance of the comb is *very* small, and this drives the voltage *very* high. Of course, the charges at the end of an antenna element change from negative to positive at the RF frequency rate, whereas the charge on the comb is "static."

Thus far we have analyzed the matched condition (where there is no reflection generated), and the two possible extreme conditions that generate the maximum possible reflected wave: an open-ended line (infinite Ω impedance) and a shorted line (0 Ω impedance). Now we describe two additional conditions: where the load impedance is lower (but not zero) than the line $Z_0$ and where the load impedance is higher (but not infinite) than the line $Z_0$. These two conditions, along with the matched condition, form the three possible conditions found for an actual functioning system.

## Wave Reflection and Scattering on Transmission Lines

We are now in a position to formalize the above discussion of a reflected wave. The term we use is the reflection coefficient which most often uses one of the terms ρ or Γ. In this text we will use ρ, the same term used in the *ARRL Antenna Book*. Since we are dealing with a ratio of voltages (reflected to incident voltage), the value of the reflection coefficient can only take a value between zero (no reflection, perfect match) and 1 (complete reflection of the incident wave. If there is no reactive part of the load impedance (pure resistance between 0 and ∞Ω), this simple equation suffices:

$$\rho = \frac{V_r}{V_f} \qquad \text{(Equation 4.24)}$$

where $V_r$ is the voltage of the reflected wave and $V_f$ is the voltage of the incident, or forward wave. The same equation also holds for the reflection of current, except, obviously we use $I$ rather than $V$ for the terms.

If the load has a reactive part, we must take the reactive and resistive (real) parts of the impedance into account and derive:

$$|\rho| = \sqrt{\frac{(R_a - R_0)^2 + X_a^{\ 2}}{(R_a + R_0)^2 + X_a^{\ 2}}} \qquad \text{(Equation 4.25)}$$

where $|\rho|$ is the complex reflection coefficient, $R_a$ is the real part of the load impedance, $X_a$ is the imaginary part of the load impedance, and $R_0 = Z_0$, the characteristic impedance of the transmission line.

Over the past several decades a new mathematical tool has come into prominence in RF engineering that greatly simplifies calculations in both circuits and transmission lines: scattering parameters, or simply S-parameters. The following *very brief* introduction is presented simply to introduce the reader to this tool which has become the preferred method of circuit analysis and has direct applicability to transmission line problems. The principle of "scattering" a signal was introduced in Chapter 2 as the "scattering aperture" of an antenna. The antenna scattering aperture, especially as it relates to transmission lines, is discussed in more detail later in this chapter. S-parameters are derived from a very powerful mathematical technique that uses a matrix of terms to solve equations. Matrices form an important part of Linear Algebra. Here we will briefly introduce S-parameters only as they relate to a 2-port "circuit". In this case, the "circuit" is the junction point between an antenna's feed point with a transmission line.

In **Figure 4.13** the left side of the matrix is the transmission line and the right side is the antenna or load. One advantage of using 2-port S-parameters is that they explicitly and simultaneously define the transfer functions of both the forward and reflected waves, (a 1-port situation would be used for a dummy load).

Most descriptions of forward and reverse waves only use a one-way system problem: like a transmitter feeding an antenna. Then they usually "assume" the antenna being used as a *receive* antenna is simply the reciprocal function of a transmit antenna. As Appendix F points out, this can lead to mistakes. For the transmit case a signal is applied to the left (a) side of the matrix. $S_{11}$ defines the portion of the incident wave that is reflected (scattered) where $S_{11} = \rho$. $S_{21}$ is the portion of the signal that is passed to the right side (b, the antenna feed point). Since there is no signal applied to the b side, $S_{12}$ and $S_{22}$ are zero. In the receive case, the antenna is the signal source, and the transmission line/receiver is the load. In the receive case, the antenna is the signal source and the transmission line and receiver form the load. Therefore $S_{22}$ is the reflected wave *from* the antenna to the line and $S_{12}$ is the portion of the received wave from the antenna that is transferred to the line. So, the S-matrix solves the two-way problem with a set of closed form equations.

**Figure 4.14** shows the mathematical form that solves the problem of determining the exact characteristics of the signal (b) appearing on the antenna terminals with the input signal (a) with the S-matrix defining the transfer function. For those who are interested in exploring S-parameters in detail, there are now a multitude of sources since the advantages of matrix analysis has become ubiquitous for radio design.

## Basic Consequences of Reflected Waves on Transmission Lines

In the last sections we described the causes of reflection (scattering) of an electromagnetic wave at an impedance discontinuity on a transmission line. We also offered simple equations that define this phenomena, and two techniques for their solution. On a transmission line, the wave is constrained to two directions, forward, and reflected. We will now investigate the consequences of the existence of reflected waves on a line and how they interact with the incident, or forward wave.

First, we will consider the interrelationships between incident and reflected waves on a lossless "perfect" transmission line. First, again, let us assume that the load resistance is equal to $Z_0$. The RF V/I ratio along the line is constant, therefore if we observe the RMS voltage along the line, it will be constant at all points along the line. Of course, the instantaneous voltage will be sinusoidal, but our meter is not

$$\begin{pmatrix} b_1 \\ b_2 \end{pmatrix} = \begin{pmatrix} S_{11} & S_{12} \\ S_{21} & S_{22} \end{pmatrix} \times \begin{pmatrix} a_1 \\ a_2 \end{pmatrix}$$

ARRL1854

Figure 4.14 — This figure shows the mathematical form that solves the problem of determining the exact characteristics of the signal (B) appearing on the antenna terminals with the input signal (A) with the S-matrix defining the transfer function. For those who are interested in exploring S-parameters in detail, there are now a multitude of sources since the advantages of matrix analysis has become ubiquitous for radio design.

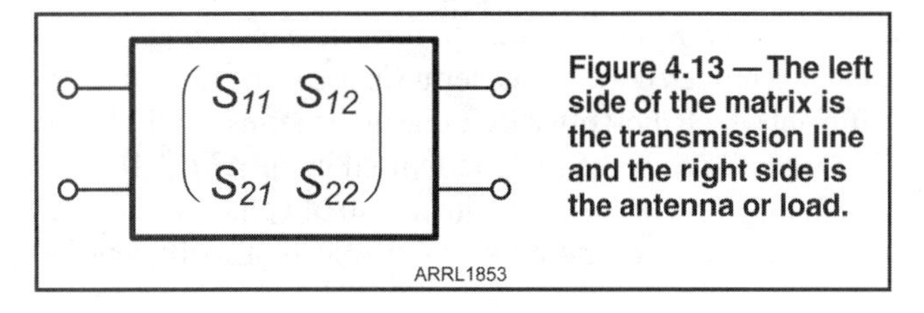


Figure 4.13 — The left side of the matrix is the transmission line and the right side is the antenna or load.

fast enough to follow millions or billions of sinusoidal cycles per second, so it *integrates* and therefore displays some "average" of the sinusoidal voltage. The statistical method for determining power in an ac (including RF) circuit is the "root mean square" of an RF signal, or "RMS." The RMS of a 1-V peak sine wave is 0.707 V, the coefficient being

$$\frac{\sqrt{2}}{2}$$

The RMS value is used for sinusoidal voltage and current to determine power. A "true" RMS voltmeter will provide the "true" RMS value by accurately performing the proper equations to determine true RMS (either through analog or digital means).

The detailed discussions on the creation of the reflected wave in the preceding sections have another consequence: the voltage and current of the reflected wave are *out of phase* whereas the incident signal's voltage and current are *in phase*. The V/I ratio of both the incident and reflected waves' *must* be defined by the $Z_0$ of the line. However, *the sum of these opposing waves and the ratios of V/I will not be defined by* $Z_0$. This is an extremely important point: Since the sum of the opposing waves create ratios of voltage to current (sums) not conforming to $Z_0$, impedances different from $Z_0$ exist at different points along the line. Therefore, a transmission line with both an incident and reflected wave(s) acts as an impedance transformer. Furthermore, the *distribution of these sums* are "standing", or not moving on the line even though both waves are moving. Therefore, at any given point on the line for a constant $Z_0$ and load impedance, the resulting impedance *at any point* on the line remains *constant*, assuming, of course, the wavelength, and thus the frequency of the signal is constant. If the frequency changes for a given line/load condition, the distribution of sums and thus the distribution of impedances on the line will also change.

### Standing Waves

It is important to remember that the incident and reflected waves are completely independent of each other on the line. Both conform to the V/I ratios defined by $Z_0$. It is the *sum* of these two waves which cause *standing waves*. This condition offers many possibilities in the use of transmission lines. But *how* do these opposing signals sum, and how does the impedance change along the line?

"Summing" simply means that you add up all the RMS voltages or currents present at a given point on a line, and that is the sum at a given point along the line. If they are equal in amplitude but out of phase, they sum to zero, but both are still there like passing ghosts in the night. An RF RMS voltmeter will "automatically" do the summing of all of the waves present at the point of measurement. This is analogous to a field strength meter which simply sums the V per m field strengths from all signals present at the point of measurement in free space.

Imagine we apply a CW signal to matched line, move the RMS voltmeter (like in **Figure 4.15**) up and down the line, and plot the measurements. In this case we will measure only the forward wave (there is no reflected wave) and thus the RMS voltage will be the same at all points along the line (again assuming no line loss). If, however, the line is mismatched, we then find a *wave shaped distribution* that doesn't move, hence the term "standing waves" applies. These waves are stationary therefore, they are *not* electromagnetic waves defined by Maxwell's equations, but again, simply the distribution of summing of all waves' RMS values up and down the line. Also, it is appropriate to repeat the *standing waves* are created by summing two or more propagating waves traveling up and down the line.

The graph in **Figure 4.16** plots the RMS voltages up and down a mismatched line using different values of ρ. This graph assumes an incident RF voltage of 1, also called the "normalized" voltage. Only one reflected wave is present in that the reflected wave is not re-reflected back toward the load. Notice that the standing wave voltage becomes a maximum voltage point at the 3π phase point on the right, which is the point on the line connected to a load with an impedance higher than $Z_0$. If the load impedance were lower than $Z_0$, we would simply define the load point as

$$\frac{5\pi}{2}$$

but still move to the left (toward the signal source). As we can see, the RMS voltages (but also the RMS currents) are changing in a repetitive manner.

We have so far defined the reflection coefficient ρ and VSWR and can formalize the relationships between these terms. If the load impedance is complex, the reflected voltage will be shifted somewhere in between. In effect the voltage at the load will not be maximum or minimum.

As described in Chapter 3, this mechanism is also present in free space, since Maxwell's equations must be satisfied in both guided waves and free space waves. At any instant in time, the sum of all electromagnetic wave values of **E**'s sum at a given point in 3D space. However, these separate waves

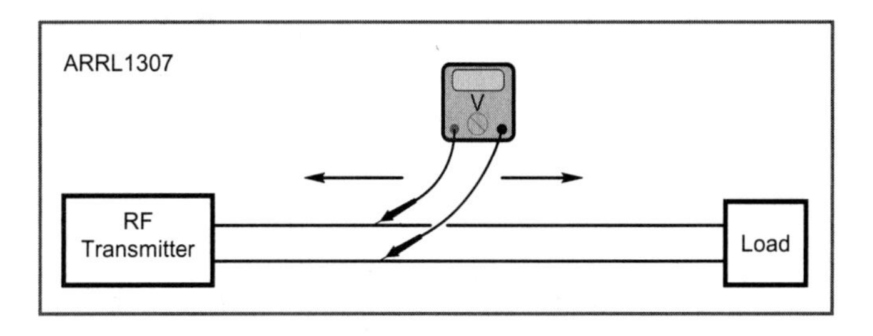


**Figure 4.15 — When the load and line impedances are equal, an RF voltmeter will read a constant voltage at all points. When the line is not matched to the load, a reflected wave will add and subtract to the forward voltage at different points along the line creating voltage maximums and minimums and thus "standing waves"The ratio of these maximum and minimum standing wave values is the *voltage standing wave ratio* or VSWR.**

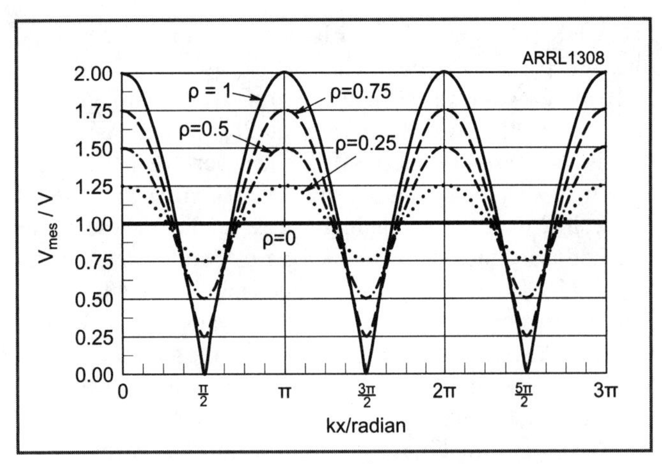


**Figure 4.16 — This graph plots the measured voltage ($V_{mes}$) as we work "back" from a load value (on the far right) for various values of ρ. "V" is the normalized voltage along the line that would be read everywhere when the VSWR is 1:1. All load impedance values are higher than the line impedance. The length of the line shown here is 3/2 λ. As we move an RMS RF voltmeter up and down the line we will observe different voltage values that vary by a sine function, or "standing waves." Several values of ρ are plotted for comparison.**

in no way are affected by each other either in a line or in free space. Yet another analogy can be explained from circuit theory. If we add two sine waves, we may see the result on an oscilloscope of the voltages adding, but the two original signals are present. In contrast, if we *multiply* the two sine waves, the original signals completely disappear, and new signals of different frequencies and amplitudes appear. This is precisely how a "mixer" works in a radio to provide an IF signal by multiplying a signal from the antenna with a local oscillator signal. In free space and in transmission lines (without metal corrosion!) there is no multiplicative transfer function, only summing of the waves present. The difference is as simple as the difference between addition and multiplication, but of trigonometric functions not simple numbers. Consequently, knowledge of simple addition and multiplication of waves (trig functions) gives you the fundamental mathematical basis for describing relatively complicated essential functions in radio theory!

### The Voltage Standing Wave Ratio

Now we gather the terms introduced above and present the essential equations relating them. First, the VSWR results from the summing of all waves present on the line: the incident wave and all reflected waves. Thus:

$$VSWR = \frac{1+\rho}{1-\rho} = \frac{V_{max}}{V_{min}} = \frac{I_{max}}{I_{min}} = \left|\frac{Z_{load}}{Z_{line}}\right| \quad \text{(Equation 4.26)}$$

This equation relates VSWR to the reflection coefficient ρ, where

$$\left|\frac{Z_{load}}{Z_{line}}\right|$$

is the absolute value of the complex impedance ratio.

And the inverse:

$$\rho = \frac{SWR-1}{SWR+1} = \frac{V_{reflected}}{V_{forward}} \quad \text{(Equation 4.27)}$$

A common source of confusion often arises when it becomes apparent that with a single reflection an "infinite" VSWR is possible, yet the maximum voltage along the line is limited to twice the value of the incident wave. Referring to Equation 4.26, note that as $V_{max}$ approaches 2 as in $V_{max}/V_{min}$, $V_{min}$ will approach zero, thus the ratio (VSWR) approaches infinity.

In practical lines, there are multiple reflections, and losses. The reflections of these waves bouncing back and forth on the line encounter ρ values at both the load and source ends that can vary between 0 and 1, and therefore loss values from near zero to many dB. Knowing the maximum voltage on a line is critical for assuring that the line will not "arc" when high VSWR conditions are encountered. When an arc occurs within the foam of a coaxial dielectric, the plastic foam turns into carbon, and carbon is a semiconductor, not an insulator. Thus, the cable (at least at that point on the line) has been destroyed. Also, when calculating voltages for preventing "arcs" it is necessary to calculate the peak-to-peak between the two conductors, which is $V_{ptp} = 2.828\ V_{RMS}$. After a detailed discussion on line reflections, the maximum possible line voltage is presented as Equation 4.32

Thus far we have only discussed real values of impedances. Even with real load impedances, any VSWR greater than 1 will result in impedances along the line with reactive components. The analytical solution to solving the resulting impedance at any point along an unmatched transmission line is both complicated and time consuming.

Perhaps the easiest way to understand the effects of a mismatched line upon line and load current, voltage and power is to consider that *power* must be conserved in a lossless system. Equations 4.26 and 4.27 are of fundamental importance and when combined with power conservation, all the parameters can be defined and emplaned.

## The Smith Chart

Fortunately, there is a very convenient tool which solves for the complex impedance at any point along a transmission line: the Smith chart (see **Figure 4.17**). It was invented by Phillip H. Smith in 1939. In effect, this chart converts the nasty equations for the transformation of voltage and current ratios along with their phase relationships into equivalent impedances by simply drawing circles on the graph.

Unfortunately, many explanations of using Smith charts discourage beginners. At first sight, the very appearance of the Smith chart appears very complicated and thus discouraging. However, it is very simple to use. It has a complicated appearance only because it is solving complicated equations! The hard part of solving the problem is already accomplished in that you can find answers to these equations by simply

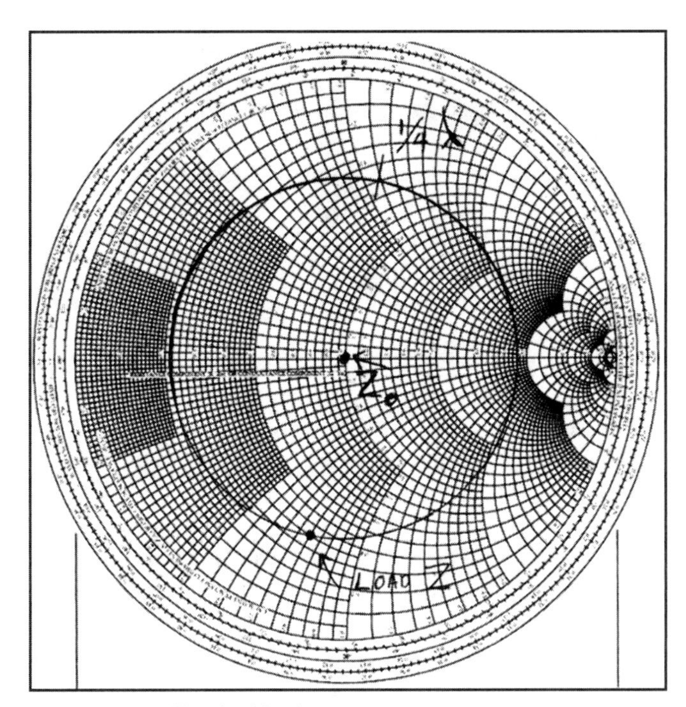

Figure 4.17 – The Smith chart.

finding two applicable points, assign a single value, draw a circle and read the resulting impedance at the desired point(s). So, following a very long sequence of authors attempting to explain the use of the Smith chart, the following very simple 3-step process is offered.

To calculate impedances on a transmission line using a Smith chart:

**Step 1:** "Normalize" the chart by referencing the center of the chart with the characteristic impedance of the transmission line $Z_0$ you wish to analyze. In this case, we're using an open wire line with $Z_0 = 450\ \Omega$. Notice the center of the chart is the intersection of a circle with a value of 1.0 and another circle (actually this one is the horizontal straight line defining a diameter of the circle) labeled 0. Since we normalized the chart for 450 Ω, we simply multiply all the line values by 450 in all subsequent calculations, so the dot at the middle of the chart is very simply 1 × 450 = 450 Ω resistive, and 0 × 450 = 0 Ω reactive.

At this point take a minute or two and see how the sequences of circles relate to both the real and reactive impedances. As a "sanity check" notice that *all* the resistance lines form complete circles on the graph, while *none* of the reactance lines form full circles. Again, the horizontal diameter line represents 0 reactance, so any impedance along this line is a pure resistance whose values are the lines on the chart multiplied by 450. Also notice that along this line all possible real impedances can be plotted (between 0 and ∞). Reactance line values *below* this straight diameter line represent positive reactance (inductive) and line values above, represent negative reactance (capacitive). Again, a non-reactive impedance can only be found on the horizontal diameter line that bisects the circle.

**Step 2:** Find the point on the graph that corresponds with the load impedance (input impedance to the antenna), in this case, it's $Z_{load} = 157.5 + j337.5\ \Omega$, or 157.5 Ω resistive, and 337.5 Ω inductive reactance. Since our graph is normalized to 450 Ω, we find the point that intersects the 0.35 resistance line (157.5/450), and the 0.75 reactance line *(lower half of the chart* 337.5/450). This is the "load Z" point.

**Step 3:** Draw a complete circle centered on the graph's center through the load impedance dot. The resulting circle now defines the impedance on the line *at any* point on the line! Imagine you have a 1/2 λ transmission line (include velocity factor when calculating the physical line length). Working *clockwise* along the circle, you can determine the impedance at any point along the transmission line compared to its *phase* distance (physical length × velocity factor) from the load by simply using inverse arithmetic. Notice that when completing the circle, you're now 1/2 λ away from the load. The complete circle represents 1/2 λ or 180 degrees, or π radians. Printed versions of the Smith chart typically have a circular degree calibration on the circumference for convenience in finding points. If the line is longer than 1/2 λ, we simply keep going around the circle to find the point as each complete rotation around the circle is 1/2 λ. To find the impedance at the point 1/4 λ from the feed point (1/2 rotation around the circle), we see that point has an R value of 0.5 and a –X (capacitive reactance) of 1.1. Multiplying by 450, we derive the actual complex impedance at this point: $225 - j495\ \Omega$. Thus, we solved a set of complex equations using a specialized graph, three points and a circle. Note: *For transmission line calculations it is necessary to move clockwise from the load impedance point back toward the source.*

If the line is perfectly lossless, the original circle defines the impedance at any length of line. You simply keep going in circles until arriving at the applicable point, i.e. to find the feed point impedance of a line 1-1/8 λ, you start at the load impedance point and travel 2-1/4 rotations clockwise from the load impedance point. In the case of losses on the line, the circle becomes a spiral, which spirals inward until converging on the center point, which of course is $Z_0$.

If the line is lossy as you move around the circle, the "circle" will spiral inward until finally will converge at the center point which is the $Z_0$ of the line. However, for lines with reasonably low losses, a simple circle will be reasonably accurate.

Step 1 should only require a minute or two of familiarization with the graph, after which you will have achieved expertise with using the Smith chart for transmission line impedance calculations. The Smith chart can do a great deal more than this simple calculator function, so this first level of competence can be a springboard for a great many more types of impedance problems without having to use all those messy complex equations.

## How a Transmission Line Achieves an Impedance Match

Thus far we have described how reflected waves are formed and how they interact with the forward wave. We made some assumptions about load impedances, but only hinted that there may be more, maybe much more occurring than described above. The description already presented and what follows is actually a chronological sequence of events, sequences that are timed in pico, nano or micro second increments (the time delay along the transmission line).

We began by launching an RF signal down the transmission line. Upon reaching the end of the line we described what happens if there is a matched or several possible unmatched conditions. Now let us return to the matched condition and discover why it's not as simple as we surmised above. Let's assume we have a 50 Ω line terminated in a 50 Ω feed point of a resistor. In this case the resistor is a lumped element, therefore the incident signal (*instantly, or almost*) sees a 50 Ω load, and thus the terminating impedance is established *instantly*. This is what we assume in basic circuit theory. On the other hand, an antenna is *not* a lumped circuit and therefore we cannot assume that the incident wave instantly sees its "normal" feed point impedance. As we described in earlier chapters, an antenna feed point is a function of several variables: a resistive part that includes $R_r$ at the feed point and $R_l$ from resistive losses on conductors and/or from ground losses. It also will have a reactive component that mirrors any phase differences between the RMS voltage and current values that *end up* at the feed point. Also, a traveling wave down a transmission line or entering an antenna element cannot "anticipate" or "see" what lies ahead.

When the incident signal first meets the antenna feed point, the impedance looks quite different from the "normal" antenna value. Since there is not yet any current on the antenna element there can be no resistance value in the feed point impedance. There can be no reactive value since there is not yet any voltage or current to be in or out of phase! However, the incident signal *does* see a continuation of the transmission line conductors, therefore there is some kind of finite and real impedance at the feed point. This initial feed point impedance is likely *not* 50 Ω, therefore in our supposed "perfect match" for the first few nano or micro seconds there is a mismatched condition.

After the initial wave encounters the feed point, RF current begins its journey down the elements along with its voltage and the resulting E and B fields associated with them. The charges travel to the ends of the elements where they are reflected, or circle around (as in loop antennas) back to the feed point. *Only* after these currents and voltages are "set up" on the antenna that the "normal" impedance of 50 Ω is "established" at the feed point. Therefore the "circuit" equivalent of a transmission line or antenna cannot be assumed until all the conditions "settle" throughout the entire system. That's just the first step! How is this "establishment" accomplished? If the initial impedance is not 50 Ω, then what happens to the reflected signal?

The answer to this question is identical to the same question involving an unmatched antenna feed point. We will now set up a typical system that involves a transmitter, tuner, transmission line and antenna and explain the chronological events that transpire within that system.

The system shown in **Figure 4.18** uses a typical transceiver system using two 50 Ω "jumpers" to connect the transceiver with the VSWR meter and then to the tuner, then an open wire line running to the antenna of some arbitrary feed point impedance. This type of system offers very flexible capabilities in that the antenna can be "tuned" over a very large range of frequencies. Also, there is no need for impedance matching at the antenna feed point. The complex impedances are simply brought down to a tuner which can be conveniently located at the operating position and eliminating weight on horizontal elements. Low losses are realized even with comparatively high VSWR since the matched losses are low on open wire or ladder lines. There are limits, however, as losses can become significant even on open wire lines with very high VSWR values. These limits will be quantified in detail later in this chapter.

In effect, a reflected wave coming back toward the transmitter sees the tuner's complex conjugate circuit. Normally we would expect that this reflected wave will continue through this match and on to the transmitter. This is exactly what happens when the transceiver switches to receive as the load for the reflected wave (or an incident received wave) is now 50 Ω. When transmitting, why doesn't the reflected wave go through the tuner and on to the transmitter circuit? A simple answer can be provided by employing some convenient capabilities of *EZNEC* (**Figure 4.19**). We construct a transmission line connecting two very short (relative to the wavelength) wires. We place a source at either end of the transmission line, one representing the forward wave and the other representing the reflected wave. The reflected wave (at the source, including the tuner) will always have a lower voltage and current than the forward wave. Therefore, we simply use a lower source current for the reflected wave source (Source 2) than the incident wave (Source 1).

In Figure 4.19 you have an *EZNEC* model using two sources at opposite sides of a transmission line. Source 1 represents the incident wave, source 2 the reflected wave.

Source 1 is the transmitter output (incident signal). Source 2 represents is the smaller current reflected wave.

In **Figure 4.20**, notice that Source 2 (the simulated

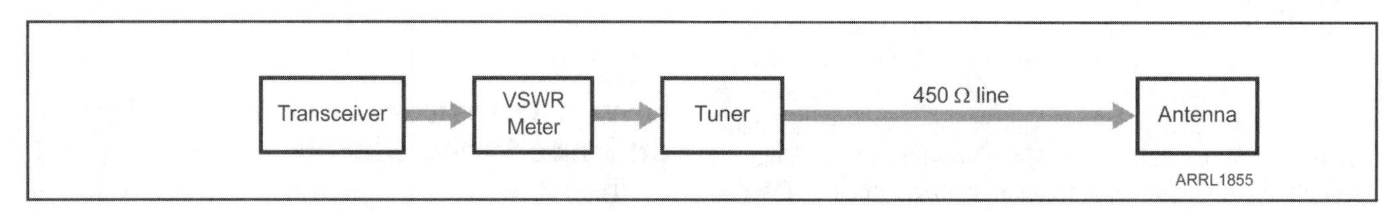


**Figure 4.18 — The system uses a typical transceiver system using two 50 Ω "jumpers" to connect the transceiver with the VSWR meter and then to the tuner, then an open wire line running to the antenna of some arbitrary feed point impedance.**

reflected wave) encounters a *negative resistance* at the transmitter output resulting from a higher power output going in the opposite direction (toward the reflected wave). In contrast, the incident wave encounters a positive resistive component. When there are more than two sources applied to an antenna and/or transmission line system, negative resistance impedances may appear (i.e. often encountered in 4-square vertical arrays). Negative resistance informs the source is a *negative source* meaning that the feed point and transmission line are *absorbing not supplying power* to this part of the system. Therefore, reflected power on a *transmitting* transmission line cannot enter the transmitter *assuming a conjugate match for the incident wave.*

If the incident wave is not matched, the reflected wave may or may not see a negative resistance, and thus may or may not be dissipated or reflected or both. This depends upon the reflecting impedance seen by the incident wave and the details of the output amplifier of the transmitter, which can become very difficult to measure much less calculate. This *EZNEC* model was used *only* to show that negative impedances can exist in antenna/transmission line systems. **Note:** *Accurate* modeling of very low impedances as well as negative resistances is beyond the scope of *EZNEC*'s capabilities but *does* serve to illustrate the point.

Sources

Source Edit Other

| | Sources | | | | | | | |
|---|---|---|---|---|---|---|---|---|
| | No. | Specified Pos. | | Actual Pos. | | Amplitude | Phase | Type |
| | | Wire # | % From E1 | % From E1 | Seg | (V, A) | (deg.) | |
| | 1 | 1 | 50 | 50 | 1 | 1 | 0 | I |
| ▶ | 2 | 2 | 50 | 50 | 1 | 0.9 | 0 | I |
| * | | | | | | | | |

**Figure 4.19 -- *EZNEC* model using two sources at opposite sides of a transmission line. Source 1 represents the incident wave, Source 2 the reflected wave. Source 1 is the transmitter output (incident signal). Source 2 represents is the smaller current reflected wave.**

Source Data

File Edit Search Format Help

```
                         EZNEC Pro/4 ver. 6.0

Open Wire Line Simulation                        3/7/2019     2:22:39 PM

          --------------- SOURCE DATA ---------------

Frequency = 28 MHz

Source 1        Voltage = 1.446 V at -90.0 deg.
                Current = 1 A at 0.0 deg.
                Impedance = 1.345E-08 - J 1.446 ohms
                Power = 1.345E-08 watts
                SWR (50 ohm system) > 100  (473 ohm system) > 100

Source 2        Voltage = 21.87 V at -90.0 deg.
                Current = 0.9 A at 0.0 deg.
                Impedance = -1.28E-08 - J 24.3 ohms
                Power = -1.037E-08 watts
                SWR is undefined (source R is negative)

                Total applied power = 3.078E-09 watts
```

**Figure 4.20 — Notice that Source 2 (the simulated reflected wave) encounters a *negative resistance* at the transmitter output resulting from a higher power output going in the opposite direction (toward the reflected wave). In contrast, the incident wave encounters a positive resistive component.**

## The Conjugate Match

In Chapter 3 we discussed differences among various forms of electric, magnetic and electromagnetic fields. In free space and on a transmission line, we encounter electromagnetic fields. In an LC circuit (usually used for complex conjugate matching), the fields are *not* propagating waves, since the fields created by the lumped components are usually dominated by either the **E** (in a capacitor) or **H** (within and around an inductor). A wave that encounters a matching unit of reactive lumped components is converted into mostly inductive fields, currents and voltages within the circuit. (There is always some small amount of radiation from lumped components such as inductors and capacitors). This energy inside the circuit then sees two impedances represented by the input and output ports of the matching circuit. The source impedances within the matching circuit are "tuned" to the corresponding impedances presented by the transmission lines connected to it. This is a definition of a conjugate match. If there are no "incoming" signals present at either port, the energy will travel in both directions. If, however, there are waves (received or transmitted) one of the ports assumes a negative resistance impedance. This creates a situation where the RF energy in the tuner can travel only toward the positive resistance load. In turn the currents, voltages and resulting fields align themselves accordingly to propagate in the correct direction, thus obeying Maxwell's equations.

The existence of a negative impedance on one of the tuner's ports also indicates a second wave's energy entering the tuner. We assume that both the reflected and incident waves incident on the tuner are ultimately from the same source (a transmitter in the transmitting mode or an antenna in the receiving mode) and therefore are identical in frequency (including sidebands produced by modulation) but different in amplitude (because of attenuation) and phase (because of time delays on the line). Once inside the tuner, the orientation of the fields while propagating disappears and we are left with a complex arrangement of new fields, current and voltages. Within the complex conjugate circuit, these "ingredients" of the energy being stored within the tuner circuit are "forced" to sum and align themselves in the direction of positive resistive impedance. An intuitive understanding can be derived by the action in Figure 3.25. Within an oscillating LC circuit, any *smaller* RF signals of the same frequency added to the circuit or already exiting in the circuit will conform to the phase of the dominant signal. In a perfect tuner, there is no loss, so power (energy) loss is impossible by cancellation. Therefore, the currents, voltages and fields add and propagate toward the load, conforming to the **S** of the incident wave.

This explanation applies to both the transmit and receive conditions; the only difference being the power levels and direction of propagation within the line. Referring to Figure 4.18, in the transmit condition, the transceiver is the power source and the antenna is the load. In the receiver case, the opposite holds true; the antenna is the power source and the transceiver is the load.

## Explanation of Conditions on a Transmission Line

The above discussion introduced the basic dynamics of a mismatched transmission line. Now we will discuss a step-by-step explanation of how a settled condition is established on a line. Using **Figure 4.21**, we will consider the transmit condition. We have tuned the tuner to match 50 Ω to the complex conjugate of whatever impedance is presented to it in the "settled" condition. We tuned the tuner to the "settled" condition by applying a much lower level signal to the input of the 50 Ω (either by greatly reducing the transmitter output or using an antenna analyzer). We tune the tuner until a matched condition is indicated by the VSWR meter, or at least a very low VSWR. The following chronological sequence happens so fast, to the human tuning the tuner circuit it appears as instantaneous but explains *how* a complex conjugate match is achieved.

### Step 1: Transmit Condition

See Figure 4.21. The tuner is tuned to the complex conjugate of the impedance appearing at its output port when the incident signal is feeding the antenna and the "normal" impedance at the antenna feed point has been established. However, the first initial signals coming from the transmitter see a non-conjugate load impedance at the input of the tuner since the tuner is not tuned to 450 Ω. In effect, the first initial signal has not had time to propagate down the line to the antenna and thus set up the beginnings of the settled condition. The tuner reflects a portion of the forward power back to the transmitter but passes some of the power down the line toward the antenna. The antenna's impedance has not yet been established since there is no current yet on the antenna. However, there is an impedance on the antenna different that 450 Ω and therefore a temporary $\rho$ value >0. A portion of the power appearing at the antenna feed point does enter the antenna, setting up the normal current, voltage and field distributions. This changes the feed point of the antenna to its "normal" value. If the tuner is tuned to the $Z_0$ of the line (in this case 450 ohms), then the line itself will act as the load for the tuner and there will be no reflection of the transmitted signal initially. However, again, this is usually not the case since the tuner will be set to whatever impedance is determined by the settled condition.

Another way to think of this problem is to imagine a *very* long transmission line being hundreds of wavelengths long. The output of the tuner can only "see" the $Z_0$ of the line until the first reflected signals from the load (way far away in distance and therefore also time) arrive back at the tuner.

### Step 2: Transmit Condition

See **Figure 4.22**. After the initial "activation" of the system, the transmitter keeps supplying power. After the time delay (round trip time on the 450 Ω line) the reflected signal from the antenna arrives at the tuner and sums with the new incident signal. We now have the first complete "round trip" on the line, thus creating the first standing waves on the line, and where again, some power is accepted by the tuner circuit and some is reflected back up the line. However, the new standing waves on the line create an impedance much closer to the complex conjugate for the tuner. New incident waves see a lower $\rho$, and thus more power is transferred to the line.

### Step 3: Transmit Condition

In **Figure 4.23** the Step 2 process keeps repeating (a "step" process of successive approximation) until the VSWR "settles" and the impedance on the line at the output of the tuner becomes the complex conjugate of the tuner's output. Therefore, $\rho$ becomes 0 at the tuner input, reflection back to the transmitter stops, and the same amount of power delivered by the transmitter is accepted by the antenna (minus transmission line losses). Also, the reflected power on the mismatched 450 Ω line settles to the standard equations for $\rho$ and VSWR, where

$$VSWR = \left|\frac{Z_{load}}{Z_{line}}\right|$$

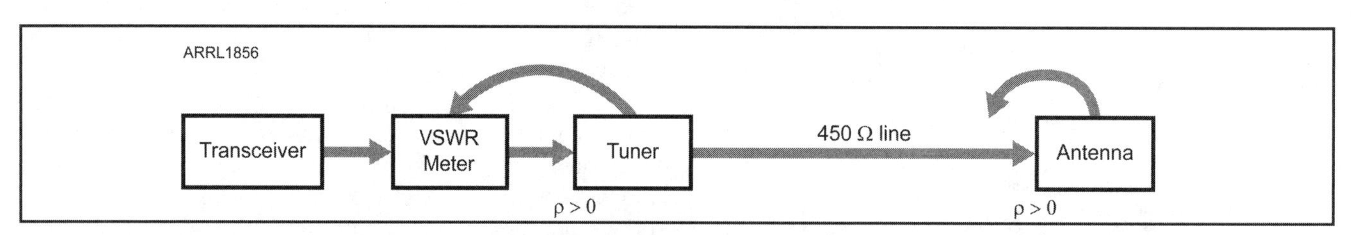


**Figure 4.21 — The initial condition on a "tuned" transmission line.**

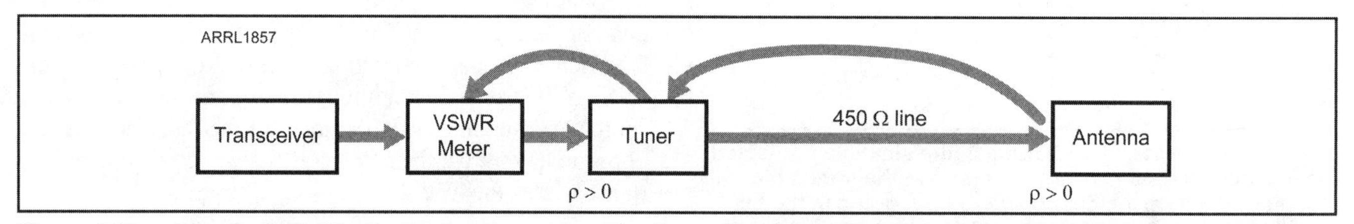


**Figure 4.22 — After the initial "activation" of the system, the transmitter keeps suppling power. After the time delay (round trip time on the 450 Ω line) the reflected signal from the antenna arrives at the tuner and sums with the new incident signal.**

In this settled condition, the power at the output of the tuner (without line loss) will be the power output of the transmitter plus re-reflected power from the antenna. Since there is now a constant supply of power (energy) on the line from the transmitter and reflected waves, power accepted by the antenna equals power output from the transmitter. The constant presence of the reflected waves creates the VSWR, which in turn provides the impedance transformation. The resulting VSWR on the line conforms to Equation 4.26 and the complex impedance at the tuner's output is found by using the Smith chart. All this is completed within nano or microseconds after the initial wave is sent, assuming relatively short transmission line lengths.

The "settling" of voltages and currents on the antenna involves a complex (but necessary) step to "settle" the over-all system parameters. If we ignore the antenna settling process by replacing the antenna with a lumped element, the system settling becomes easier to understand.

Any "keying" of the transmitter involves finite "attack and release times." Thus, the initial wave is of a much lower power level and gradually (usually measured in *milliseconds*) adds adequate buffer to the successive approximation "build-up" to the settled condition. Also, the settled condition will only last for the period of a "dot" or "dash" on CW, a syllable of voice, or a symbol of data.

A common mistake is assuming once an antenna is "tuned" by a remote conjugate matched tuner, that the VSWR drops to 1 on the line. It is true that the VSWR may be unity, or close to it, on the jumper line between the transceiver or source and the conjugate match, that's usually the method to establish a conjugate match! The VSWR on the line directly feeding the antenna will conform to Equation 4.26. In effect, the reflected wave is constant once the settled condition is established.

In most texts, the problem with an antenna settling is ignored as the closed form equations describing the process become very complex and worse, time dependent. Fortunately, the process shown in Figure 4.23 in most instances becomes a close approximation (but never exact) even when non-lumped impedances like antennas are used. The process in Figure 4.23 is *much* easier to understand compared to factoring in the effects of tuners, amplifiers and line losses.

In Chapter 2 we described the basics of how reciprocity occurs on an antenna between transmission and reception. Using the same system as above, we now review the reciprocal operation: the antenna is now the signal source and the transceiver the load.

### Step 1: Receive Condition

See Figure 4.24. In a receive orientation, the antenna is now the signal source and the transceiver (receiver) is now the load. Let us assume the antenna feed point impedance is complex, as in the transmit case. (The feed point impedance is nearly always the same for a given antenna for both transmit and receive conditions.). If the feed point impedance of the antenna is not 450 Ω, ρ will be $> 0$, and thus a part of the received power will be reflected back onto the antenna and re-radiated by the antenna's scattering aperture. If the antenna provides a 450 Ω source, then the system is matched and there are no reflected waves, as in a matched transmit system. In an unmatched system, some of the power is accepted by the line and travels toward the tuner. As in the initial transmit condition, the tuner is not matched to the line, it is tuned to the complex conjugate of the line with a VSWR, which does not yet exist on the line. Therefore, the waves from the antenna see a mismatch and a portion of the incident waves are reflected.

### Step 2: Receive Condition

As shown in **Figure 4.25**, at this point the process of establishing the settled condition is identical to the transmit condition with a few caveats. In the receive case, the identical tuner complex conjugate tuning is used as with transmitting. The discontinuity of impedances: 450 Ω vs. the complex conjugate *creates* the reflected wave back to the antenna. In effect, identical VSWR conditions exist on the line for both the transmit and receive conditions. The power (energy) distribution on the line is total line energy = received power + reflected power = load power (receiver), again, assuming no losses in the line or tuner. In effect the same successive approximation process applies to the receive and the transmit

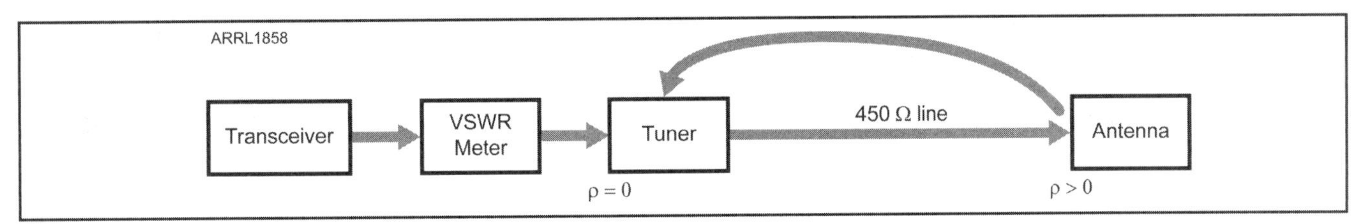


**Figure 4.23 — The settled transmit condition.**

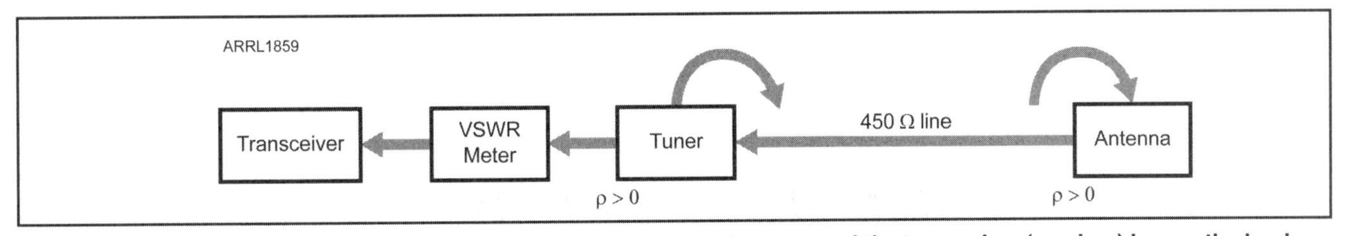


**Figure 4.24 — In a receive orientation, the antenna is now the signal source and the transceiver (receiver) is now the load.**

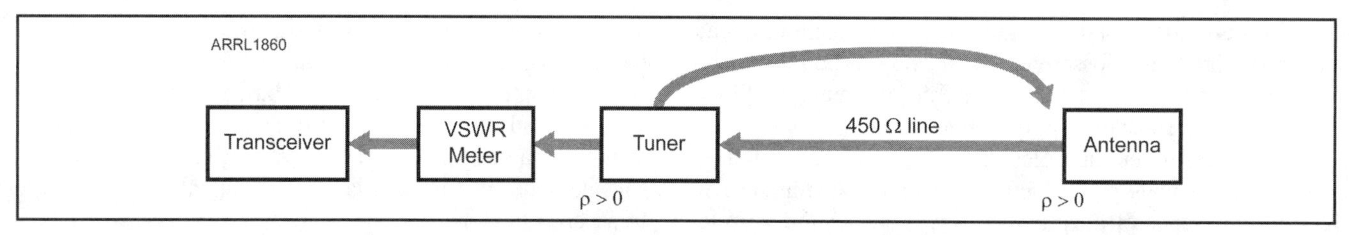


**Figure 4.25 — At this point, the process of establishing the settled condition is identical to the transmit condition with a few caveats. In the receive case, the identical tuner complex conjugate tuning is used as with transmitting. The discontinuity of impedances: 450 Ω vs. the complex conjugate *creates* the reflected wave back to the antenna. In effect, identical VSWR conditions exist on the line for both the transmit and receive conditions. The power (energy) distribution on the line is total line energy = received power + reflected power, again, assuming no losses in the line or tuner.**

condition, but the directions of incident and reflected waves have reversed.

## Two System Examples

We now present two system examples. The first (Figure 4.26) is a solid state transmitter feeding an umatched load. The second example (Figures 4.27-4.30) shows the details of how a conjugate match achieves the settled condition.

**Figure 4.26** is a diagram illustrating the conditions on a system comprising of a 50-W transmitter feeding a 50 Ω line that terminates into a 150 Ω resistive load. The line is exactly 1/2 λ electrical length so that the load impedance is the same as the line impedance where it is attached to the source. Here we can easily derive the VSWR as

$$\frac{150}{50}\Omega = 3, \textit{ and } \rho = 0.5.$$

The forward signal is shown as 50 V at 1 A which conforms to the requirement for a 50 Ω line. Since ρ = 0.5, the reflected voltage *and* current *must be* half of the incident voltage and current. Therefore, the reflected power is 12.5 W, which corresponds to $P_{reflected} = P_{forward}\,\rho^2$. The product is 37.5 W dissipated by the load thus satisfying system power. In this case, we assume a resistive power source (such as the case with a solid-state amplifier), the reflected 12.5 W will be dissipated (as heat) by the amplifier's transistors. In an actual system, the line will accept *less* power than 50 W since the line impedance at the transmitter will be 150 Ω. However, the forward and reverse signals V/I ratio *must* conform to characteristic line impedance of 50 W

### Step 1

We now place a conjugate match into the system (**Figure 4.27**) where the reflected power will be *combined* and therefore for *added to* the forward power in the conjugate match. The conditions shown in Figure 4.27 represent the instant in time *before* the first reflected wave reaches the conjugate match and is combined with next forward wave.

In reality, the forward wave is *not* as it is shown here since the load impedance has not yet been "seen" by the initial wave. Let's imagine that the conjugate match automatically "sees" the 450 Ω load and that in the further steps that follow, it matches automatically to the changes to a conjugate condition. Gradually (as we will see) the impedance at the

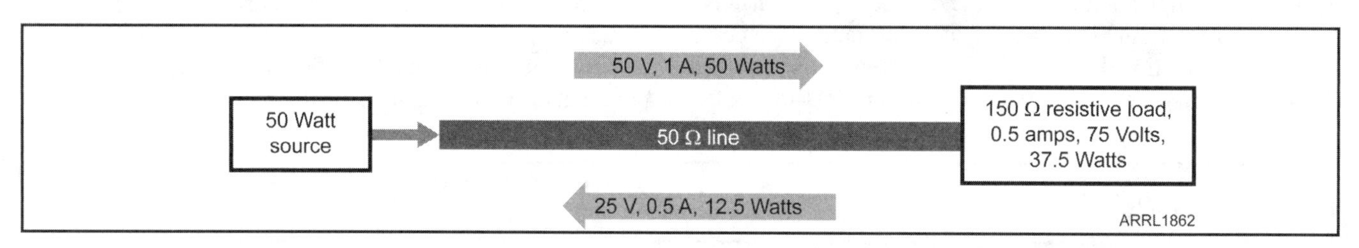


**Figure 4.26 — This is a diagram illustrating the conditions on a system comprising of a 50-watt transmitter feeding a 50 Ω line that terminates into a 150 Ω resistive load with no conjugate match.**

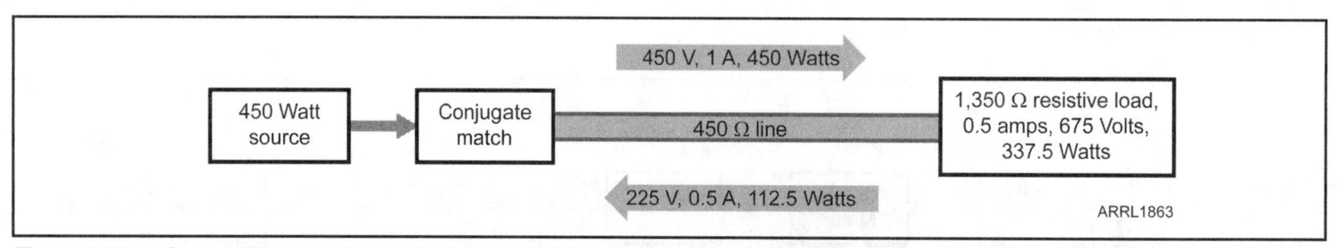


**Figure 4.27 — Step 1: We now place a conjugate match into the system where the reflected power will be *combined* and therefore for *added to* the forward power in the conjugate match. The conditions shown above represent the instant in time *before* the first reflected wave reaches the conjugate match and is combined with next forward wave.**

conjugate match will converge to the load impedance of 1350 Ω since the 450 Ω line is 1/2 λ electrical length. The line connecting the 450-W source (transmitter) to the conjugate match is (presumably a 50 Ω coaxial line) of any convenient length. However, these steps do provide an accurate conceptual sequence of steps to establishing a settled condition when using a mismatched line tuned with a conjugate match.

### Step 2

See **Figure 4.28**. After the first summing of power in the conjugate match, the new forward power is 562.5 W (450 + 112.5). Notice that the reflected voltage and current conforms to the system's $\rho = 0.5$. This is the condition before the second reflected wave reaches the conjugate match. (add reflected power from Step 1 to the 450-W incident power as this condition is *before* the 140.6 W reflected power combines with the 450-W incident power.)

### Step 3

**Figure 4.29** includes the data for the third reflected wave Notice that all currents, voltages and powers are increasing. Also notice that the *rate* of these increases has slowed. This points to a very gradual increase until we reach the *limit*. The limit is defined where the load power is equal to the transmitted power (450 W). Of course, this assumes there are zero losses in the system. With losses, the load power limit will be somewhat less than the input power of 450 W. (add reflected power from Step 2 to the 450-W incident power). Again, this is the condition *before* this new reflected power is added to the incident power.

### Step 4

See **Figure 4.30**. In principle there are an *infinite* number of steps to reach the settled condition since the solution is achieved through successive approximation to reach the limit. With each new reflection, the limit is approached in smaller and smaller intervals of all the terms. It is interesting to do the computations for the settled condition for the critical terms: VSWR, ρ, and power conservation. We begin with the 450 W from the transmitter. In the conjugate match we add the 150 W reflected power resulting in 600 W forward power on the 450 Ω line. The load absorbs (through heat or for an antenna radiates) 450 W and the remaining 150 W is reflected back to the conjugate match and the process repeats.

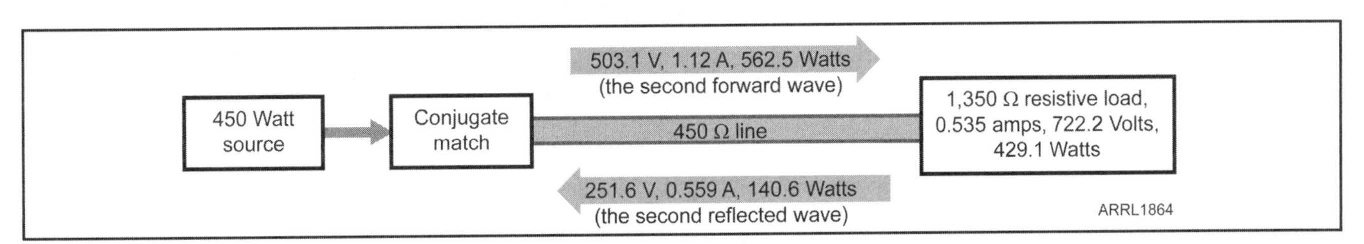


**Figure 4.28 — Step 2: After the first summing of power in the conjugate match, the new forward power is 562.5 watts (450+112.5). Notice that the reflected voltage and current conforms to the system's ρ = 0.5.**

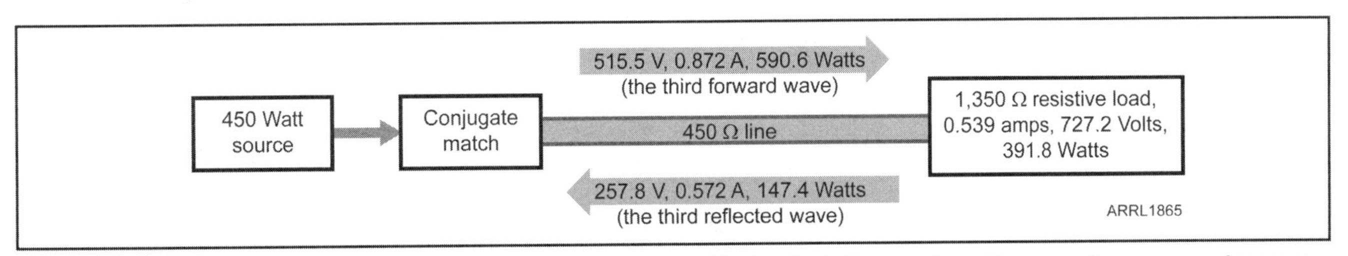


**Figure 4.29 — Step 3: Here is the data for the third reflected wave Notice that all currents, voltages and powers are increasing. Also notice that the *rate* of these increases has slowed. This points to a very gradual increase until we reach the *limit*.**

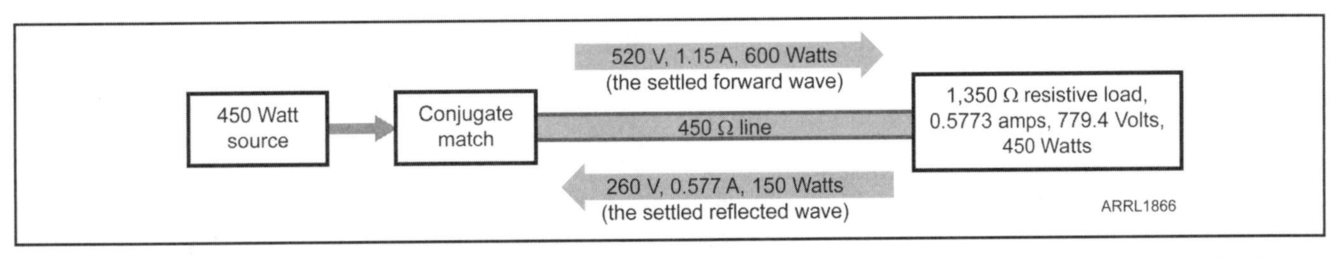


**Figure 4.30 — Step 4: In principle there are an *infinite* number of steps to reach the settled condition since the solution is achieved through successive approximation to reach the limit. With each new reflection, the limit is approached in smaller and smaller intervals of all the terms. It is interesting to do the computations for the settled condition for the critical terms: VSWR, ρ, and power conservation. We begin with the 450 watts from the transmitter. In the conjugate match we add the 150 watts reflected power resulting in 600 watts forward power on the 450 Ω line. The load absorbs (through heat or for an antenna radiates) 450 watts and the remaining 150 watts is reflected back to the conjugate match and the process repeats.**

The condition is stable and conforms to all critical terms.

In this settled condition we can see that 150 W is constantly moving up and down the line in both directions in addition to the constant application of the incident 450 W from the transmitter and the 450 W constantly being removed from the system by the load. The two 150-W signals are like two ping-pong balls being reflected back and forth at the same time. In effect, we can calculate the *total power* on the line simply by adding up the new forward power (600 W + the reflected power 150 W). In effect, it is this "stored" (but propagating) power on the line which creates the voltage and current peaks and nulls along the line and thus also provides different impedances at different points along the line.

## Defining Voltage and Current Maximums and Minimums

As discussed before, the existence of voltage and current minimums and maximums along a line with a single reflection are formed by simple addition and subtraction of the forward and reverse signals propagating on the line. However, on a line with a conjugate match and multiple reflections, and using the settled condition, the peak RMS voltage in this example requires the use of both the VSWR and the *total* forward power or,

$$V_p = \sqrt{VSWR(PZ_0)} \quad \text{(Equation 4.28)}$$

where

$$\sqrt{(PZ_0)}$$

is the RMS voltage of the incident wave using the power and line impedance for the calculation. However, if we are calculating true "peak" voltage (better for considering arcing on lines, insulators, passive components, switches, etc.) we need to multiply the RMS voltage by 1.414, or in this example, 1,003 V. Now that we have these terms, we can easily compute voltage, and current maximums and minimums. See Table 4.2.

The above calculations assume that all the reflected power is re-reflected back to the load by virtue of the conjugate match.

In the first example we assumed that all the reflected power was absorbed by the source (usually the transmitter's final amplifier. This is a close approximation in solid state transmitters, where the output circuit is usually fixed. In the old tube transmitters, the adjustable pi-net output circuit could permit some if not all reflected power to be re-reflected (either a partial or near-perfect conjugate match). From the first examples we see that the maximum voltage will be 50 + 25 = 75 and the minimum 50 – 25 = 25, which also gives the same VSWR of 3:1. However, the equations for minimum and maximum RMS values change. See Table 4.3.

**Table 4.2**

Maximum and minimum RMS voltages and currents along a transmission line with conjugate matching (all power is dissipated by the load). For peak values multiply the RMS values by 1.414.

| *Term* | *Equation for RMS Values* |
|---|---|
| Voltage maximum | $V_{max} = \sqrt{VSWR\ (PZ_0)}$ |
| Voltage minimum | $V_{min} = \sqrt{\frac{(PZ_0)}{VSWR}}$ |
| Current Maximum | $I_{max} = \sqrt{\frac{(P)(VSWR)}{Z_0}}$ |
| Current Minimum | $I_{min} = \sqrt{\frac{P}{Z_0(VSWR)}}$ |

**Table 4.3**

Maximum and Minimum RMS voltages and currents along a transmission line with no matching (all reflected power is dissipated at the source). For peak values multiply the RMS values by 1.414.

| *Term* | *Equation for RMS Values* |
|---|---|
| Voltage maximum | $V_{max=\sqrt{PZ_0}} + \rho\sqrt{PZ_0}$ |
| Voltage minimum | $V_{min=\sqrt{PZ_0}} - \rho\sqrt{PZ_0}$ |
| Current maximum | $I_{max=\sqrt{PZ_0}} + \rho\sqrt{\frac{P}{Z_0}}$ |
| Current minimum | $I_{min=\sqrt{PZ_0}} - \rho\sqrt{\frac{P}{Z_0}}$ |

## Power Conservation

A common source of confusion arises from a simple question: Using the example in Figure 4.27 in the initial condition we have a 450 volt initial wave and 225 volts is reflected, but 675 volts appears across the load. How can this be? It is important to note three key points: First, ρ defines the voltage and current reflection coefficients *on the line*. Second, the voltage and current at the load must conform to Ohm's Law. Third, power conservation must also be obeyed so we get a voltage transformation due to the $IR_{load}$ voltage drop across the load where $I$ is set by ρ. This is the case for a load impedance *higher* than the line impedance. In the case where the load impedance is *lower* than the line impedance, the load voltage is set by ρ and the current is defined by

$$\frac{V}{R_{load}}$$

Of course, these simple relationships imply a pure resistive load. We can conclude this discussion by simply pointing out that ρ defined the *reflection* coefficients for line voltage and current, while $\rho^2$ defines the coefficient for reflected power, while power conservation must be obeyed where

$$\rho^2 = \frac{P_{reflected}}{P_{incident}} \quad \text{(Equation 4.29)}$$

And

The proportion of forward power is dissipated by the load in a single reflection condition is:

$$1-|\rho^2| \quad \text{(Equation 4.30)}$$

On a transmission line with a $Z_0$ that has no reactance (most practical transmission lines have a very small reactive component so are usually assumed to have a real impedance only), there can be only a purely "real" power in both the incident and reflected waves. The voltage and current are always in-phase for waves moving toward the load and always 180 degrees out-of-phase for wave moving toward the source. The proportion of "real" power in a wave is defined by the power factor,

$$Power\ factor = |cos\theta| \quad \text{(Equation 4.31)}$$

where θ is the phase difference between voltage and current on a transmission line or a circuit. Since on a transmission line this phase difference is only 0 or 180 degrees, only real power exists. The percentage of real power is simply 100 × the power factor. A phase difference of 90 or 270 degrees indicates only "imaginary" or reactive power.

If the load impedance is complex, the reflected wave is "real". However, the location of the standing waves will shift. Whereas when the load is a pure resistance, the maximum voltage or maximum current (VSWR) will appear at the load but with a reactive load the voltage and/or current maximums and/or minimums will not appear at the load. A load impedance with a reactive component will produce a purely "real" reflected wave but will have the effect of phase shifting the distribution of standing waves on the line.

In **Figure 4.31** you'll find the plots the percentage of power reflected vs. the VSWR where:

$$\%\frac{Reflected\ power}{Incident\ Power} = 100\left(\frac{VSWR-1}{VSWR+1}\right)^2 \quad \text{(Equation 4.32)}$$

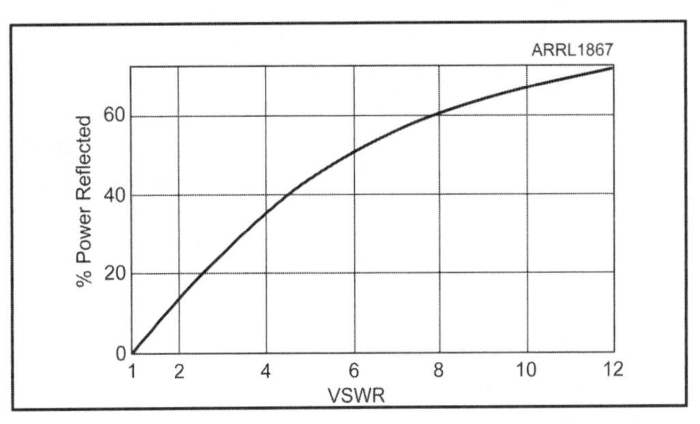


**Figure 4.31 – A plot of the percentage of power reflected vs. the VSWR. See text.**

## Resistive and Radiation Losses in Transmission Lines

Earlier in this Chapter we introduced the basic factors that lead to line loss. The two sources of resistive losses in coaxial transmission lines are the resistance of the line's conductors and dielectric losses. These losses also will modify the characteristic line impedance as shown in Equation 4.23. In open wire lines, loss through line radiation also must be considered (Equation 4.39) whereas in coaxial lines radiation loss is usually assumed to be zero. In many cases this can be a mistaken assumption. The advantages of coaxial lines over open wire lines have already been discussed. However, most common coaxial lines use "braided" metal wire for the outer conductor rather than a solid metal tube. Any variation from a solid outer conductor will permit line radiation, and through reciprocity, introduction of unwanted signals entering the coaxial line. Many techniques have been devised to lessen this problem: higher density mesh, multiple mesh layers, and the use of solid foil in addition to mesh, to name a few.

A well-known design minimizes dielectric losses in coaxial lines by using a helix shaped insulating separator between the inner conductor and the shield. In this type of "heliax" configuration, the distance between the center conductor and shield must be increased to provide adequate physical integrity to the line and to keep the center conductor from shorting to the outer conductor when the cable is bent in installation. Another common approach is to use insulation "beads" to stabilize the separation but also requires the size of the line to increase. The increase in size necessitates the use of considerably more metal and insulating material thus increasing the cost and weight of the line. However, it also further reduces the resistive loss of the line.

In **Figure 4.32** we see helictical coaxial cable (heliax) with the insulator and conductors configured as helixes. This configuration provides for very low dielectric loss since most of the dielectric is air. A solid

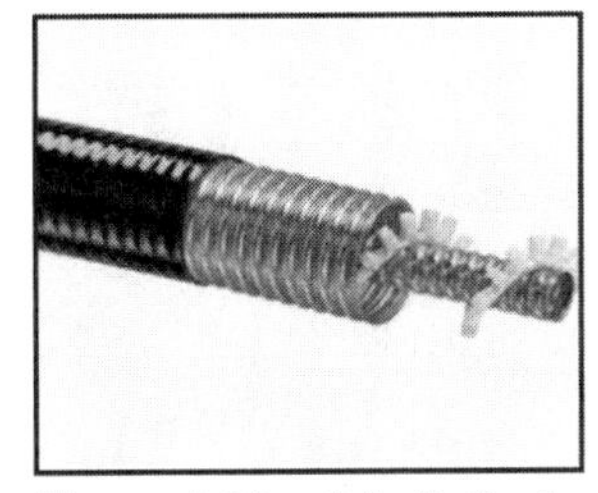

**Figure 4.32 -- A helictical coaxial cable.**

cylindrical outer conductor also minimizes cable radiation. The corrugated conductors also allow for bending of the cable (within a given minimum radius). The conductors can be either helictical shaped or simply corrugated to permit bending without kinking the conductive tubing. To keep moisture from entering inside the air dielectric, dry nitrogen gas can be used to fill the inside at a few pounds per square-inch pressure.

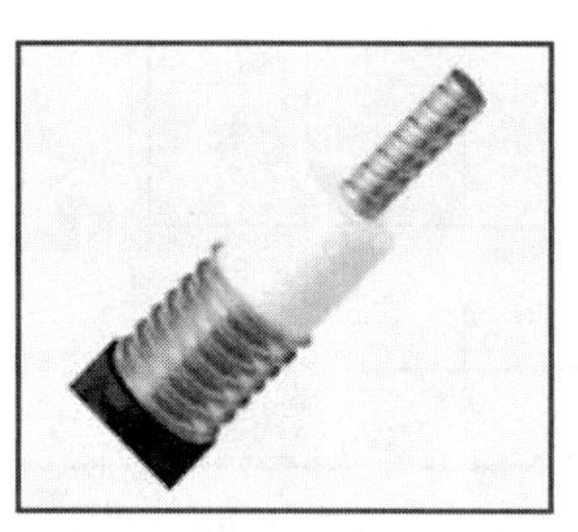

**Figure 4.33 -- A configuration of helictical conductors with a foam**

In **Figure 4.33** you'll see a configuration of helictical conductors using a foam dielectric. This permits the coax to bend (to a limited radius) without risk of distorting the two conductors but has the disadvantage of using a foam dielectric vs. air. The use of foam dielectric heliax is more common in smaller diameter cables. Radiation from heliax cables is minimum and resistive losses are typically low due to the larger size of the cable, and the associated conductors.

Earlier in this chapter, in Table 4.1 and Figure 4.3, it is assumed that radiation from a two-wire transmission line is canceled since $R_r$ is zero. A caveat was also stated that some radiation will occur in practical lines since the spacing between the two conductors is not zero, and therefore the spacing is some very small fraction of the signal's wavelength. Therefore, we conclude there must be some high frequency limit for a given spacing of the conductors and thus intuition informs that using higher frequencies requires closer spacings

**Table 4.4**
**Summary of Terms and Equations for Reflected Waves on a Transmission Line**

| Term | definition | Equation(s) |
|---|---|---|
| $V_{forward}$ | Forward or incident voltage | $V_{forward} = \frac{P_{forward}}{I_{forward}}$ |
| $V_{reflected}$ | Reflected voltage | $V_{reflected} = \frac{P_{reflected}}{I_{reflected}} = \rho V_{forward}$ |
| $I_{forward}$ | Forward or incident current | $I_{forward} = \frac{V_{forward}}{P_{forward}}$ |
| $I_{reflected}$ | Reflected current | $I_{reflected} = \frac{V_{reflected}}{P_{reflected}} = \rho I_{forward}$ |
| $Z_0$ | Transmission line characteristic impedance | Equations 4.5 to 4.18 |
| $P_{forward}$ | Forward or incident power | $P_{forward} = \frac{V_{forward}{}^2}{Z_0}$ |
| $P_{reflected}$ | Reflected power | $P_{reflected} = P_{forward}\,\rho^2$ |
| ρ | Reflection coefficient | $\rho = \frac{V_{reflected}}{V_{forward}} = \frac{SWR - 1}{SWR + 1}$ |
| VSWR | Voltage standing-wave ratio | $VSWR = \frac{V_{max}}{V_{min}} = \frac{1 + \rho}{1 - \rho}$ |
| VF | Velocity factor | $VF = \frac{v_{line}}{c}$ |
| $P_{load}$ | Power dissipated by load | $P_{load} = P_{forward}(1 - \rho^2)$ |
| $V_{maxref}$ | Maximum rms voltage on a line (reflected power re-re-reflected at the source) | $V_{maxref} = VSWR\sqrt{P_{out}Z_0}$ |

for similar non-radiating performance. Equations 4.11 and 4.18 show that spacing (S) of the conductors is also a major determinate of the line's characteristic impedance, therefore building lines with closer spacing will result in a lower characteristic line impedance as well as lower radiation at higher frequencies. Before coaxial cable was widely available, television antennas typically used 300 Ω "twin lead" with wire spacings typically less than 1/2 inch. The very small spacing is critical when considering VHF and UHF feed lines. When considering open wire lines, a key question is: what is the relationship between frequency and loss to radiation for a given spacing? The "loss" also results in susceptibility to unwanted signals being received thus both transmit and receive performance must be considered.

In Table 4.1 and Figure 4.3 we assumed 0 Ω $R_r$ for a transmission line whose spacing is sufficiently small relative to a wavelength. In coaxial cables the electric and magnetic fields are constrained between the shield and the inner conductor. Therefore, in principle, no radiation can emerge from such a geometry. However, in open wire and ladder lines, there is a finite spacing. Therefore, some radiation will occur, unless the "open" wire line is contained within a solid shield.

Also, the conductors that form the open wire line will also exhibit resistive loss. In "ladder line" about 50% of a given length will have a layer of plastic (polyethylene) as a dielectric. However, this plastic is typically wide (about 1 inch) and thin (about 1/16 inch), reducing the dielectric loss compared to "twin lead".

The usual loss specifications for transmission lines show loss for a given type of line for a perfectly matched condition (ML) as in **Figure 4.34** taken from the *ARRL Antenna Book*. The graph shows frequency on the X axis and dB loss per100 feet on the Y axis. Log-log graph formatting typically yields a liner relationship. In this type of presentation, loss is dependent upon resistive losses in the conductors and the dielectric material separating them.

I have devised a method for accurately modeling open wire lines using *EZNEC* for my own needs and will now share this method for those that may find this technique useful. I have been using "ladder lines" for many years, but never really knew what to expect under varying conditions of use. I also was curious to know if many of the detailed specifications

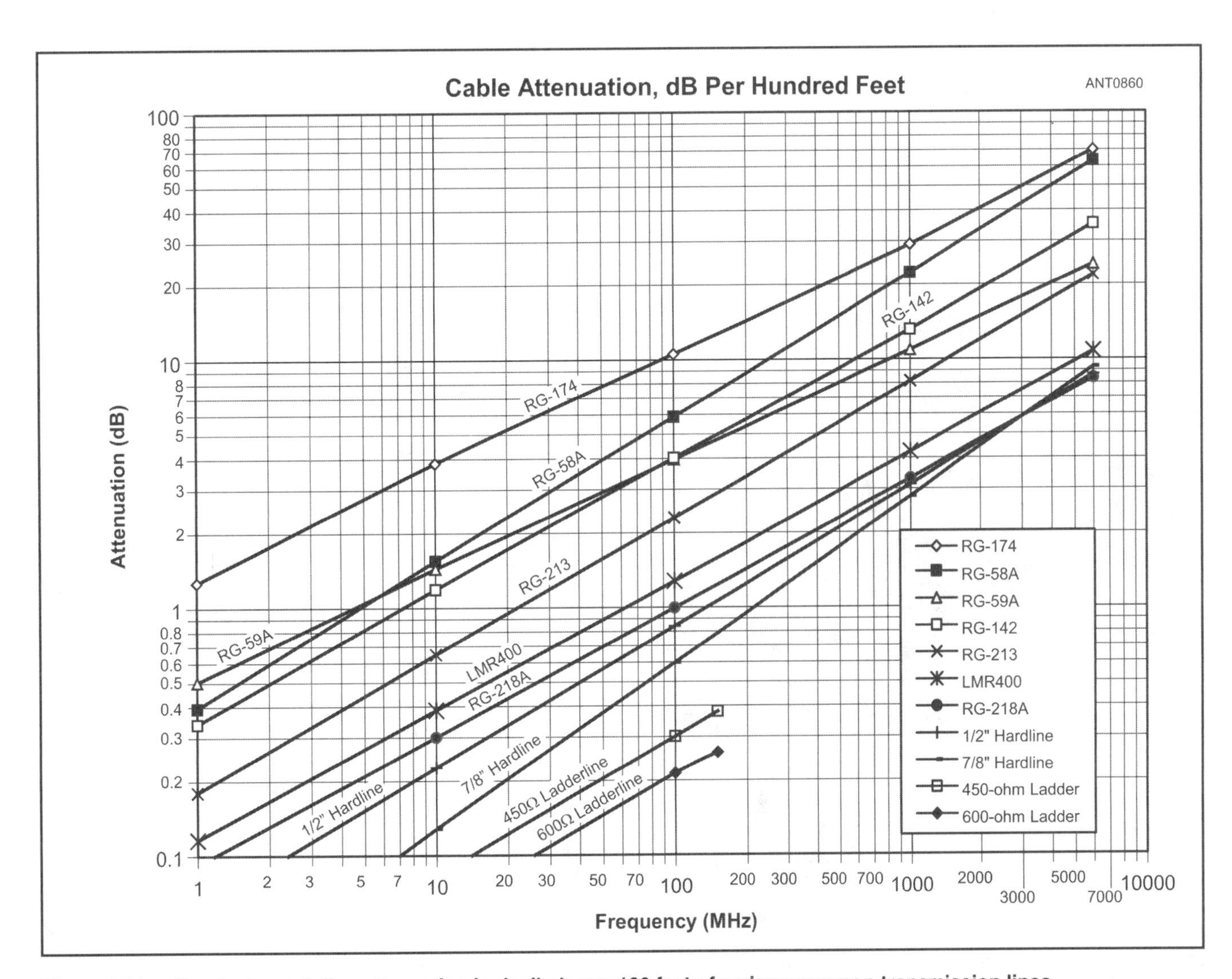


**Figure 4.34 — Nominal match-line attenuation in decibels per 100 feet of various common transmission lines.**

for coaxial cables published in the *ARRL Antenna Book* and the multiple other sources available would be applicable also to open wire and ladder lines alike. This section provides answers that were satisfactory for my own needs and hopefully for the readers of this section.

For those not interested in performing detailed models, I include an extensive amount of data and conclusions, mainly in graph form, that may be helpful. The reader may skip the part describing the *EZNEC* method and go right to the results.

Among the topics covered from data taken from *EZNEC* models are:

1. Matched (VSWR=1) resistive line losses vs
   - Line wire size
   - Wire spacing (and thus also line impedance)
   - Different line wire metals
   - Proximity to ground
   - Ladder vs. open wire lines
2. Unmatched losses (increased loss vs. VSWR)
   - Losses from line radiation vs
   - 2-wire line vs. 4-wire line
   - Wire separation
   - Line length

All this data includes plots over frequency of operation and applicable variables.

## Open Wire Transmission Lines

Thus far we have addressed transmission lines generally and focused only on coaxial type lines. In today's amateur literature there is little detailed information on open wire and so-called "ladder lines". Before the ready availability of coaxial cables (many decades ago) open wire lines such as those shown in **Figure 4.35** were the main type of transmission line used, and in most cases the only type of line available. The use of open wire lines extended to nearly every radio service. AM broadcast systems used them extensively.

Figure 4.35 — A 70-foot run of homebrew open wire line at the author's station. A 500-foot run connects the shack (200 feet below this point and hidden to the right) and the terrain peak behind the camera.

Over the decades coaxial cable has become the de-facto standard line used in amateur installations. Today, detailed information regarding open wire lines is anemic. This lack of readily available information has caused considerable misunderstanding regarding the characteristics of open wire lines especially when compared to coaxial lines. This section is an attempt to re-introduce readers to these advantages and disadvantages as well as dismiss some oversimplifications and errors.

## Modeling Open Wire Line Using EZNEC

The following introduces and explains in detail, how to model open wire lines using *EZNEC*. The following section "Data from *EZNEC* for Open Wire Lines" is included for readers who may only need answers to some basic questions on open wire lines. The inclusion of this section is to provide details of this method of using *EZNEC* to enable the serious reader with this tool for his/her special modeling needs for station design and/or additional research. For those that do not need this comprehensive analysis, there is an excellent supplement to the *ARRL Antenna Book* (23rd edition) in the form of a Windows calculator called *TLW, Transmission Line Program for Windows* by N6BV.

Comparisons of simulations using *NEC2* and *NEC4* engines does produce slight differences in results, but these differences are typically less than 1% for free space modeling. However, for ground proximity models, *NEC4* must be used (a license for using *NEC4* is required from the Lawrence Livermore Labs), and the more advanced versions of *EZNEC* (*EZNECpro*) are required to run these more advanced software engines. The free space simulations presented here use *NEC2* on the *EZNEC* Pro4 v. 6.0 platform. As this program was developed, the modeling results were carefully compared with readily available information on open wire lines. The results conformed very well with established equations and their derivative data. This provided confidence in the validity of the program and to pursue new areas of interest in open wire research.

Transmission lines are an included option within *EZNEC* (**Figure 4.36**), however, as shown in Figure 4.36 it is necessary for the user to *know* beforehand the critical parameters of $Z_0$, VF and loss per 100 feet. It is also possible to model open wire lines directly onto antenna models, but these "custom" transmission line models do not yield critical line parameters *per se*. Isolating and characterizing lines as subsystems requires some extra steps using *EZNEC*.

Figure 4.45 uses six arrows on the *EZNEC* Control Center to point out the critical *EZNEC* tools used in this technique. Therefore, all references to "arrow #" refer to this Figure which shows the *EZNEC* Control Center.

Transmission Lines

Trans Line Edit

| Transmission Lines | | | | | | | | | | | |
|---|---|---|---|---|---|---|---|---|---|---|---|
| No. | End 1 Specified Pos. | | End 1 Act | End 2 Specified Pos. | | End 2 Act | Length | Z0 | VF | Rev/Norm | Loss | Loss Freq |
| | Wire # | % From E1 | % From E1 | Wire # | % From E1 | % From E1 | (ft) | (ohms) | | | (dB/100 ft) | (MHz) |
| * | | | | | | | | | | | | |

**Figure 4.36 – Transmission lines are an included option within *EZNEC*. However, the antenna engine *ignores* the effect of the transmission line on the antenna's pattern and it is necessary for the user to *know* beforehand the critical parameters of $Z_0$, VF, R loss, and radiation loss.**

## Wire Modeling a Two-Wire Line

For a two-wire line this process involves the use of 4 wires in the *EZNEC* model: two wires for the line proper (wires 1 and 2), one short wire (wire 4) contains the signal source while another short wire (wire 3) contains the load (**Figure 4.37**). It should be pointed out at this point that this modeling technique can be combined with the more typical antenna models using NEC. The combination should be intuitive for users with competent *EZNEC* skills.

The wire definitions for a typical two wire line are shown in **Figure 4.38**. The length of the line is 100 feet., shown along the X axis. The wire spacing is 6 inches, shown on the Y axis. Therefore 6 inches is also the length of wires 3 and 4 which contain the load and source.

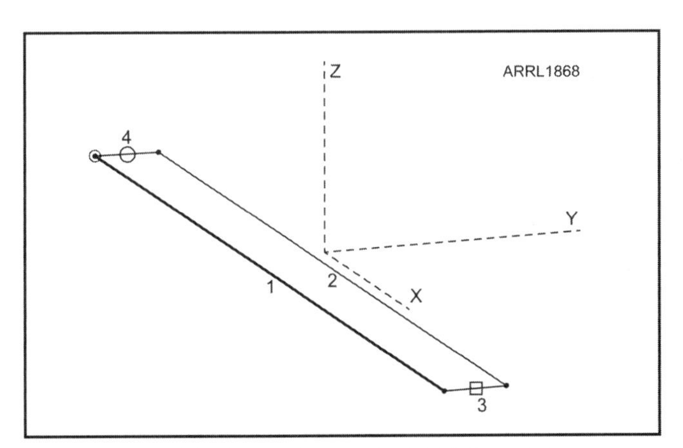


**Figure 4.37 -- This figure shows the *EZNEC* model used in this technique. The line shown here is very short to show the details. To maximize accuracy the same number of segments should be used on both wires comprising the transmission line itself (wires 1 and 2).**

The diameter .081" corresponds to wire gauge #12. Notice that all four wires use the same wire thickness. The NEC engine may make mistakes if different thicknesses are used, including the end wires, especially for $Z_0$ calculations.

**Table 4.5**
**AWG vs. Wire Diameter**

| *AWG* | *Diameter (Inches)* |
|---|---|
| #6 | 0.162 |
| #12 | 0.081 |
| #14 | 0.064 |
| #16 | 0.051 |
| #18 | 0.040 |

Wire thickness in inches can be found in Table 4.5 from AWG gauge numbers. This information is also readily available on-line from multiple sources.

## Wire Insulation Material

In Figure 4.38 the wire prompt box includes provision for wire insulation. *EZNEC* can only model insulation thicknesses commonly found on "real" wires. The two key parameters for line spreaders (which also apply to RF insulators generally) are the dielectric constant $(\varepsilon)$ and the dielectric tangent loss. These two parameters are also shown to specify the type of insulation on the wire. This data can be found for some common electrical wire insulators in Table 4.6.

Insulation on the wires will have near-zero effects on loss, even when using PVC insulation. However, the insulation will affect the velocity factor by a few percent.

## Source placement, Load placement and Determining the Line's $Z_0$

The next step (see **Figure 4.39**) will be to place the source at one end of the line and the load at the other end.

Wires - (2)

Wire Create Edit Other

☐ Coord Entry Mode ☐ Preserve Connections ☑ Show Wire Insulation

| Wires | | | | | | | | | | | | | |
|---|---|---|---|---|---|---|---|---|---|---|---|---|---|
| No. | End 1 | | | | End 2 | | | | Diameter | Segs | Insulation | | |
| | X (ft) | Y (ft) | Z (ft) | Conn | X (ft) | Y (ft) | Z (ft) | Conn | (in) | | Diel C | Thk (in) | Loss Tan |
| 1 | 0 | 0 | 0 | W4E1 | 100 | 0 | 0 | W3E2 | 0.0808 | 57 | 1 | 0 | 0 |
| 2 | 0 | 0.5 | 0 | W4E2 | 100 | 0.5 | 0 | W3E1 | 0.0808 | 57 | 1 | 0 | 0 |
| 3 | 100 | 0.5 | 0 | W2E2 | 100 | 0 | 0 | W1E2 | 0.0808 | 1 | 1 | 0 | 0 |
| 4 | 0 | 0 | 0 | W1E1 | 0 | 0.5 | 0 | W2E1 | 0.0808 | 1 | 1 | 0 | 0 |
| * | | | | | | | | | | | | | |

**Figure 4.38 — The wire definitions for a typical two wire line in *EZNEC*.**

**Table 4.6**

| *Insulation (wire type designation)* | $e_m$ *1 MHz* | $e_m$ *1 GHz* | *Δ 1 MHz* | *Δ 1 GHz* |
|---|---|---|---|---|
| Air | 1 | 1 | 0 | 0 |
| Polyvinylchloride PVC (THHN) | 2.7 | 2.8 | 0.006 | 0.019 |
| Polyethylene PE (XHHW) | 2.2 | 2.2 | 0.0003 | 0.0003 |
| Teflon PTFE | 2.0 | 2.0 | 0.0002 | 0.0002 |

**Figure 4.40** shows the boxes for the load data entries. The next step is to determine the line's $Z_0$. This will involve simultaneous changing for the values of the load resistance and the Alt SWR Z0 (Arrow 6). But first, we will get an approximate value for $Z_0$ from **Figure 4.41**.

We are using #12 wire spaced at 6 inches, therefore Figure 4.41 gives a $Z_0$ of about 600 Ω. We set the load resistance value to 600 Ω and the **Alt SWR Z0** also to 600 Ω. The rest of the Control Center settings should be the following:

**Frequency**: 1.8 MHz (but not critical at this point)

**Transmission Lines**: 0 (we're not using the *EZNEC* option)

**Ground type**: free space

**Wire loss**: copper (or whatever metal is used)

**Plot type**: 3D (typically used for this program, details later)

## Segmentation Check

In **Figure 4.38** the number of segments for each wire is shown. The load and source wires (wires 3 and 4) may be always left with a single segment. There are two critical rules that must be followed for segments on the transmission line wires (wires 1 and 2). First, these wires must be identical in length. Second, make certain that there are the proper number of segments for the frequency being modeled and the length of the line. I found it most useful when modeling multiple frequencies to begin with the lowest frequency and set the segment numbers for wires 1 and 2 to 1 segment each. When you do this an error message will appear. See Figure 4.42. To get the proper number of segments on wires 1 and 2, press "segmentation" and then "conservative". *EZNEC* will then auto set the number of segments up to four for wires 1 and 2.

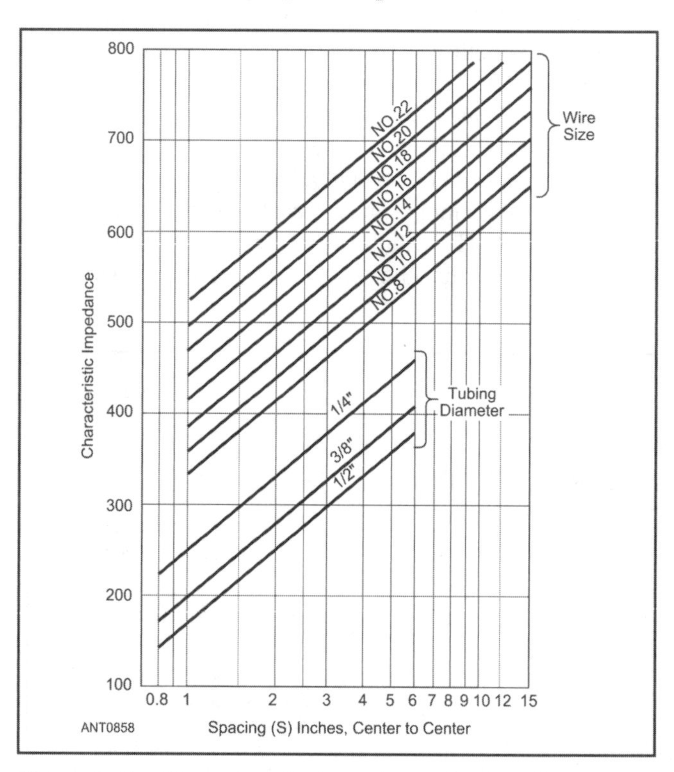


**Figure 4.41 – Characteristic impedance as a function of conductor spacing and size for parallel conductor lines.**

Sources - (2)

Source Edit Other

| | Sources | | | | | | |
|---|---|---|---|---|---|---|---|
| No. | Specified Pos. | | Actual Pos. | | Amplitude | Phase | Type |
| | Wire # | % From E1 | % From E1 | Seg | (V, A) | (deg.) | |
| ▶ 1 | 4 | 50 | 50 | 1 | 1 | 0 | I |
| * | | | | | | | |

**Figure 4.39 — Within *EZNEC*, we place the source at one end of the line and the load at the other end. See text.**

Loads (R + j X) - (2)

Load Edit Other

| | Loads | | | | | | |
|---|---|---|---|---|---|---|---|
| No. | Specified Pos. | | Actual Pos. | | R | X | Ext Conn |
| | Wire # | % From E1 | % From E1 | Seg | (ohms) | (ohms) | |
| 1 | 3 | 50 | 50 | 1 | 551 | 0 | Ser |
| * | | | | | | | |

**Figure 4.40 – This shows the boxes for the source and load data entries. See text.**

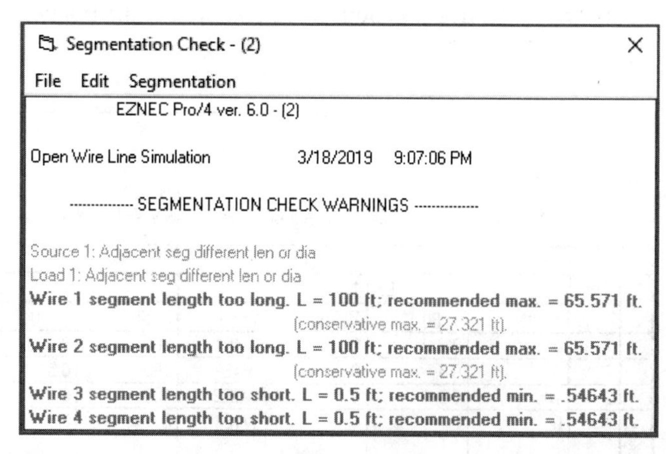


**Figure 4.42 — After selecting "segmentation" and then "conservative", the problem with wires 1 and 2 disappear.**

Now we are left with "too short" for wires 3 and 4 (see **Figure 4.43**). These warnings, either in red or orange can be ignored. Six-inch wires are indeed too short for 1.8 MHz. These end wires are not being used for radiators but may cause some very small errors in the calculation. Six inches is also too short at 3.5 MHz, but the color turns to orange indicating "less of a problem" from *EZNEC*. I have seen no unexpected results from simply ignoring these errors for the two very short end wires. However, again, make certain that the thickness (gauge) of all 4 wires are identical.

Now you are ready to run the simulation. Hit FFPlot.

See **Figure 4.44**. Notice in the 600-ohm system SWR shows 1.006, almost a perfect match! If it is farther off than this approximate value, change the R values slightly up or down in *both* the load R value *and* the **Alt SWR ZO** until the minimum SWR is found.

These steps have modeled the line and determined the line's $Z_0$. This simulated $Z_0$, measurement should coincide very well with figure 4.41, thus giving us confidence that this modeling method is indeed valid.

**Note:** The actual $Z_0$ determined by this method is not usually critical for building the line proper. In a practical sense, after a line is actually built, the actual $Z_0$ can be easily measured by the above method and then slight adjustments in spacing made, if necessary. However, the extra effort to determine the *modeled* $Z_0$ is necessary to derive accurate efficiency results from this *EZNEC* simulation. It also is required to achieve a desired $Z_0$.

When moving to higher frequencies the segmentation error may return. When it does, simply repeat "Segmentation" then "conservative". It may be helpful to write down the number of segments for each frequency simulated. Pay attention to segment warnings when changing frequency.

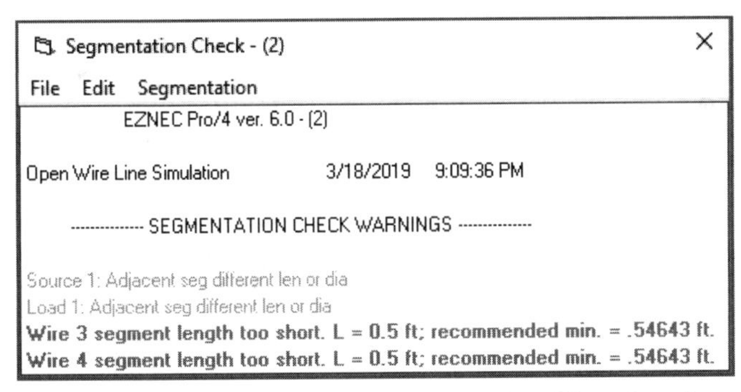


**Figure 4.43 — Now we are left with "too short" for wires 3 and 4. These warnings, either in red or orange can be ignored.**

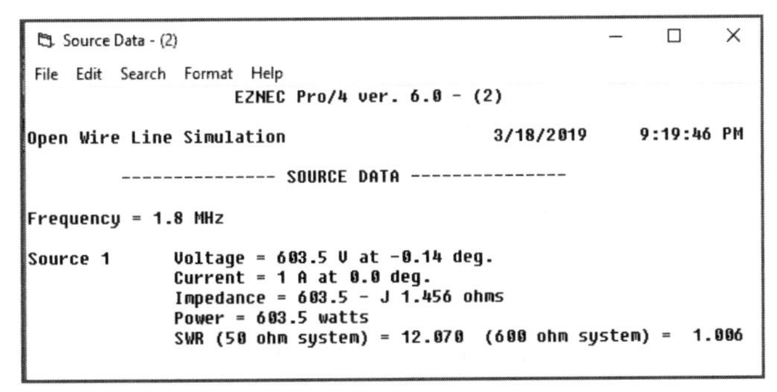


**Figure 4.44 — Running the simulation and seeing the results.**

## Interpreting the Results

See **Figure 4.45**. Beside determining the $Z_0$ of the line, we can derive a wide range of additional information. First, we can determine the line's total loss by accessing load data (Arrow 2).

In **Figure 4.46** we can see that our line provides 600.1 W for a line source applied input of 603.5 W, or 99.4% efficient, or 3.4 W lost by:

$$efficiency = \frac{P_{load}}{P_{source}} \qquad \text{(Equation 4.33)}$$

$$P_{loss} = P_{in} - P_{load} \qquad \text{(Equation 4.34)}$$

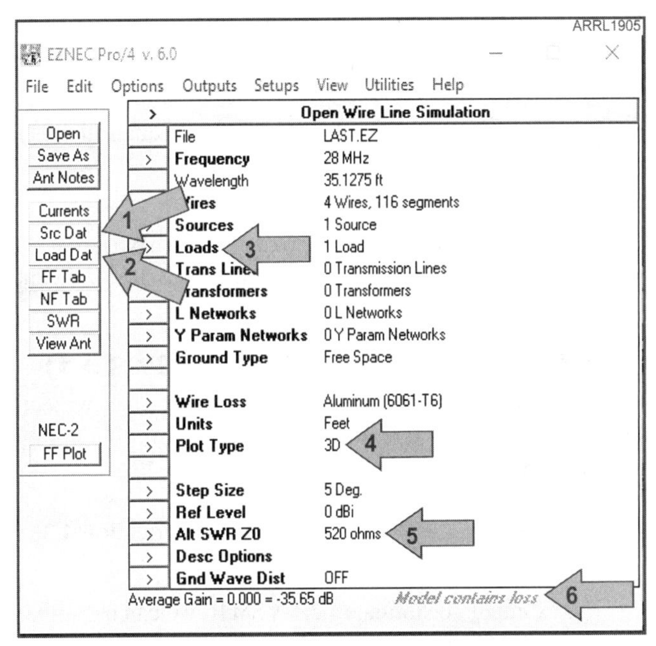


**Figure 4.45 — Location of arrows referenced in the text.**

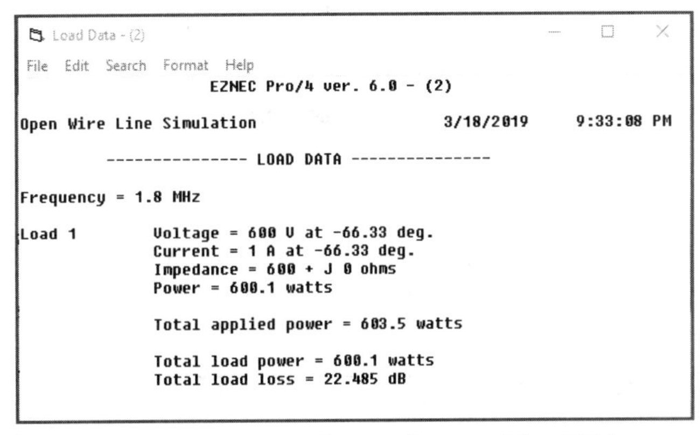


**Figure 4.46 — We can see that our line provides 600.1 watts for a line source applied input of 603.5 watts, or 99.4% efficient.**

The ML (dB) at 1.8 MHz is:

$$ML(dB) = 10\log\left[\frac{P_{load}}{P_{in}}\right] = 0.025 \text{ dB} \qquad \text{(Equation 4.35)}$$

Since this is a 100-foot line, we can assume that our result is also applicable to a dB per 100 feet specification (0.025 dB per 100feet).

See **Figure 4.47**. At the bottom of the control center (Arrow 6) we find the Average gain to be -56.52 dB. This means that all radiated power (integrated over a sphere surrounding the entire line) from this line is 56.52 dB below the input power to the line, or, in this case,

$$\left[10^{-.1dB}\right]P_{in} = P_{radiated} \qquad \text{(Equation 4.36)}$$

Which in this case is 1.34 mW radiated for 603 W input. In other words, *very* small loss from radiation. NOTE: it is necessary to use the 3D Plot Type (#4) to activate the Average Gain calculation.

$$R_{total} = R_l + R_r \qquad \text{(Equation 4.37)}$$

The loss from $R_r$ is therefore 1.34 mw.

The loss from $R_l$ is: 3.4 W (total) – 1.34 mW (radiated) ≈ 3.4 W (resistive). Thus, the radiated loss is insignificant compared to the resistive loss.

### Modeling Mis-matched Lines (VSWR > 1)

Since we have determined the $Z_0$ of the line, we can set the load to any value and thus simulate any line at any VSWR, where

$$VSWR = \left|\frac{Z_0}{Z_{load}}\right| \qquad \text{(Equation 4.38)}$$

For example, to simulate a 2:1 VSWR, we can use either a 300 or 1200 Ω resistor for a 600 Ω line. The results of this program are in very good agreement with the figure from the *ARRL Antenna Book* that graphs additional line loss vs. frequency and VSWR. This graph holds for open wire line as well as coaxial lines.

### Finding the Velocity Factor of a Line

To make calculations easy, set the frequency to 9.8352 MHz using a 100-foot line. In free space 100 feet is almost exactly 1 λ. Use 1000 segments for both wires of the line. Remove the end line with the load on it thus configuring an open-ended line. Run the simulation and click "currents" (arrow 1). Look for the two segments with the maximum "local" currents. They should be about 1/2 λ apart. 500 segments represents 1/2 λ in free space, thus the number difference in segment numbers divided by 500 x 100 will give the velocity factor in %. For example, we find the local current maximums at segments 760 and 279. Therefore, they are 481 segments apart. Divide by 500 and multiply by 100, and you get VF= 96%. *EZNEC* does not permit the use of spacers, thus the VF will likely read a few percent higher. Typical calculated values are closer to 99%, but it is possible to measure the effect of insulated lines and subtract the delta from a more typical uninsulated VF of about 95%. If more exact measurements are needed, empirical measurements of the actual line are most reliable since VF may change with age and/or slight manufacturing variations.

Average Gain = 0.000 = -56.52 dB    *Model contains loss*

**Figure 4.47 — At the bottom of the control center we find the Average gain to be -56.52 dB.**

### Tapered Line Calculations

Tapered lines can be used as impedance transformers much like the well-known 1/4 λ transformer. This is a relatively simple modeling exercise using this *EZNEC* technique and can yield lower loss characteristics. Since open wire line is usually "home brew" there is little extra effort in building a tapered line directly into the design for specific applications.

### Conjugate Matching

It is also possible to model LC networks and transformers using the *EZNEC* tool. This can be useful when designing more extensive systems.

### Using Modeled Lines with Antennas

*EZNEC* was designed to model antennas. However, using this program together with antenna design offers much expanded capabilities. For example, when feeding a non-resonant antenna with a long open wire feed, the *system* efficiency can be directly computed using the Average Gain function.

## Using Excel Spread Sheets for Quick Calculations

For calculating, saving and comparing multiple models, I simply use *Excel* spread sheets. More advanced use of *Excel* could be used for optimization etc., etc.

**Figure 4.48** shows a typical spread sheet to which the data is entered, and critical repetitive calculations are performed automatically. Here are the *f(x)* entries for the "B" column:

Cell B13: =B11-B12

Cell B14: =B12/B11*100

Cell B16: =10^(B15/10)*B11

Cell B17: =B13-B16

Cell B18: =10*log(B11/B12)

The "B" column corresponds to the 1.8 MHz column, so the same formulas must also be provided for the corresponding C, D, E, F, G, H and I columns for the other frequencies. Watch out for cell B16 where B15 is divided by 10 not B10!

There are ways to import and/or export data between *EZNEC* and *Excel*, thus it may be possible to build a "calculator" using this program.

|  | A | B | C | D | E | F | G | H | I |
|---|---|---|---|---|---|---|---|---|---|
| 1 | **LINE PHYSICAL DATA** |  |  |  |  |  |  |  |  |
| 2 | Wire diameter (AWG) | #12 |  |  |  |  |  |  |  |
| 3 | Line length (feet) | 100 |  |  |  |  |  |  |  |
| 4 | Wire type | copper |  |  |  |  |  |  |  |
| 5 | wire insulation ε δ type |  |  | none |  |  |  |  |  |
| 6 | insulation thickness (inches) |  |  |  |  |  |  |  |  |
| 7 | wire spacing (inches) | 6 |  |  |  |  |  |  |  |
| 8 | **EZNEC CALCULATIONS** |  |  |  |  |  |  |  |  |
| 9 | **MHz** | **1.8** | **3.5** | **7** | **14** | **28** | **50** | **144** | **432** |
| 10 | RtΩ=Zo for 1:1 VSWR | 601 | 601 | 601 | 601 | 601 | 620 | 620 | 474 |
| 11 | Power input W | 602.8 | 599.2 | 603.3 | 599.5 | 609.3 | 628.7 | 577.1 | 667.8 |
| 12 | Power at load W | 599.4 | 594.5 | 596.7 | 590.1 | 595.7 | 609.1 | 545.2 | 436.5 |
| 13 | total power lost W | 3.4 | 4.7 | 6.6 | 9.4 | 13.6 | 19.6 | 31.9 | 231.3 |
| 14 | line efficiency % | 99.436 | 99.21562 | 98.906 | 98.432 | 97.7679 | 96.8825 | 94.4724 | 65.3639 |
| 15 | dBi average gainW | -56.52 | -48.21 | -43.23 | -36.87 | -31.01 | -26.25 | -16.09 | -5.59 |
| 16 | power lost to Rr W | 0.00134 | 0.009048 | 0.02868 | 0.12325 | 0.48287 | 1.49088 | 14.1988 | 184.351 |
| 17 | power lost to Rl W | 3.39866 | 4.690952 | 6.57132 | 9.27675 | 13.1171 | 18.1091 | 17.7012 | 46.9486 |
| 18 | line attenuation (dB) | 0.02457 | 0.034199 | 0.04777 | 0.06864 | 0.09804 | 0.13755 | 0.24695 | 1.84662 |

**Figure 4.48 -- A typical spread sheet to which the data is entered, and critical repetitive calculations are performed automatically.**

### Modeling Noise Pick-up Rejection on Open Wire Lines

In the receive mode, the degree of rejection from external signal (including noise) sources will depend upon the field strength of the undesired signal at the line, its polarization, and what portion of the line is being subjected to the field.

A simple procedure can accurately simulate an undesired signal's effect upon a transmission line and thus a receiving system.

1. Using your model for a transmission line, create an external "noise source" by adding wire (s) and move the source to that new structure.

2. Replace the two short end wires on the transmission line with resistor values equal to $Z_0$ of the line. At one end (corresponding to the receiver) a complex conjugate match is usually assumed, therefore the resistor value of $Z_0$ is appropriate. You will be measuring the equivalent noise power input to the receiver at this resistor. This is the end used for "source" on the transmit more. A good approximation of the noise level can be obtained by simply terminating both ends of the line with $Z_0$. In cases where more exact measurements are needed, the actual antenna complex feed point, together with a complex conjugate match would be required.

3. The placement (distance from the line) of the noise source antenna will automatically determine if the source is in the near or far field.

4. Determining the actual geometry of the noise source wire(s) can be difficult. However, if you have a suspect power line noise source, in effect, you can make a good approximation by estimating the dimensions of the power line. A noise "point source" can be modeled as a very short single wire.

5. Run a simulation of the new noise source antenna at the desired frequency. Check the real part of the input impedance on the noise wire (*EZNEC* Src Dat). Now adjust the power of the noise source by adjusting the current amplitude of the *EZNEC* source (Sources, Amplitude),

where

$$P_{noise} = \frac{I^2}{Z_{real}},$$

where I is the current of the source and $Z_{real}$ is the resistance part of the feed point impedance taken from Src Dat. Notice that the calculated current value for low noise power can be very low.

6. Simply run the simulation and observe the power at the receiver input (Load Dat). This is the real power value. To convert to dBm: $dBm = -30 - 10 log P_{noise}$. The standard S-unit (S-9) is -73 dBm, and S-units are in 6 dB steps, (i.e. S-8 is -79 dBm).

**Note:** S-unit calibration can vary considerably among commercial receivers. For empirical measurements, a spectrum analyzer (if available) can offer a calibrated wattmeter for these types of measurements. Accurate *relative* signal strength measurements can be taken by turning the receiver's AGC off and simply measuring the *relative* audio output levels.

## Data from EZNEC for Open Wire Lines

The pervious section is presented as the procedure used here to model any open wire line using cylindrical conductors. This section provides data and interpretations using the above procedure to investigate multiple properties of lines and present both data and relevant conclusion based on that data.

### Effect of Wire Size on Loss

The first analysis will be the effect on matched line loss (ML) on the size of the wire comprising the line. We will use a constant spacing between the line wires for this comparison. However, when using constant spacing the $Z_0$ of the line will decrease as the wire size increasing. As we show a bit later, ML *increases* as $Z_0$ decreases thus we will also show ML for different wire sizes with corrected spacings to provide equal $Z_0$ for each different wire size. However, it is also very useful to plot ML for different wire sizes with constant spacing.

In **Figure 4.49** we see the total line loss per 100 feet, 2-inch spacing matched load for various copper wire with different wire diameters and one aluminum wire. Heavier electrical wire is commonly available in aluminum which provides for a much lighter line and is thus included in this example. Notice that #6 aluminum wire has nearly identical loss per 100 feet as #12 copper wire except at UHF frequencies. At 432 MHz, losses from line radiation become significant but #6 wire has greater diameter than #12 wire, therefore, there is better radiation cancellation (lower loss) with #6 wire. In practice, the simple 4-wire straight model for open wire line give excellent results over the HF band. However, as operating frequencies increase, bends in the line and proximity issues begin to distort the modeling results. For example, at 50 MHz a 500 foot open wire line with multiple "sags" in the line (for cable flexibility) significant anomalies are present, where at low HF frequencies, they are insignificant.

The graph in **Figure 4.50** shows total line loss (dB) for #12 copper wire at various spacings per 100 feet. At VHF, losses due to radiation become noticeable. Notice that the 2-inch spacing retains lower loss than 1 inch spacing since radiation loss has not yet overtaken the loss advantage of higher impedance line. This forms a fundamental trade-off at higher frequencies.

The graph in **Figure 4.51** is based on identical data as in Figure 4.47 but the higher frequencies are removed for better resolution at HF. Again, this plots dB loss per 100 feet as functions of frequency and wire spacing using #12 copper wire. At these frequencies, the effect of line radiation compared to total loss is minimal so wider spacings provide for lower losses. At wider spacings the line impedance is higher, thus the line voltage is higher, and the line current is lower.

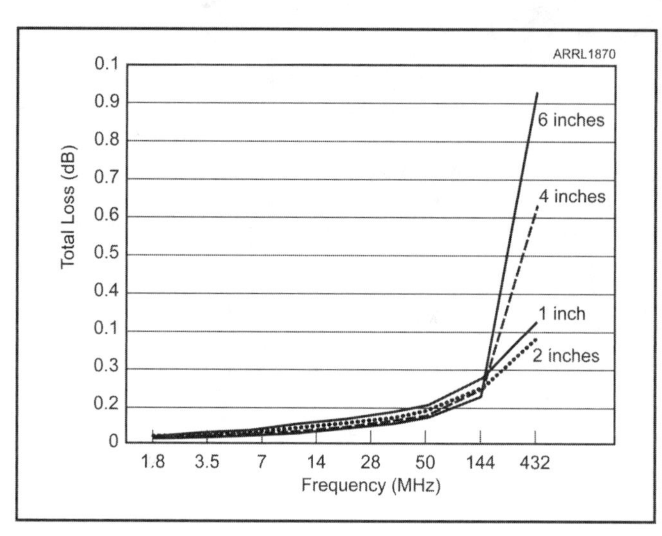


**Figure 4.50 — Total line loss (dB) for #12 copper wire at various spacings per 100 feet.**

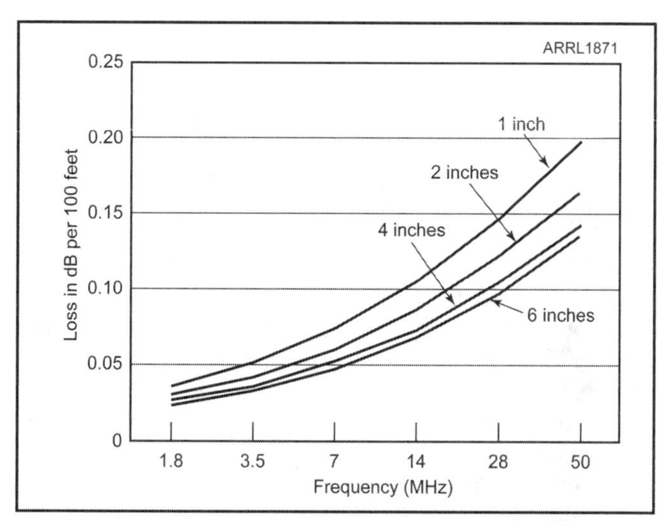


**Figure 4.51 – Loss in dB per 100 feet as functions of frequency and wire spacing using #12 copper wire. At these frequencies, the effect of line radiation compared to total loss is minimal so wider spacings provide for lower losses.**

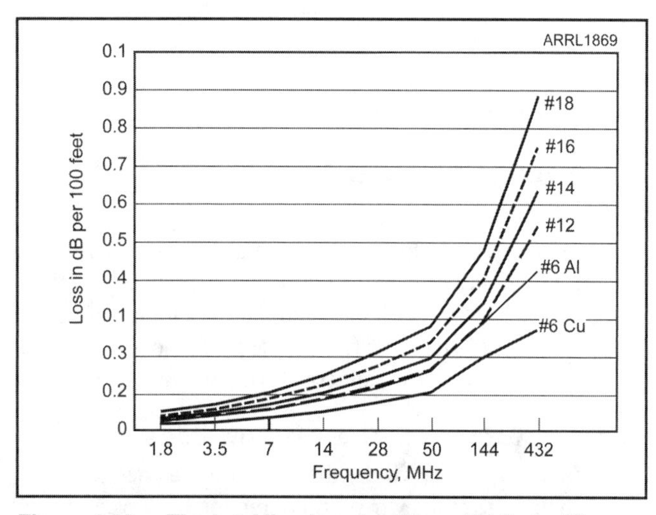


**Figure 4.49 — The total line loss in dB per 100 feet with a 2-inch spacing and a matched load for various copper wire with different wire diameters and one aluminum wire.**

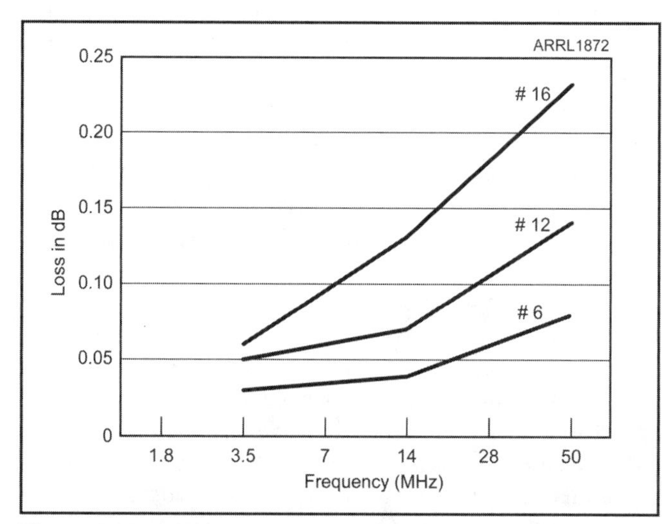


**Figure 4.52 — ML line losses for different size copper wires with spacing correction to maintain 525 Ω .**

This results in lower $I^2R$ losses, exactly what we see in high voltage power transmission lines. This points to a fundamental trade-off in selecting the spacing of an open wire line. The wider the spacing, the lower the resistive losses, but at the expense of greater line radiation and response from undesired signals, and of course the need for longer and heavier spacers.

In **Figure 4.52** we see ML line losses for different size copper wires with spacing correction to maintain 525 Ω $Z_0$. Notice there is a very small variation for 3.5 MHz in Figure 4.51 vs. 4.52 for #12 wire. The difference is about .02 dB per 100 feet. The difference used two different models and the resulting VSWR was very small resulting in a very small difference in the load current, but significant enough to yield this very small difference. This small difference shows both the accuracy of this model but also the degree of error being very small if used carefully.

The results from Figures 4.50, 4.51, and 4.52 show that for wire spacings larger than about 3 inches loses are almost insignificant for changes in separation. However, wire size is critical for *any* separation since it effects line loss more than spacing more than about 3 inches.

## Open Wire Line Loss for Different Metal Wire Types

Perhaps the most important consideration is the differences between copper and aluminum wire, since they represent the most common materials for electrical wire.

See Table 4.7. Notice that the loss for steel wire is very similar to the loss for tin. These resistance values (except for steel) are the default values within the *EZNEC* tool. The aluminum alloy is 6061-T6, commonly used for antenna element tubing. These values are for DC resistance. For equivalent resistance at an RF frequency, skin effects must be considered. Skin effect is automatically calculated within the *NEC* engine.

The chart in **Figure 4.53** shows total line loss at 3.5, 14, 28 and 50 MHz for #12 and #18 copper, aluminum, zinc and tin wire at 2-inch spacings in dB per 100 feet. **Figure 4.54** offers a close-up of how wires are oriented for a 4-wire source or load.

As you can see in **Figure 4.55**, the total loss of 4-wire line is nearly identical to the 2-wire line with comparable spacings and identical wire diameter. Also see Table 4.8. Both the 2 and 4 wire lines use #14 wire. The table is included since the graph places both traces on nearly identical paths. The four-wire line suggests lower resistive loss since there are four wires rather than two. However, when using the same wire spacing four-wire spacing results in lower $Z_0$ so losses increase resulting in almost perfect correlation in losses between the two configurations. Four wires lower resistive loss (more conductors in parallel), but lower $Z_0$

**Table 4.7**
**Wire Resistivity**

| *Wire Material* | *Resistance Ω Meter* |
|---|---|
| Copper | 0.7E-8 |
| Aluminum | 4E-8 |
| Zinc | 6E-8 |
| Tin | 1.14E-7 |
| Steel Piano wire | 1.18 E-7 |

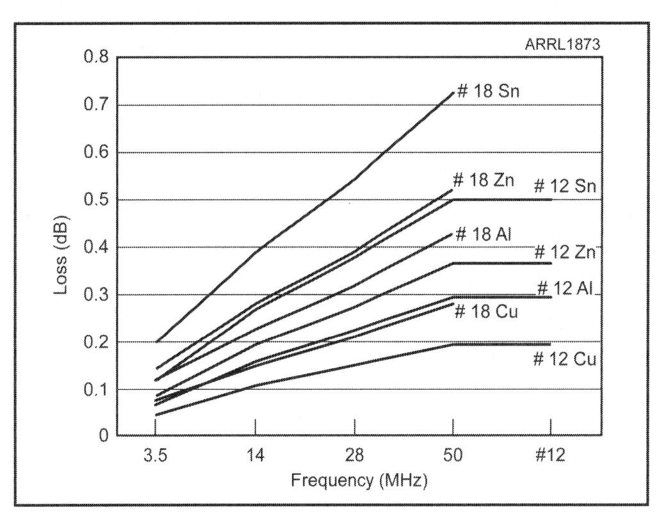


**Figure 4.53 -- Total line loss at 3.5, 14, 28 and 50 MHz for #12 and #18 copper, aluminum, zinc and tin wire at 2-inch spacing in dB per 100 feet.**

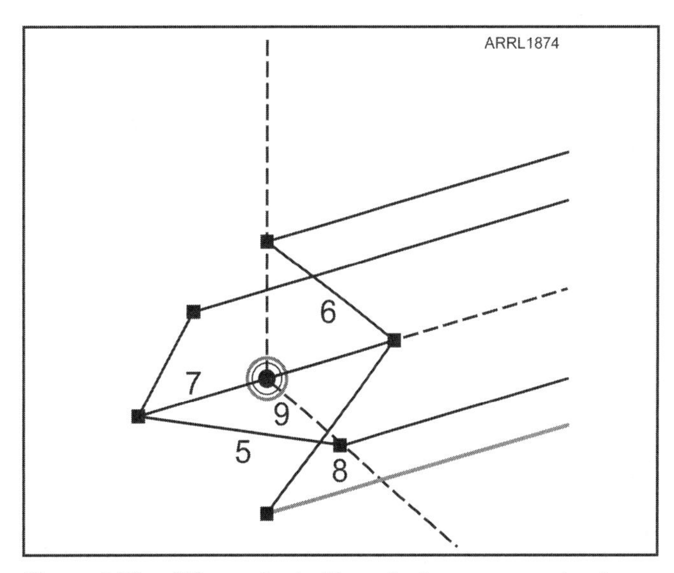


**Figure 4.54 — Wires oriented for a 4-wire source or load.**

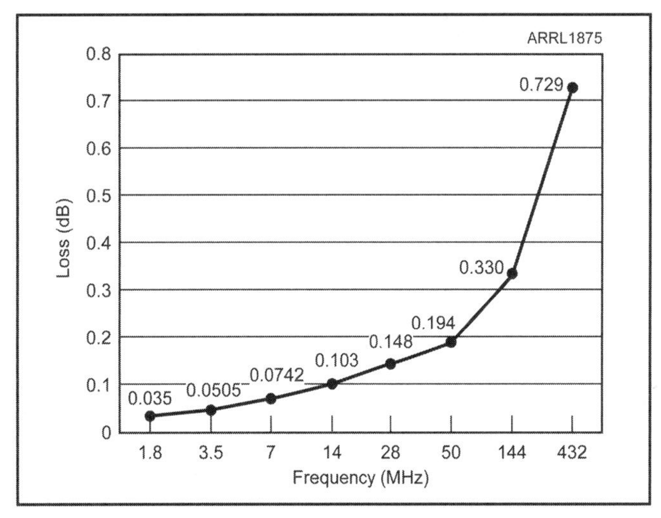


**Figure 4.55 — The total loss of 4-wire line is nearly identical to the 2-wire line with comparable spacings and identical wire diameter.**

(lower voltage and higher current on the line).

In both cases the radiation losses are considerably less than the resistive losses, even up to UHF frequencies. However, as the wire spacing increases radiation losses becomes far more important.

**Figure 4.56** illustrates the dB loss for 100-foot lines through radiation compared with power input to line, 2-wire line is #14 wire, 2 inch spacing, 4-wire line is #14 with 2-inch spacing (a 2-inch cube structure). This graph illustrates a key advantage to 4-wire line: at lower frequencies 4-wire line will attenuate noise pick-up and therefore line radiation by more than 10 dB over a similar 2-wire line making it a considerably better choice for receiver applications, but with considerable mechanical complexity.

**Figure 4.57** shows the total radiated power (dB below power applied to the line) as a function of line length. The line uses #12 copper with 6-inch spacing. For lines shorter than about 1/4 λ losses decrease until at 0 length there theoretically should be no losses from any mechanism (there can be no resistive or radiated loss from a line of zero length!). A line with a length longer than about 1/4 λ will lose power by radiation at a constant rate no matter the added length and conform to Equation 4.39. However, notice the slightly negative slope for lengths longer than 1/4 λ. Radiation decreases along the line due to less current at longer distances which in, turn is caused by resistive loss. Resistive loss decreases radiated loss!

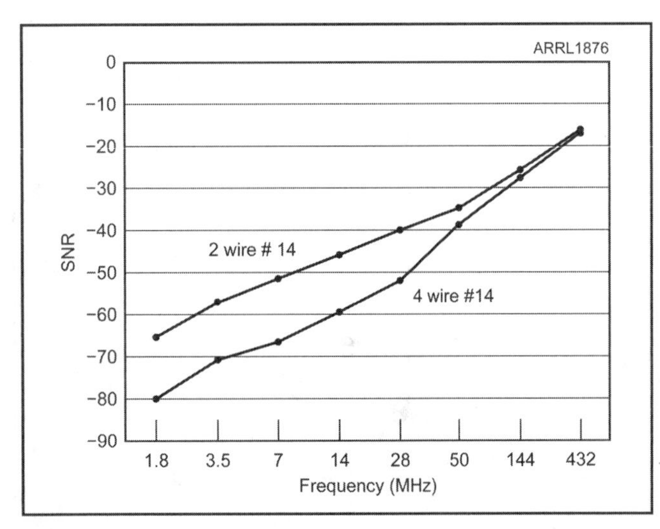


**Figure 4.56 — This graph illustrates a key advantage to 4-wire line: at lower frequencies 4-wire line will attenuate noise pick-up and therefore line radiation by more than 10 dB over a similar 2-wire line.**

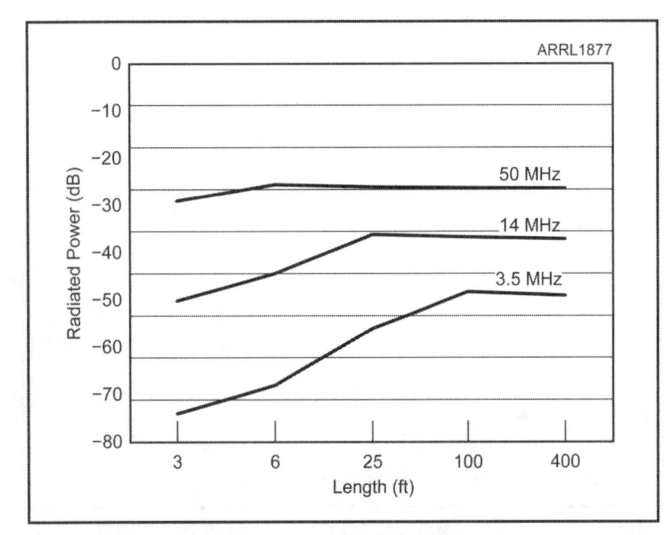


**Figure 4.57 — The total radiated power (dB below power applied to the line) as a function of line length.**

$$\frac{P_{radiated}}{P_{input}}(dB) \approx 20log\frac{wire\ spacing}{\frac{1}{2}\lambda} \quad \text{(Equation 4.39)}$$

where , wire spacing << λ, line length is longer than about 1/4 λ, and wire spacing and λ are in same units.

This surprisingly simple empirically derived (by the author) equation will give a good approximation for two-wire line power loss through radiation vs. line input power in dB and/or a figure of merit to its susceptibility to noise pick-up on receive. This equation is for perfectly matched two-wire lines that is at least 1/4 λ long, but longer total line lengths do not matter, except for a gradual reduction of radiation due to increased resistive loss in longer lines. Wire spacing and λ use the same units. For commonly used wire sizes (#18 to #6 AWG) derivations from this equation are minimal. The 20 log term indicates that the mechanism is indeed the degree of cancellation of **E** and **B** fields as a function of wire spacing.

### Additional Line Loss Due to VSWR

One of the biggest advantages of open wire line is the ability to minimize line losses even if the line is subjected to considerably high VSWR. There are limitations, however, to this sweeping statement as losses do increase with increasing VSWR. Open wire line is also inherently much less likely to "arc over" in the presence of high VSWR with the associated high voltages present. Through extensive *EZNEC* modeling it was determined that Equation 4.40 is valid for both coaxial and open wire lines and thus the total line loss is determined by both the ML and the additional VSWR losses shown in Figures 4.60 and 4.61.

The previous sections have assumed near perfect matches (where ML can be assumed with VSWR < 1.01). We can model high VSWR conditions on *EZNEC* by simply changing the load value on the line with a pure resistance. This will provide the condition usually associated with VSWR effects

**Table 4.8**

| *Frequency (MHz)* | *Resistive Loss in dB 4-Wire* | *2-Wire* |
|---|---|---|
| 1.8 | 0.035 | 0.037 |
| 3.5 | 0.0505 | 0.052 |
| 7 | 0.0742 | 0.073 |
| 14 | 0.103 | 0.104 |
| 28 | 0.148 | 0.146 |
| 50 | 0.194 | 0.194 |
| 144 | 0.339 | 0.34 |
| 432 | 0.729 | 0.698 |

on line performance: the VSWR *at the load.* A major advantage of open wire and ladder line compared with foam dielectric coaxial cables is their inherent low loss characteristics. When subjected to high VSWR conditions the losses will increase, as in any transmission line, but since the "starting points" for wire lines are so low, the total losses often remain significantly lower even with very high VSWR conditions, but there *are* limitations. As with any line there is a trade-off between VSWR values and total line loss, so the often-used assumption that any arbitrarily high VSWR can be used with open wire lines is not correct. Calculating the additional loss of an open wire line due to an expected VSWR value should be made to assure the design will not have excessive total loss.

Once a reasonable added loss is derived, we can ignore matching an antenna to the line and perform matching at a far more convenient point, i.e. on the ground. Also, since the tuning is more conveniently performed on the ground, it is usually much easier to tune, i.e. a wire antenna, to any frequency.

Figure 4.61 plots (from the ARRL *Antenna Book*) added line loss in the presence of VSWR's higher than 1. However, does this plot hold for open wire lines as well as coaxial cables? I often wondered if this graph and the equation that was used to derive the data also applied to open wire lines. As will be shown, this equation does indeed hold for both coaxial and open wire lines, again shown with good correlation to the *EZNEC* model.

To test this similarity, we start with two simple open wire lines and see if the *EZNEC* data agrees with Equation 4.40.

In **Figure 4.58** we see the additional loss due to VSWR per 100 feet for #18 wire at 1-inch spacing higher impedance at load vs. $Z_0$ of the line. For example, when $Z_0 = 450\ \Omega$ the load resistance used is 900 Ω for a VSWR = 2.

See **Figure 4.59**. The data shows that open wire line conforms as well as coaxial lines to Figures 4.60 and 4.61.

However, when we try to use the *Antenna Book's* Figure 4.61, we find that the range on the plot is too short. Open wire lines commonly exhibit *very* low ML and thus I include Figure 4.60 which is simply and extension of Figure 4.61. Also, since the ML is so low with open wire lines, I also greatly expand the VSWR to demonstrate where the desired (or undesired) limits occur.

In **Figure 4.60**, the graph plots additional line losses (dB) as a function of ML and the VSWR. Total line loss is simply ML + the additional line loss. Open wire lines often have considerably lower matched loss (ML) than coaxial lines. Figure 4.61 is published also in the *ARRL Antenna Book*, but the minimum matched loss shown is 0.2 dB. This graph, in effect, extends Figure 4.61 using equation 4.40 down to 0 dB ML. It also shows extended added loss for up to 100:1 VSWR, not unusual for open wire applications. This graph represents data taken from the *EZNEC* model but also coincides very well with Equation 4.40

$$Total\ resistive\ loss\ (dB) = 10 log\left(\frac{a^2 - |\rho|^2}{a\left(1-|\rho|^2\right)}\right)$$

(Equation 4.40)

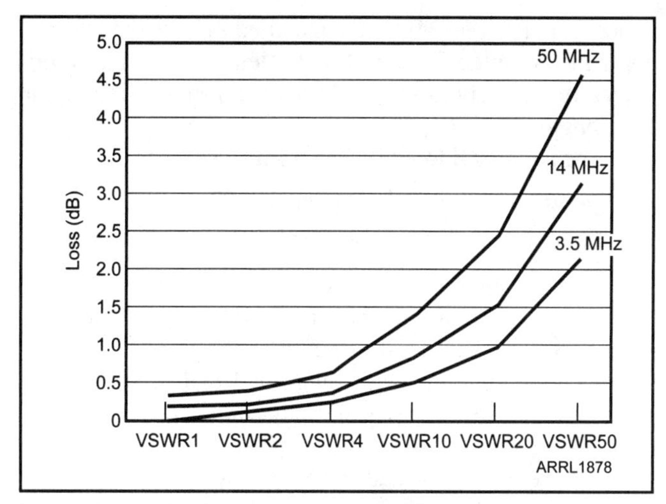


**Figure 4.58 -- Additional loss due to VSWR per 100 feet for #18 wire at 1-inch spacing higher impedance at load vs. of the line.**

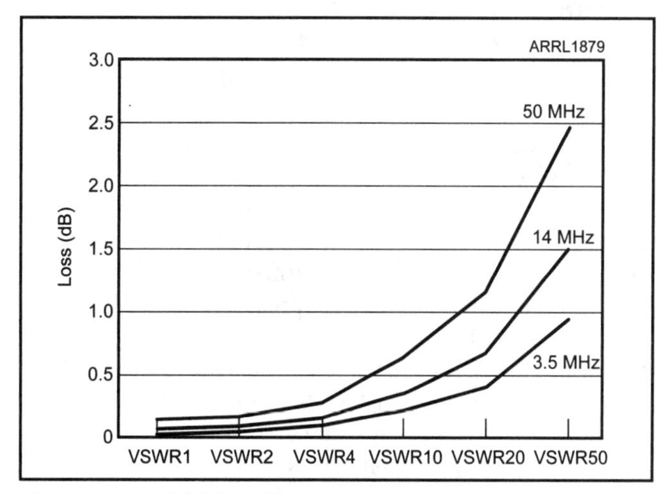


**Figure 4.59 -- Additional loss due to VSWR per 100 feet using #12 with 6-inch wire spacing.**

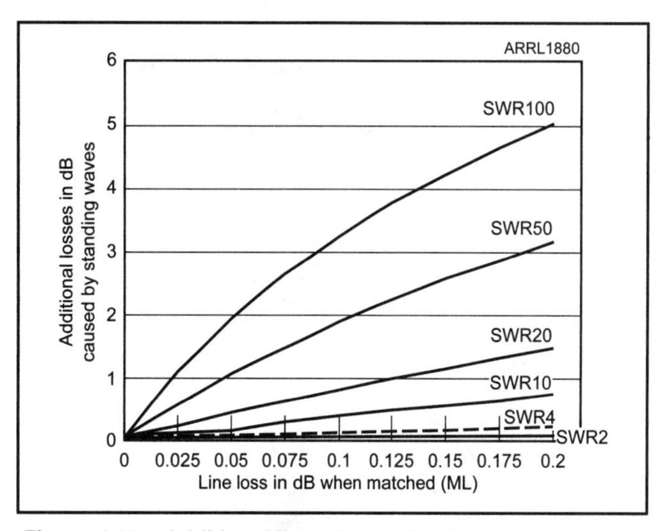


**Figure 4.60 – Additional lines losses (in dB) as a function of ML and the VSWR. The plot is a continuation of Figure 4.61 (from the ARRL Antenna Book). Here ML is continued down to 0 dB since open wire lines can have ML's much less than 0.2 dB.**

where $a = 10^{0.1ML}$, ML is the matched line loss in dB, and $|\rho|$ is the complex reflection coefficient. Notice that in this equation, ML represents *total* resistive matched line loss at any length.

The additional loss of a line from loss resistance is

*Additional loss from VSWR* (Figures 4.60 and 4.61) =

*Total Resistive Loss* (Equation 4.40) – *ML*

(Figure 4.29 and other sources)

These equations hold for both coaxial and wire transmission lines. For added losses due to VSWR greater than 1, calculate the total line loss (using line length × ML per 100 feet). That is the number you use on the X axis for Figures 4.60 and/or 4.61.

**Figure 4.61** plots added VSWR losses vs. ML, as depicted in the *ARRL Antenna Book* and taken from Equation 4.40. Notice that this graph is not dependent upon the line length.

## Effects Upon Open Wire Line from Nearby Objects

All of the previous discussions and results have modeled open wire lines in free space. One of the advantages of coaxial lines over open wire lines is that no consideration need be given to proximity of materials since the fields are completely contained within the structure. Coaxial cables may be directly buried, run through walls, equipment chassis, etc. with no concern for performance degradation. Open wire line installations *must* take proximity to materials into consideration. This section will deal with this issue, again, using the *EZNEC* model. Close proximity to ground requires the more sophisticated *NEC4* than *NEC2* engine and use of the more advanced *EZNEC* interface.

The **Figure 4.62** graph plots line efficiency per 100-foot lengths for various line elevations above average ground, frequencies of 1.8, 14 and 50 MHz at VSWR values of 1, 10, and 20. Notice the strong dependence on frequency and VSWR. Also, this graph shows efficiency in percent, not dB loss. This line uses #12 copper wire with 6-inch spacing.

Look at **Figure 4.63**. Beside operating frequency and VSWR, we see a dramatic decline in losses due to ground proximity when a lower line impedance is used (closer wire spacing). This graph plots lines with #12 copper wire but with 2-inch spacing. (Compare this graph with the 6-inch spacing in Figure 4.56.) In **Figure 4.64** we can see the effect on total losses vs. line length.

**Figure 4.65** shows the efficiency of 4-wire line vs. height above average ground. This graph shows a key advantage of 4-wire lines: much less effected by nearby objects. These lines are 100 feet long at fixed heights above ground. Also,

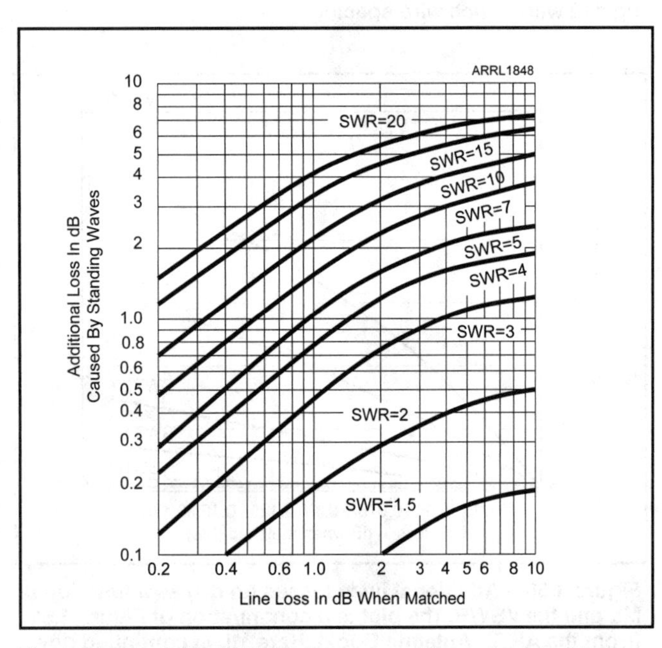


**Figure 4.61 – Added VSWR losses vs. ML as depicted in the ARRL Antenna Book and taken from equation 4.40.**

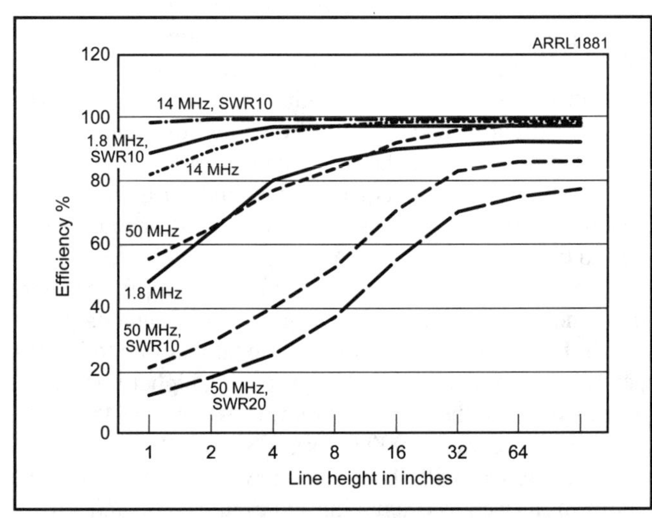


**Figure 4.62 — Line efficiency per 100-foot length for various line elevations above average ground, frequencies of 1.8, 14 and 50 MHz at VSWR values of 1, 10, and 20. Line is #12 wire at 6" spacing.**

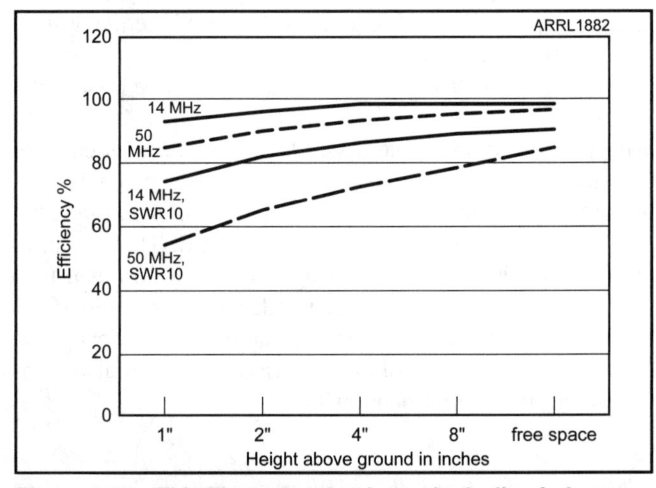


**Figure 4.63 – This illustrates the dramatic decline in losses due to ground proximity (compared to Figure 4.62) when a lower line impedance is used (closer wire spacing). Here the line used #12 wire at 2" spacing.**

this graph shows efficiency in %, not dB loss. Note that the 8-inch height is very close to free space loss even at 50 MHz.

### Ladder Line vs. Open Wire Line

Ladder line has become increasingly popular since it may be installed "out of the box". Open wire line typically requires often tedious construction techniques especially the installation and mechanical stabilization of the spreaders.

Table 4.9 was provided by Wire Man, a distributor of ladder line. These products all use polyethylene insulation material and use Copper clad steel wire, except for part number 561. Both of these materials are excellent choices (electrically, mechanically and economically) for their intended functions.

The graph in **Figure 4.66** plots ML values for Wire Man's ladder line #551 (#18 wire), #554 (#14 wire), and #14 wire, 1-inch spacing open wire line. The two ladder lines' data comes from "Transmission Line Program for Windows" a supplement in the ARRL Antenna Book, 23rd edition, by N6BV. Product number 554 has identical spacing and wire size as the open wire line. The additional ML is due to the added dielectric loss in the ladder line.

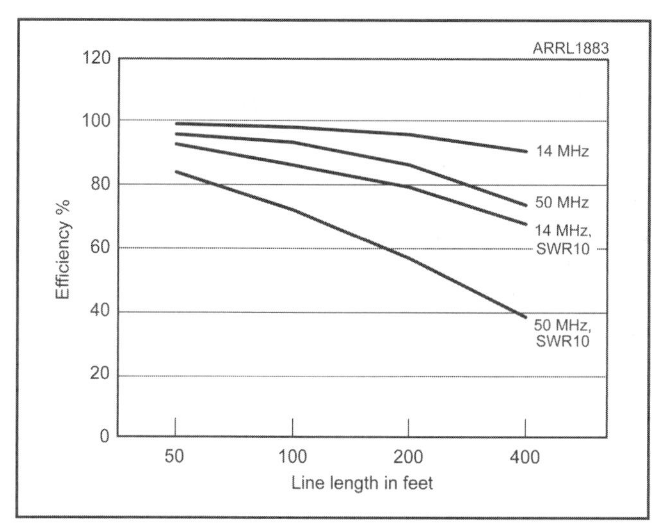


**Figure 4.64 – This graph clearly illustrates the effect on total losses vs. line length.**

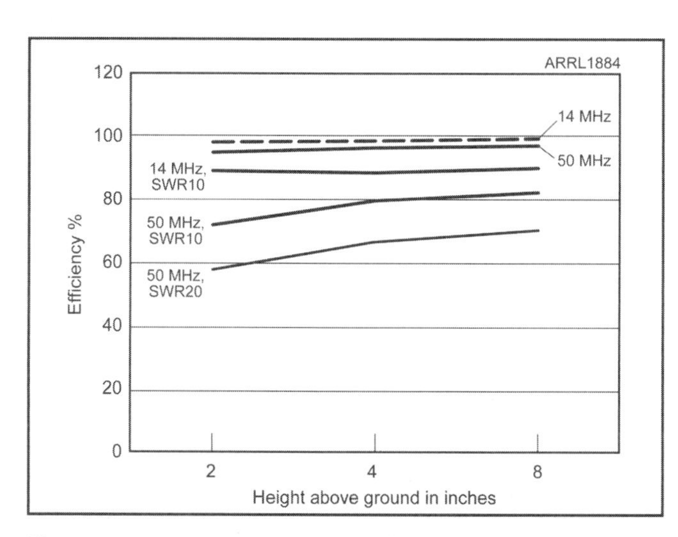


**Figure 4.65 -- The efficiency of 4-wire line vs. height above average ground.**

**Table 4.9**

| *Part Number* | *Gauge/ Strands* | *Insulation diameter and Width (Inches)* | $Z_0$ (Ω) | *Velocity Factor* |
|---|---|---|---|---|
| 554 | 14/19 | 0.065 × 0.930 | 370 | 0.928 |
| 552 | 16/19 | 0.065 × 0.930 | 400 | 0.917 |
| 553 | 18/19 | 0.065 × 0.930 | 450 | 0.898 |
| 551 | 18/1 | 0.065 × 0.930 | 450 | 0.902 |
| 561 | 20/7 BC | 0.150 × 0.400 | 300 | 0.800 |
| 562 | 18/19 | 0.150 × 0.400 | 300 | 0.820 |

#### Summary of Conclusions for Ladder and Open Wire Lines

Here are some basic conclusions regarding open wire and ladder lines.

1. The standard method of determining dB loss of a coaxial line is to state the loss (in dB) as dB per 100 feet, or dB per 100 meters vs. frequency. The relationship of dB per coaxial length is assumed to be linear. Two simulations (at 7 and 50 MHz) were run with a 200-foot line in Simulation 4.2. The comparison of line attenuation at these two frequencies shows a nearly perfect doubling of dB attenuation from a 100-foot line simulation, exactly what we would expect from a coaxial line: double the line length, double the attenuation in dB. Consequently, we can conclude that this same rule also holds for open wire line.

2. The added loss equation and graphs due to VSWR published in the *ARRL Antenna Book* and also published here are applicable to open wire and ladder lines as well as coaxial lines.

3. The near-average of the total power radiated from open wire lines increases at 6 dB per octave of frequency (Equation 4.42). As the frequency rises, the spacing between the wires becomes a greater percentage of the wavelength, therefore the transmission line gradually begins to increase its radiation. The line radiates twice the field intensity for a given halving of wavelength or doubling the wire spacing thus the power quadruples (6 dB). However, the overwhelming majority of line loss is due to resistive loss compared to radiated loss, even up to 50 MHz and relatively wide spacing.

4. The amount of power radiated by a line does *not* increase as the line is made longer. This is analogous to a comparison between a 1/2 wave dipole to a very long wire: the directivity patterns may be quite different, but the total radiated power will be very similar. This holds for lines ≥ about 1/4 λ.

5. Line radiation from open wire lines does *not* increase from higher VSWR, but again, resistive losses do increase obeying the same equation as for coaxial lines.

6. Ladder line has increased loss compared with open wire line of similar spacing and wire. This is due to the polyethylene flat spacers. However, ladder line has considerably less loss than coaxial cables and is much more convenient

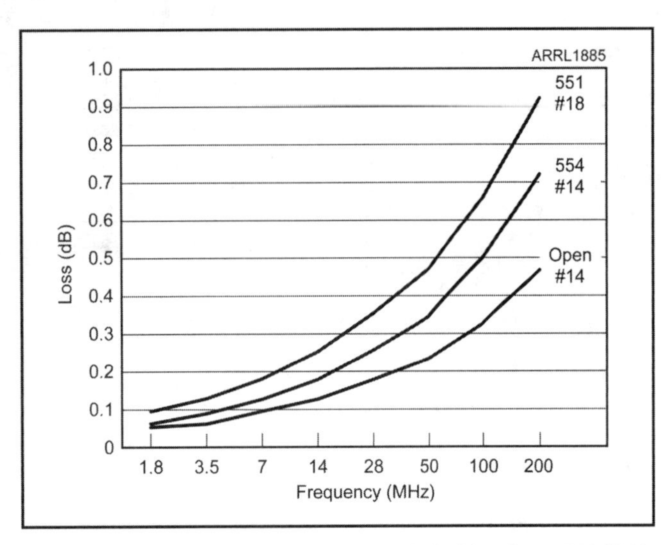


**Figure 4.66 -- ML values for Wire Man's ladder line #551 (#18 wire), #554 (#14 wire), and #14 wire, 1-inch spacing open wire line.**

than open wire line to install. Also, line construction is not required.

7. Four-wire line offers considerably less line radiation and therefore considerably less noise pick-up than two-wire line. This holds true through the HF region for wire spacing of 2 to 3 inches, Above HF the line radiation characteristics become similar. The other main advantage of 4-wire line over 2-wire line is its considerably lower adverse effects from close proximity to materials, including close proximity to the ground. ML losses are about the same in 4-wire vs. 2-wire lines of comparable wire spacing. The doubling of conductivity for 4-wire lines vs. 2-wire lines is nearly cancelled by the lowering of the line impedance in the 4-wire line.

8. ML will decrease as wire spacing increases since the $Z_0$ increases and therefore the line voltage increases. However, at higher frequencies, line radiation and noise pickup also increase. This creates a trade-off between the two issues.

9. Much higher VSWR conditions can be typically tolerated on both ladder and open wire lines than coaxial lines since the ML of the wire lines is typically much lower than coaxial lines. Also, since the physical separation between the two conductors is much greater than typical coaxial cables, there is less risk of arcing along an open wire line. However, *very* high VSWRs can result in very high line losses. This is especially true in long line lengths.

## References

There are many excellent treatments of transmission lines. I can recommend three as especially helpful for the preparation of this text:

• *ARRL Antenna Book*, 23rd edition, ARRL, 2015: The common symbols used in this chapter were derived from the *Antenna Book*. In addition, this text provides a wealth of information on practical issues related to measuring, designing and installation of transmission lines.

• *Reflections Transmission Lines and Antennas,* 1990, M Walter Maxwell, ARRL: This text provides an in-depth discussion regarding the mechanics of transmission lines and focuses on erroneous conclusions common among amateurs and professional RF engineers alike.

• *Transmission Lines*, 2013, Rickard Collier, Cambridge University Press: This is an advanced, in-depth theoretical treatment of transmission lines. The first part examines the more traditional approach of treating transmission lines as circuits (and differences thereof). The second part explores, again in depth, transmission lines using electromagnetic theory, i.e. Maxwell's equations. The final part is a fascinating take on photon behavior in transmission lines.

# Antennas Using Multiple Sources

Thus far we have considered only pattern formations from a single antenna element. The technique for determining antenna patterns is described in Chapter 3 using small segments (or infinitesimal dipoles). When we move into more complex antennas (here exemplified by multielement arrays), the utility of the PADL technique (Phase, Amplitude, Direction, and Location — introduced in Chapter 3) becomes indispensable. We may extend the PADL principle to include a rather sweeping statement: It does not matter if the PADL parameters are distributed along a single antenna element or multiple elements distributed in 3D space. They form the basis of the antenna pattern. It also does not matter if these segments are excited directly via a *conducted wave* (transmission line or wire) or through a *radiated wave* coupled by a *mutual impedance*. These two mechanisms define the two basic forms of multielement arrays: "driven arrays" and "parasitic arrays." Sometimes *both* techniques are used in a single antenna (intentionally or unintentionally). However, unlike the single element antennas thus far discussed, the actual values of the phase, amplitude and direction for close-spaced elements become *interdependent* among all the elements.

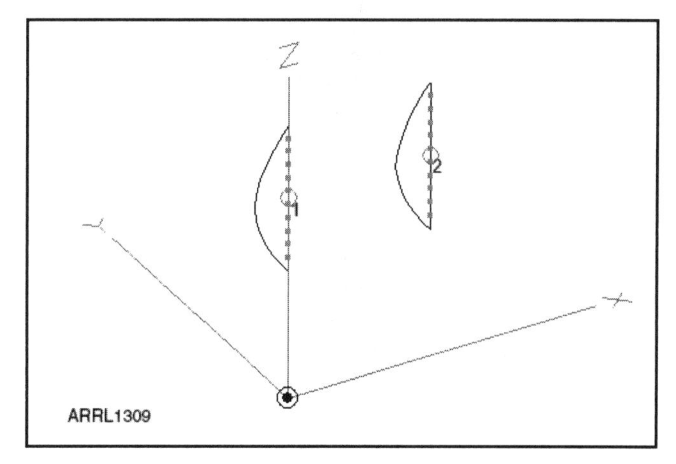


**Figure 5.1 — This figure shows two ½ λ dipoles in free space separated by ½ λ.**

## Driven Arrays

We will begin by using two simple examples that can form a more intuitive understanding of how PADL parameters create antenna patterns. Both examples use two parallel ½ λ dipoles in free space and are separated by ½ λ (**Figure 5.1**). Both dipoles are center fed from the same RF generator (transmitter). Since both of the elements that constitute the array are directly fed from the signal source, we call this a driven array. In this first example the currents being fed to the two dipoles are identical but we will use two phase relationships: in phase (0 degrees) and out-of-phase (180 degrees).

The pair of dipoles in **Figure 5.2** are fed in-phase. If we observe the array from one of the two broadside directions, we will see two fields adding in-phase. Consequently radiation is maximum "broadside" to the elements, in the

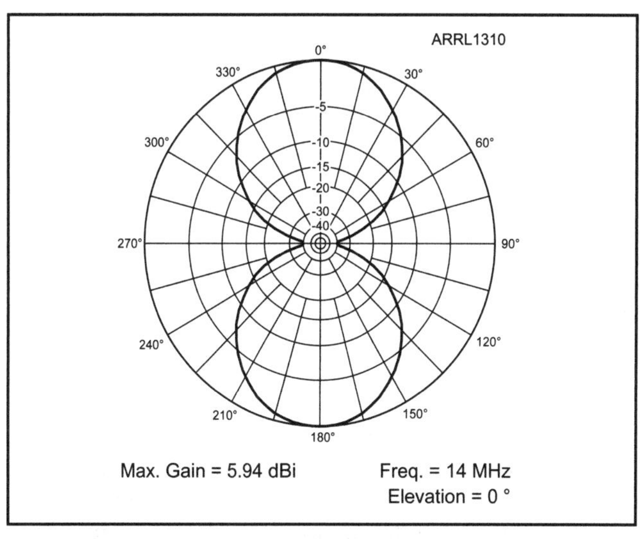


**Figure 5.2 — This figure shows the bidirectional "figure 8" pattern of the Figure 5.1 array. Fed in-phase, the radiated patterns add broadside to the array, in this case along the Y-axis.**

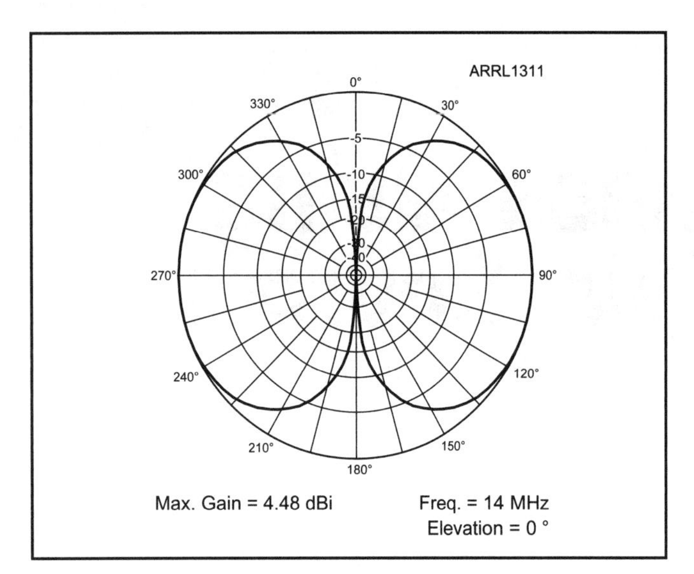


**Figure 5.3 — This figure shows the same array from Figure 5.1, but the elements are now fed 180 degrees out-of-phase creating another "figure 8" pattern. However, in this case the broadside pattern is cancelled and the end-fire (along the X-axis) sum produces the maximum gain. At other angles, there is imperfect cancellation or summing, creating intermediate gain values because of partial cancellation.**

directions normal to the plane of the dipoles. Conversely, the pair of dipoles in **Figure 5.3** are fed out-of-phase. Therefore, observing them from a broadside direction will reveal two waves cancelling. Now imagine the field from the left dipole in Figure 5.1 propagating toward the right dipole. By the time it propagates to the location of the right dipole, the phase of the right dipole has reversed, since the sine wave of the generator has changed 180 degrees. Thus the fields cancel (off the ends), or in-line with the plane of the dipoles in Figure 5.2. However, in Figure 5.3 by the time the field travels to the other dipole, the field has changed 180 degrees, but the fields are now in-phase in the plane of the dipoles. This is referred to as an "end fire" configuration.

The result is that we can provide switchable bidirectional "figure 8" patterns by simply changing the phase relationship between the two elements, again as shown in Figures 5.2 and 5.3.

The gain parameters in Figure 5.2 may appear a bit conflicting upon careful investigation. The maximum gain is 5.94 dBi, or 3.82 dBd. This is counter-intuitive if we assume a doubling of a ½ λ dipole aperture, the gain should be 5.15 dBi, or 3 dB higher than that of the ½ λ dipole. Thus we must explain: where did the extra gain and thus aperture come from?

Simple addition of apertures assumes completely independent apertures and the added assumption that there is no "duplication" of aperture coverage, like the shading of one solar panel over another. If, however, antenna elements are placed in close proximity, they become "coupled" and all their independent characteristics change, including phase, amplitude, and direction of the charge acceleration. Since these terms directly relate to aperture, then aperture can change as a result of the terms changing in a non-independent manner.

The intuitive process above serves as a general explanation of how directivity is formed, but it ignores the critical subtlety hinted above by inter-element "coupling" in multi-element arrays.

## Mutual Impedance

Now we will consider this more complicated coupling problem. Although multielement arrays can constitute driven elements spaced at distances well into their far fields, we will first consider near-field interactions. The far field and very large aperture antennas will be considered in Chapter 9. In the case of far-field coupling (through free space) there is no measurable effect from one antenna to the other, only the induced currents and voltages the transmitting antenna imparts to the receiving antenna through free space. In effect, a single antenna suspended in free space is immersed in a medium defined by the intrinsic impedance of free space ($120\pi\Omega$). When we move the two antennas closer, approaching the near fields of both, significant changes result in the new "array." In effect we have formed a new antenna with a new set of parameters. The two elements become coupled not by a conductor, but by virtue of their respective apertures.

In our example in Figure 5.1, some of the power radiated by one element will be received by the other element; they are not mutually independent. Because of the received power, the element now has a modified voltage and current value: it is the sum of the initial fed power and the received power from the other element. This is called "mutual coupling" between the two elements. Since the affected voltages and currents have been affected in both elements, there must be some sort of impedance involved.

From Figure 5.1 we now define the two antennas as antenna 1 and antenna 2 (it doesn't matter which one is which, as they are equal for this analysis). We now define the self-impedance, which is also the feed point impedance, and also the $R_r$ of both the antennas independently. Therefore the self-impedance of the two antennas is $Z_{11}$ and $Z_{22}$ respectively.

However, antenna 1 will induce power into antenna 2 and vice-versa. The current, voltages and their phase relationships define a *mutual impedance* as $Z_{12}$ and $Z_{21}$ respectively.

We can now determine the feed point impedances for both antenna elements where

$$V_1 = I_1 Z_{11} + I_2 Z_{21} \quad \text{(Equation 5.1)}$$

In other words, the term $V_1 = I_1 Z_{11}$ is the voltage on antenna 1 as if it were the only element involved, and $I_2 Z_{21}$ accounts for the additional voltage term introduced by the close proximity of fed element 2.

Since the two elements are identical and fed in-phase, we can assume $I_1 = I_2$. Also, for the same reason, we can assume that $Z_{11} = Z_{22}$ and $Z_{12} = Z_{21}$.

Therefore the terminal impedance of element 1 is

$$Z_1 = V_1 / I_1 = Z_{11} + Z_{12} \quad \text{(Equation 5.2)}$$

Also, because of the inherent symmetry of this simple array, $Z_1 = Z_2$.

Thus for this simple case, using two identical elements with identical feeds (either 0 or 180 degree phase difference and equal amplitude), we derive a simple equation for the new feed point impedance, and, in this case, also the new $R_r$. We know that the self-impedance ($Z_{11}$) of a ½ λ dipole is 73 Ω. After some rather rigorous trigonometry we derive $Z_{12}$ at $-13 + j14$ Ω. Thus

$$Z_1 = 73 - 13 + j14 = 60 + j14\ \Omega \quad \text{(Equation 5.3)}$$

Since there are two elements in parallel, the real part of the feed point is 30 Ω where the two 60 Ω feeds are connected in parallel. Again, for this simple case we can directly calculate $R_r$ as 30 Ω. We now provide a simple calculation that verifies this value. We have already provided a general equation for $R_r$. In Chapter 7 we will derive a more convenient form for calculating the $R_r$ of a ground-mounted vertical antenna.

$$R_r = 94.25 \frac{h_e^2}{A_e}$$

for a vertical antenna. The equation for a dipole in free space is of the same form. Of particular importance to this discussion is that $R_r \propto 1/A_e$. We simply take the ratio of the values of $R_r$ for a ½ λ dipole and our new array, and the reciprocal of the numeric gain numbers. Therefore, in this simple case,

$$\frac{73}{30}\Omega \cong \frac{G_{array}}{G_{dipole}} \quad \text{(Equation 5.4)}$$

Indeed, 2.433 ≅ 3.96/1.64 = 2.419, or 3.836 dBd, or 5.99 dBi. This is very close to the modeled 5.94 dBi shown in Figure 5.2. This is yet another indication of the reciprocal relationship of gain (aperture) and $R_r$.

From this discussion there are two points which should be made:

1) Often when modeling in *EZNEC* or other programs, *exact* correlations do not occur with theory. This is mainly due to not providing *exact* dimensions when modeling the antenna. However,

$$2.433 \cong \frac{3.96}{1.64} = 2.419 \quad \text{(Equation 5.5)}$$

is close enough!

2) $Z_{12}$ is taken to be $-13 + j14$ Ω. This is not an error. *Negative resistances* are commonly encountered when modeling multielement arrays. It simply implies that more current is flowing *toward* the source rather than flowing *toward* the antenna due to the mutual coupling. Therefore in this case the sum of the real part of the feed points is 60 Ω.

The array in Figure 5.3 uses a similar method of calculation. As arrays become more complex, we can no longer make the assumption of symmetry as in the simple case of Figure 5.1's array. Therefore the computation becomes more complicated, but the principle shown here is the same. A natural progression is to the quadrature array.

**Note:** The calculation of multielement arrays is usually more complex than this simple analysis which is presented to show conceptually how $R_r$ relates to the other critical terms.

## Quadrature Array

In the quadrature array (**Figure 5.4**), we use the same two dipoles, as above. However, the element spacing is reduced to ¼ λ (90 degrees).

The two dipoles retain center feed points, but now we will define the phase relationship as 90 degrees rather than 0 or 180 degrees, with the fed amplitude (current) remaining identical in both feeds. We can follow the same intuition as in the examples using ½ λ spacing. In this case, element 2's

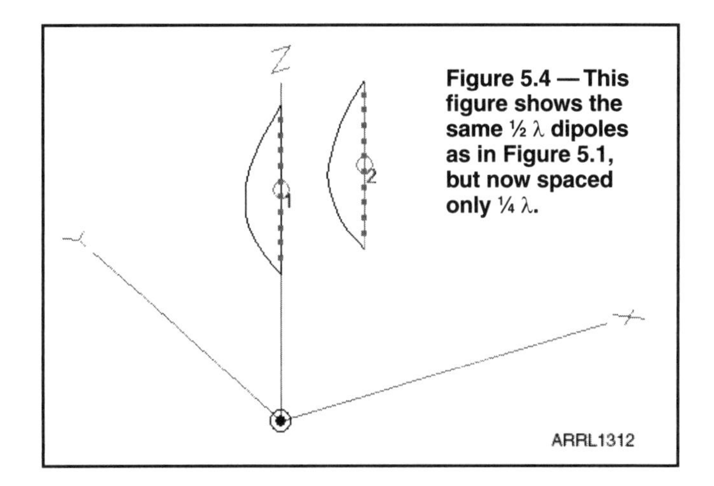

**Figure 5.4 — This figure shows the same ½ λ dipoles as in Figure 5.1, but now spaced only ¼ λ.**

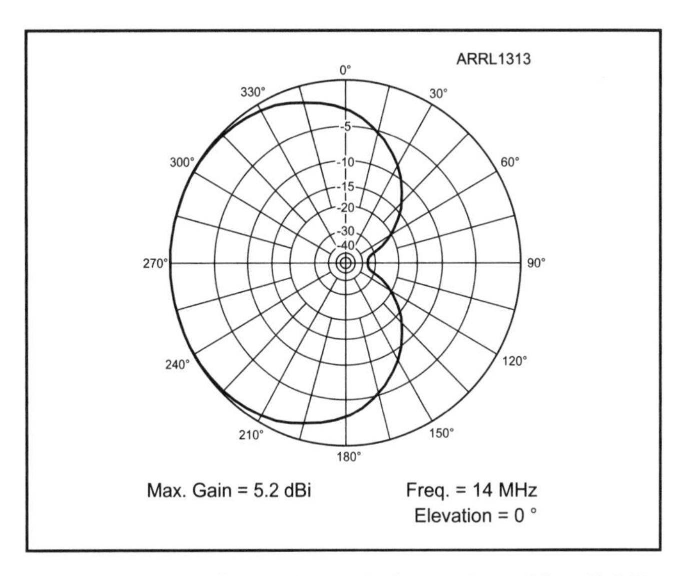


**Figure 5.5 — This figure shows the heart-shaped "cardioid" pattern when the left element phase "lags" the right element by 90 degrees. When the field radiated by the right element reaches the left element, the right element is now in phase with the arriving field and thus the two fields sum in the "left" direction along the X-axis. However, when the left element's field reaches the right element, the two fields are now 180 degrees out-of-phase and therefore there is no radiation in the "right" direction.**

phase is *leading* element 1's phase by 90 degrees. By the time element 2's field reaches element 1, the phase of the current delivered to element 1's feed point has advanced by the same 90 degrees. Thus, in the negative X direction (**Figure 5.5**) the fields add completely and we derive maximum gain.

However, where element 1's field propagates toward element 2, the phase of element 2's current has advanced another 90 degrees, for a total phase shift of 180 degrees, and there is complete cancellation in the positive X direction. Between these directions of maximum and minimum radiation, there is partial addition or subtraction of the fields, resulting in the overall pattern shown in Figure 5.5.

Unlike the ½ λ spacing examples in Figures 5.1 to 5.3, this array is non-symmetrical, in that $Z_{12} \neq Z_{21}$. Since the self-impedances of elements 1 and 2 are identical at 73 Ω pure resistive we can avoid very difficult computations by allowing *EZNEC* to perform the necessary calculus by simply reading the resulting $Z_1$ and $Z_2$ values, which are 120.3 + *j*76.95 and 52.85 – *j*9.92 Ω respectively. Therefore we can now directly compute the two different mutual impedances.

$$Z_{12} = (120.3 + j76.95) - 73 = 47.3 + j76.95\ \Omega$$

(Equation 5.6)

and

$$Z_{21} = (52.85 + j9.92) - 73 = -20.15 - j9.92\ \Omega$$

(Equation 5.7)

where Z11 = Z22 = 73 Ω.

We can see that the gain of this array is almost exactly 3 dBd or 2× the power gain over a dipole. The two real parts of the feed point impedances are 120.3 and 52.85 Ω. We calculate these two parallel resistances the same way we calculate the value of two resistors in parallel:

$$R_r = \frac{120.3 \times 52.85}{120.3 + 52.85} = 36.72\ \Omega$$

(Equation 5.8)

As expected, $R_r$ for the array is almost exactly ½ $R_r$ for the single dipole, according to Equation 5.4

**Note:** For the quadrature array we must use a slightly more complicated technique for the calculation of $R_r$. Once again, as arrays become more complex, the equivalent calculations become also far more complicated.

## Parasitic Arrays

The usual configuration for a parasitic array (**Figure 5.6**) is to employ one driven element and rely completely upon the mutual impedance(s) between or among the driven element and the parasitic element(s). In the case of multiple fed and multiple parasitic elements forming one antenna, a combination of these two analytical techniques becomes necessary. The parasitic element usually employs the scattering aperture, where the termination (center of a ½ λ linear element) is a short, or 0 Ω.

The parasitic array analysis becomes quite different from

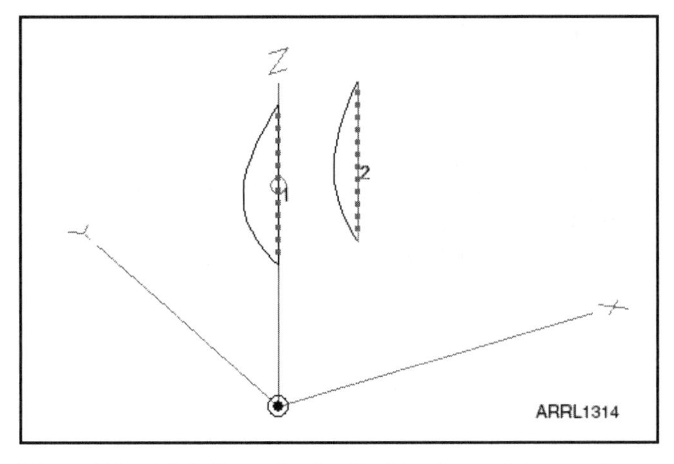


**Figure 5.6 — This figure is identical to Figure 5.4 except now only element 1 is center fed. Element 2 is the parasitic element and becomes part of the array only by virtue of its mutual impedance with element 1.**

the driven array analysis. The first step is understanding that there is a phase reversal from the driven to parasitic element. Since like charges repel, then we can assume that accelerating like charges will repel. Therefore the induced voltage and current in element 2 will be 180 degrees out-of-phase from the inducing field striking it. With this in mind, we can begin to quantify the terms

$$V_1 = I_1 Z_{11} + I_2 Z_{12}$$ (Equation 5.9)

In other words the voltage on the driven element will be the sum of the fed current/self-impedance (as if it were alone) plus the current/mutual impedance from the parasitic element. Therefore the voltage and current on the driven element is also a function of what the parasitic element "received" from and retransmitted back to the driven element! Since the parasitic element current is 180 degrees out-of-phase with its induced current,

$$I_2 Z_{22} + I_1 Z_{12} = 0$$ (Equation 5.10)

In other words, the current flowing on the parasitic element due to its own impedance is the negative value of the current flowing on the driven element that is induced into the parasitic element, so they sum to zero. Now we can describe the fundamental equation for the action of a parasitic array, and its important practical configurations including the Yagi-Uda.

$$I_2 = -I_1 \frac{Z_{12}}{Z_{22}}$$

(Equation 5.11)

In other words, the current on the parasitic element *and its phase* is the negative value of the current incident upon it. The amplitude and phase of the parasitic element is determined by the complex coefficient of the mutual impedance and the self-impedance of the parasitic element. However, the *instantaneous* phase value of the parasitic element's current will also be a function of the time (and thus phase differ-

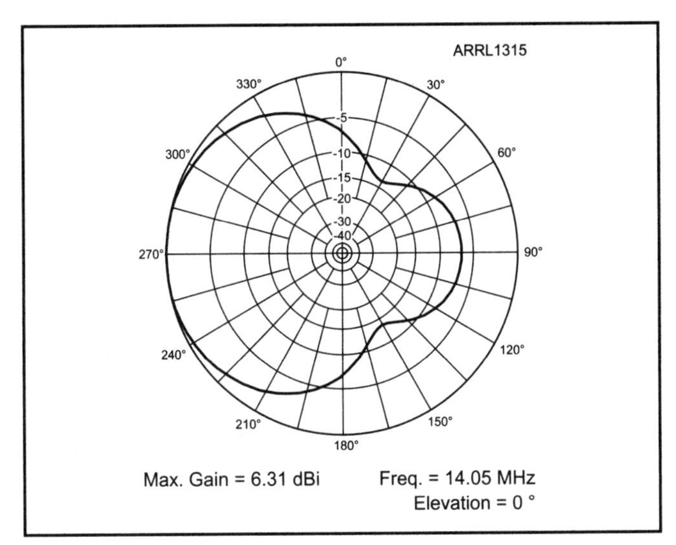


**Figure 5.7 — This figure shows the pattern of the parasitic array in Figure 5.6. The optimum spacing for such a "two element Yagi-Uda" is closer to 0.2 λ rather than the shown ¼ λ. A truly optimized array of this kind can yield about 5 dBd, much better than the mere 3 dBd gain from a two-element quadrature driven array shown in Figure 5.5. We used ¼ λ spacing for easier description of the basic principles.**

ence) due to the spacing of the driven and parasitic element. For example, if the spacing is ¼ λ there will be a 90 degree phase lag between the two elements. Therefore, the phase on the parasitic element will be –90° – 180°, or –270° or +90° relative to the driven element. Then the complex coefficient further modifies the phase difference. Finally through PADL, we can derive the parasitic antenna's pattern. Of critical importance in determining the phase of the parasitic element is $Z_{22}$. A very slight change in the complex value (slightly inductive to slightly capacitive reactance) can "flip" the phase response dramatically.

This idea should be familiar to anyone with a good understanding of ac electronics: current leads in capacitive reactive circuits and lags in inductive reactive circuits. If we extend that intuition to a current flowing up and down an antenna element it is easy to see how the phase of the current for the entire element can change with a change in the reactive impedance of the element. This, or course, leads to a change in "P" in PADL and therefore affects the directivity of the array. **Figure 5.7** shows the array's pattern.

To reiterate, the basic PADL principle applies to both driven and parasitic arrays. Driven arrays are inherently more stable since the characteristics are largely driven by the self-impedances of the elements. Parasitic arrays, in contrast, depend completely upon mutual impedances for their excitation and very careful tuning of the self-impedances for proper phase response. Finally, greater gain can be obtained from parasitic arrays because of the much larger scattering apertures inherent to parasitic elements. This is in contrast to the quadrature driven array where the antenna apertures simply add, thus providing only about 3 dBd gain. An optimized two-element Yagi-Uda provides about 5 dBd.

## Phased-Array Antennas

Although any multi-element antenna array could rightly be called a "phased array," this term has come to define a more specific type of configuration. Today phased arrays define a multi-element array that uses fixed (relative to each other) elements whose directivity is steered by adjusting the phase and amplitude relationships among the individual antennas. When used at UHF frequencies and higher, a large number of elements can be employed, even on mobile platforms.

With the advent of computer control, high directivity patterns can track satellites and mobile users, and be used as radar systems with no moving parts. In effect the steering of directivity is all performed electronically. Furthermore, the array itself can be mounted on moving platforms such as aircraft, satellites, ships, vehicles, etc. Phased arrays allow moving platforms to "lock on" to other moving vehicles by using large numbers of elements. This represents a complex problem of phase adjustment and amplitude of large numbers of elements in real time. Furthermore, by using multiple phase combining in digital processing, multiple directions can be tracked simultaneously using the same elements.

The problem extends beyond just the relative phasing of the elements. If the elements are mutually coupled, there is the added problem of tuning the array in real time. Consequently, phased array design has become an important sub-discipline unto itself, and one of high importance for commercial as well as military applications. Antenna phasing can take place either by phasing the RF signals from the individual elements or by using separate receiver front ends and analog-to-digital (ADC) converters for each element. Once all the antenna signals have been digitized, phasing (and all other signal processing), can be accomplished in DSP (digital signal processing).

A detailed explanation of the above complex systems is not possible within this text. However, a much simpler array can illustrate the basics of phased array systems and provide some added useful qualities for the specific application (low band receiving).

### The 4-Square Array

The 4-square array is a well-known configuration to amateurs using the "low bands", most typically the 160- and 80-meter bands. The 4-square array is typically set up to simply switch among four azimuth directions using coaxial cable lengths to provide for the phasing of the elements. By approaching this simple 4-element array as an opportunity for additional flexibility (a key advantage to "phased arrays") additional useful functionality can be achieved.

First, the 4-square will be described, but then we will begin a discussion on maximizing the potential of this phased array. The system described here is designed as a receive-only antenna for use at 160 and 80/75 meters. It uses element spacing of 80 feet using 20-foot-high vertical elements. Four-square and other multi-element vertical arrays are treated from a different standpoint in Chapter 6. The mutual coupling at these frequencies, spacing and heights is minimal, thus can be ignored (a huge advantage!). The result is a relatively low efficient antenna system, but that is not a problem at these frequencies as will be described in later chapters. This holds true for receive-only arrays, as transmit antennas *do* require higher efficiency, which requires longer (higher) elements and careful attention to mutual coupling as well as phasing.

**Figure 5.8** shows an *EZNEC* view of the 4-square for 160 and 80 meters with 20-foot pipe elements and 80 foot spacing among the elements. The low height of the elements reduces mutual coupling to a degree that it can be ignored from a practical view. Thus, the problem reduces to phasing/combining as the amplitude of each element is the same. Also, for the intended application, only 2-dimensional steering is needed, the target "directions" and the array itself are fixed. Of course, the directivity is steerable, in this case, 360 degrees of azimuth divided into four azimuth "directions".

In the addressed receive case, the absence of significant mutual coupling permits high impedance voltage sensing, by using high impedance input amplifiers (typically a JFET) which provide 50 or 75 Ω output sources from the four antennas. Voltage sensing does not provide for a conjugate match but does allow elimination of any front-end tuning. The active JFET amplifier provides for a high impedance voltage sampler while simultaneously providing the active impedance match to the coaxial feed as well as needed gain from the very inefficient antenna elements. 75 Ω coaxial cable has become comparatively inexpensive thanks to the massive amounts of RG6 being manufactured for cable and satellite television. No matching or phasing is necessary at the feed points, only the above functions are needed. The elimination of conjugate matching also permits uniform phase and amplitude response from the antenna element, thereby greatly simplifying the phasing and amplitude balancing necessary to create the desired directivity.

It is important to note that each of the four elements are always independently acting as omni- directional antennas (azimuthal). The technique shown here takes advantage of this in that the phasing and combining of the four signals is typically performed completely independent of the individual antennas' performance, in effect, the perfect situation for a phased array. The above description applies to existing array configurations which are very effective for their intended use. However, with some added functionality, more flexibility and better receive performance can be achieved.

## Enhancing 4-Square Array Utility

There is no reason that multiple and independent phasing and combining cannot be simultaneously performed using the same antenna array. With some additional capability, all four received directions can be *simultaneously* provided which offers additional opportunities for receive performance.

Transmission line phase delay requires three lines, since you cannot build a phase advance transmission line (unless you use very long lines). Since the spacing is small relative to the wavelengths involved, an "end fire" arrangement is typically used. In effect, the 4-square becomes a 3-element endfire array since the two middle elements contribute very little "broadside" gain. These two middle elements simply provide directivity balance and a slight broadside gain response. The usual practice is to feed the two broadside opposite elements in-phase, as shown in **Figure 5.9**. The signals from the "leading" and "lagging" (directional) elements are properly phased with a phase delay and a phase lag respectively. The phase (in degrees) is calculated by taking the free-space separation difference between the line that intersects the in-phase elements and the two phased elements, in this case 56.56 feet.

The free-space phase difference (θ) in degrees is thus

$$\theta = 180 - \frac{56.56'}{\lambda}(360°) \quad \text{(Equation 5.12)}$$

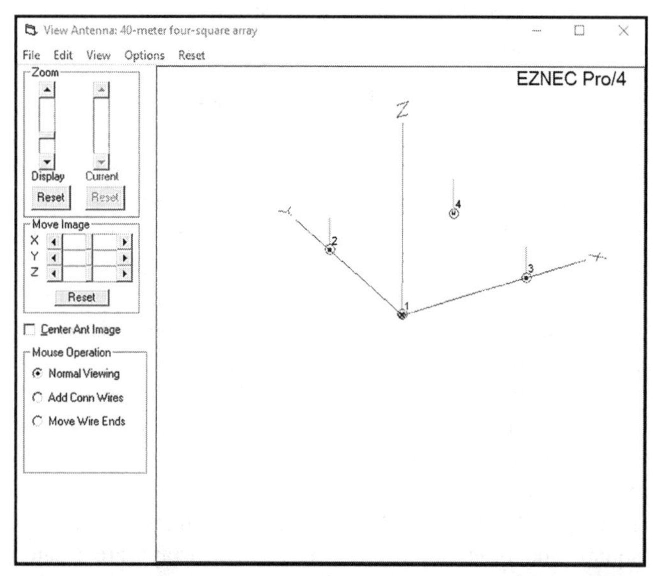


**Figure 5.8 — An *EZNEC* view of the 4-square for 160 and 80 meters with 20-foot pipe elements and 80 foot spacing among the elements.**

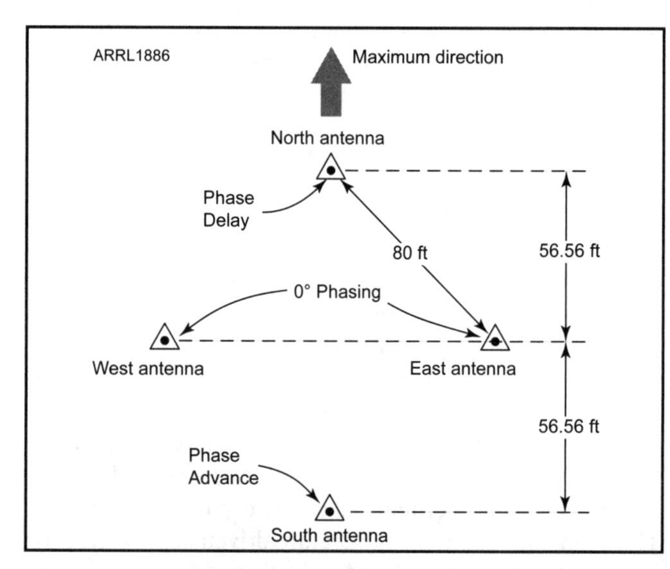


**Figure 5.9 — Top view of the basic geometry and phasing for an 80 foot 4-square. Using passive phase *advance* circuits eliminates the need for phasing delays of the middle elements.**

where λ is the wavelength in feet.

To achieve a unidirectional pattern from an *end-fire* configuration (which we are designing here), we must subtract the phase difference which is

$$\frac{56.56}{\lambda}(360°) \quad \text{(Equation 5.13)}$$

From 180 degrees to arrive at θ which is the proper *feed* phase difference.

**Table 5.1** shows the phase values required for two desired frequencies. On 80/75 meters, a mid-point frequency was chosen between 3.5 and 3.8 MHz, where most DX operation takes place. There is insignificant change is directivity, even up to 4 MHz.

**Figure 5.10** is a block diagram of the "enhanced" 4-square phased array. The four 20-foot verticals use FET voltage amplifiers at their bases. Four *identical* length feed lines (typically RG6/U) connect the antennas with the switching/phasing unit. The actual lengths of these feeders is not critical, but they must be identical in length. Each antenna feed line is split into 8 outputs to provide inputs to the 8 band/direction phasors. This provides for 8 separate band/direction outputs. (four separate directions for both 160 and 80/75 meters which provides for simultaneous reception of all 8 band-directions.

**Table 5.1**
**Phase Values Required for Two Desired Frequencies**

| *Frequency (MHz)* | *Wavelength (Feet)* | *θ (Degrees)* |
|---|---|---|
| 1.83 | 537 | 142 |
| 3.65 | 270 | 105 |

DC power is required for the remote antenna FET amplifiers. The addition of a milli-amp meter that can be selectively switched in among the four amplifiers provides a very quick check to see if the line is connected properly and the remote amplifier working properly. In effect, most of the remote system can be checked through this simple addition of a switch and meter and if there is a problem, the line/amplifier in question is identified.

**Figure 5.11** shows a phased arrangement for a "north" directivity. A 4PDT relay (preferred) or switch can be used. The relay allows more flexible control configurations. Notice that two of the elements have 0-degree phase shift.

## Advantages of a Phased Array

Since each direction has its own simultaneous output, any direction can be set for the desired direction and/or a *noise antenna*. A separate phase/amplitude combiner can be

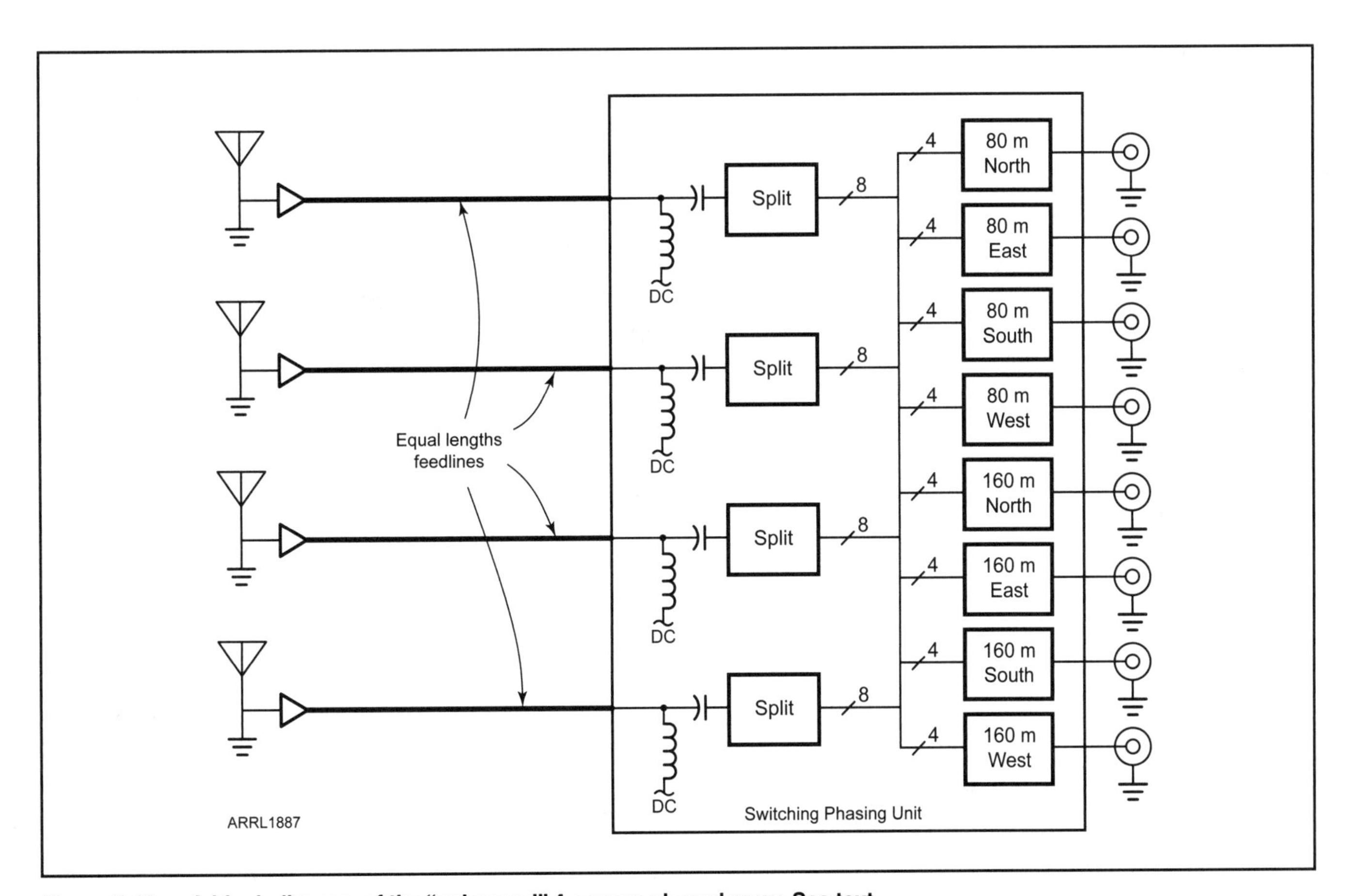


**Figure 5.10 — A block diagram of the "enhanced" 4-square phased array. See text.**

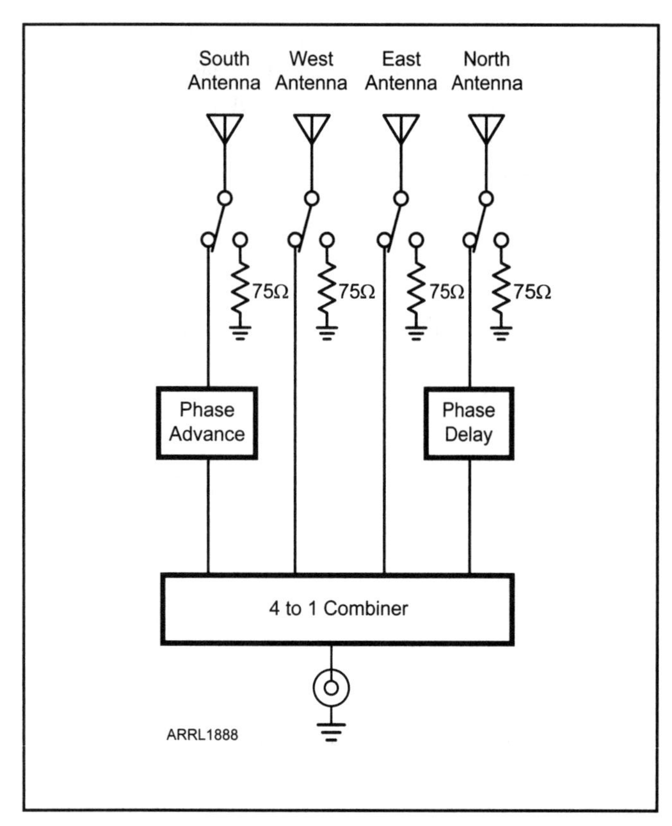


**Figure 5.11 — A phased arrangement for a "north" directivity.**

used to null undesired noise and/or interfering signals from another direction. If an undesirable signal (noise or other undesired signal) is coming from one of the three unused directions, that undesired signal can be used as a noise source. By using these two signals from different directions the undesired signal can be nulled through additional phasing and amplitude balance.

By using simultaneous twin directionality from the 4-square, we can null an interfering signal coming from the east out of the desired signal's bandpass from the north. See **Figure 5.12**. This simplified functional diagram shows that the undesired signal's level from the north and east directions must be adjusted to be identical, while the phase difference must be 180 degrees. By summing the two signals we can null the interfering signal. This nulling technique is currently commonly used, but with a second antenna array (typically a single element). By using a directive antenna for the undesired signal (noise antenna) as well as for the desired direction, this technique becomes much more effective.

Eighty- and 160-meter outputs can be used to feed separate receivers simultaneously. This can be very useful for spotting a signal of known frequency on another band while simultaneously operating on another. Also, for multi-station use (i.e. in contests) simultaneous 160- and 80/75-meter reception can be had using the same array. Of course, the second receiver can also simultaneously use another direction in the same band to monitor a known frequency of a DX-pedition while operating on other frequencies.

Using four identical receivers with their local oscillators' frequency locked, a "surround sound" effect can be implemented with 4 speakers "surrounding" the operator and give an audio directional simulation for the arrival of multiple signals. For example, the "north" speaker can be located in front of the operator, the east to the right, south to the rear and west to the left.

Any combination of the above can also be set up for great flexibility and improved reception by maximizing the utility of four simple short monopoles.

As shown above, this phased array system uses lumped circuits for phase shifting. Transmission lines could be employed, but the resulting coils of coaxial cables would be four times the number required for the existing single directional array designs. There are multiple options for phase shifter circuit configurations. The 4-square array has a huge advantage: the impedance output from the voltage sense amplifiers is 75 Ω resistive, and the load impedance can also be set to 75 Ω. This includes the impedances for the splitters and combiners. The equations for determining the C and L values for "T" and "π" phase shifters are simplified since there are no reactive (complex) terms to deal with and the load and source impedances are equal (75 Ω being typical, but 50 Ω is also used). The choice of which type of circuit to use is arbitrary for performance and one circuit may make a "T" more pragmatic while another a π. Usually the use of two capacitors and one inductor configurations is more desirable since capacitors are more convenient to use than inductors. However, sometimes an inductor value may become significantly higher at a given frequency for a given phase and that may make a π or TEE more desirable. It is important to keep the losses of these circuits low, which is fairly easy at these frequencies using silver mica capacitors and ferrite or iron core toroid inductors.

From Terman[2] page 212, we derive the equations using complex notation for the following eight equations that define phase advance and phase delay networks using both "T" and π networks.

## Phase Shifters Using "TEE" or π Networks

The TEE-network has some very desirable characteristics: it can be configured as a phase advance or phase delay circuit while not affecting the input and output impedances and, except for the small injection loss, there is no attenuation (except for inductor losses) through a tee network with identical input and output impedances. It can also simultaneously match two resistive impedances independently of the desired phase shift. However, for this circuit, we only need the simplest form which also results in the simplest mathematics to determine the L and C values needed for a specific operating frequency and phase shift. For more complex designs, I include the relevant equations since Reference 1, is out of print! Otherwise you will need to find a suitable computer design program or actually sit down and calculate it by hand, like we did in the not-so-good old days. For this antenna technique we only need to use 75 Ω in and 75 Ω out, the desired phase shift and the operating frequency. However, I also provide the needed 50 Ω values. The tee network can also be used in

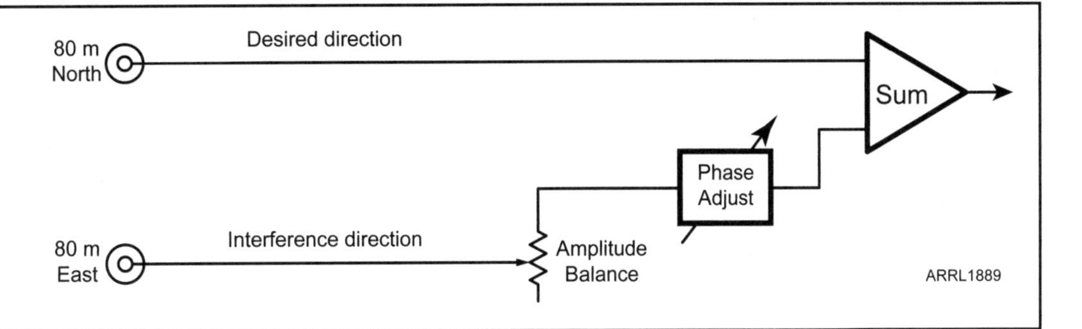


**Figure 5.12 — A simplified functional diagram shows that the undesired signal's level from the north and east directions must be adjusted to be identical, while the phase difference must be 180 degrees.**

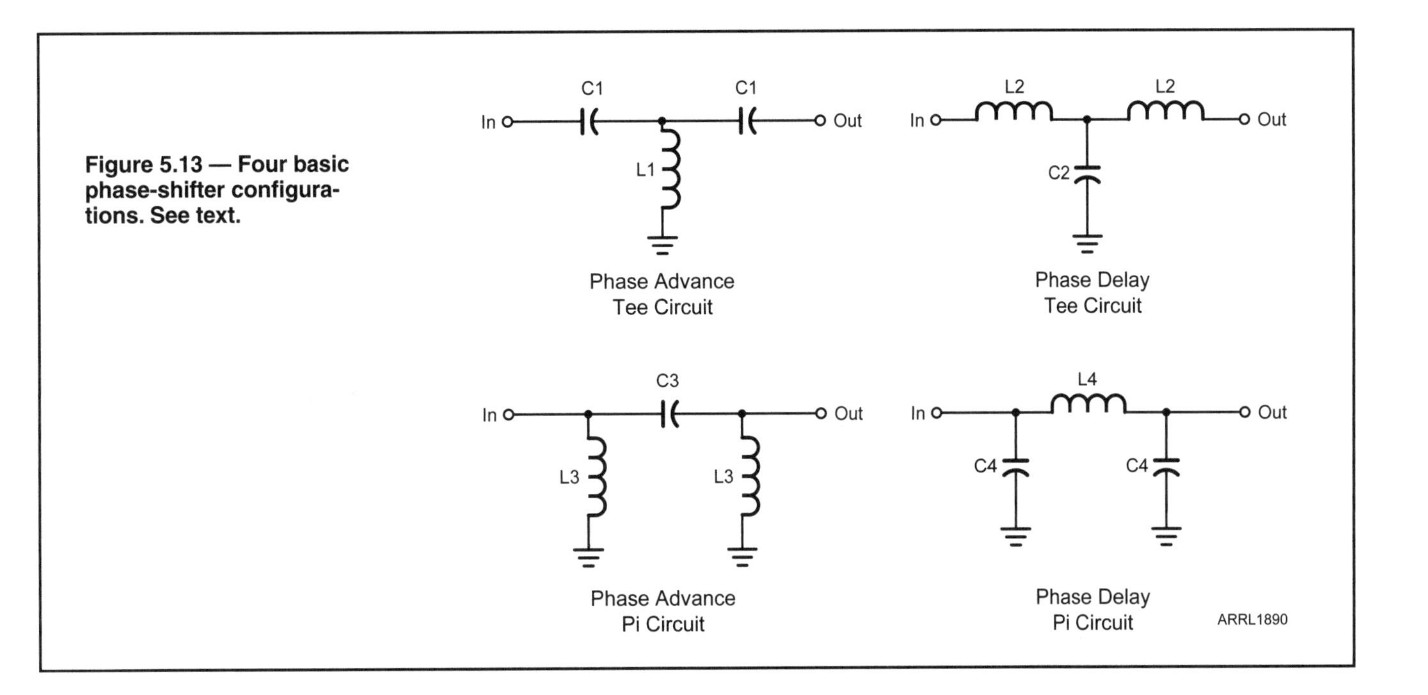


**Figure 5.13 — Four basic phase-shifter configurations. See text.**

2-element receive antennas or in 8-circle arrays.

The π network is also useful as a phase shifter. In some instances, the π will allow fewer inductors and/or lower inductor values simplifying inductor construction and lowering inductor losses.

## TEE and π Network Phase Shifters

In **Figure 5.13** we see four basic phase-shifter configurations. The equations for the values of the respective capacitors and inductors follow.

I will present only general equations that are used for calculating the frequency and phase shift for *identical* input and output impedances. The Terman[2] and Layton[1] texts also provide the more general equations for using the same circuits for simultaneous impedance matching and defined phase change. (Note: There are several typos in the Layton text for equations using different input and output impedances. However, he offers an example which is correct) I include the Layton text because the discussion on tuning and phasing AM broadcast antennas is excellent and has value for low band amateur designs as well. This text I believe is out of print, but Terman (also out of print but easier to find) states the same essential equations for this topic without error.

The eight graphs are presented for convenience of determining approximate capacitor and inductor vales and also serve to show practical possible phase shift values obtainable with both TEE and π networks.

For the following eight equations, $\omega = 2\pi f$ where f is the frequency in Hz, $Z_0$ is the input *and* output impedances, and θ is the desired phase shift in *radians,* where 180 degrees = π radians.

$$C_1 = \frac{sin\theta}{\omega Z_0 (1 - cos\theta)} \quad \text{(Equation 5.14)}$$

Equation for C1 in the TEE network phase advance circuit.

$$L_1 = \frac{Z_0}{\omega sin\theta} \quad \text{(Equation 5.15)}$$

Equation for L1 in the TEE phase advance circuit.

$$C_2 = \frac{sin\theta}{\omega Z_0} \quad \text{(Equation 5.16)}$$

Equation for C2 in the TEE phase delay circuit.

$$L_2 = \frac{Z_0 (1 - cos\theta)}{\omega sin\theta} \quad \text{(Equation 5.17)}$$

Equation for L2 in the TEE phase delay circuit.

$$C_3 = \frac{1}{\omega Z_0 sin\theta} \quad \text{(Equation 5.18)}$$

Equation for C3 in the π phase advance circuit.

$$L_3 = \frac{Z_0 sin\theta}{\omega\left(1 - cos\theta\right)} \quad \text{(Equation 5.19)}$$

Equation for L3 in the π phase advance circuit.

$$C_4 = \frac{1 - cos\theta}{\omega Z_0 sin\theta} \quad \text{(Equation 5.20)}$$

Equation for C4 in the π network phase delay circuit.

$$L_4 = \frac{Z_0 sin\theta}{\omega} \quad \text{(Equation 5.21)}$$

Equation for L4 in the π network phase delay circuit.

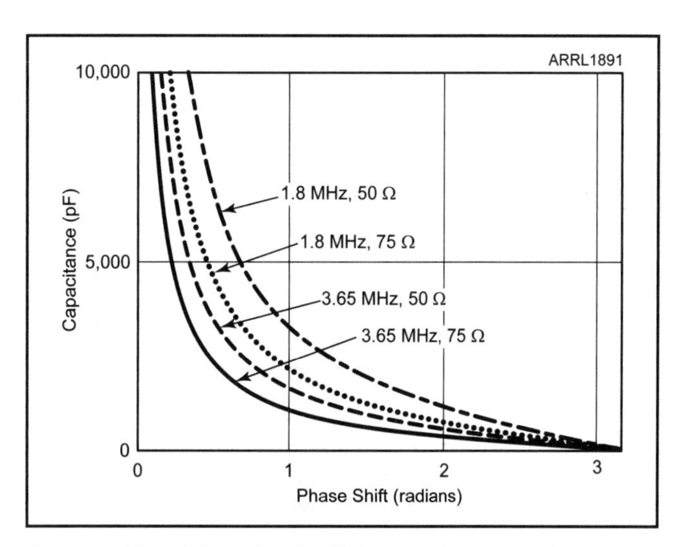


**Figure 5.14 — Values for C1 TEE network phase advance.**

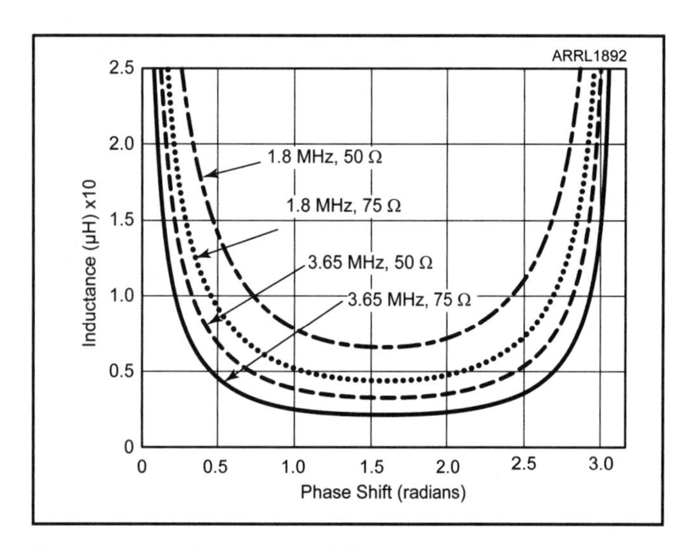


**Figure 5.15 — Values for L1, TEE network phase advance.**

**Note:** There is an interesting short cut for calculations involving the TEE network. For quadrature (90-degree phase shifts, advance or delay) the inductive and capacitive values of reactance are simply equal to the values of the identical input and output impedances. For example, for 90-degree phase shifts in a 75 Ω system, C1, L2, C2, and L2, all require 75 Ω reactive values (of course calculated at the operating frequency). This simple equation can also be useful for sanity checking your calculation for other phase shift values.

The plots shown in **Figures 5.14** through **5.21** are given for specific frequencies at both 75 and 50 Ω for convenience. These eight plots correspond to the eight previous equations.

## Phased Arrays of Small Loop Antennas

In Chapter 9 we will describe loop antennas in detail. Here we will present a brief set of designs for creating phased arrays using small loops for receiving. To configure a unidirectional loop, the typical technique is to use a resistor on the loop as in the EWE and K9AY versions. This creates the unidirectional pattern but also severely limits the forward gain. This gain reduction is a fair trade-off for the great reduction in size a single unidirectional loop affords.

To increase gain and also increase directivity it is possible

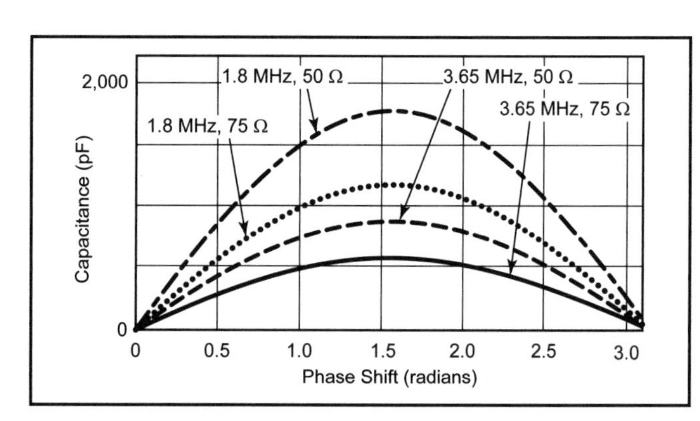


**Figure 5.16 — Values for C2, "T" Network phase delay.**

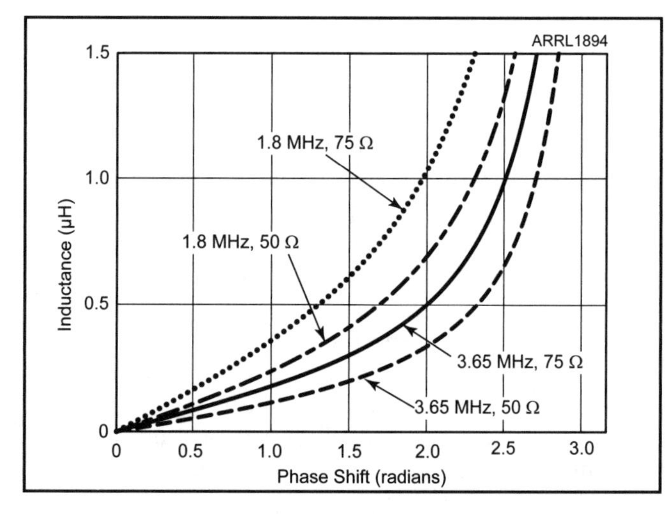


**Figure 5.17 — Values for L2, "T" Network phase delay.**

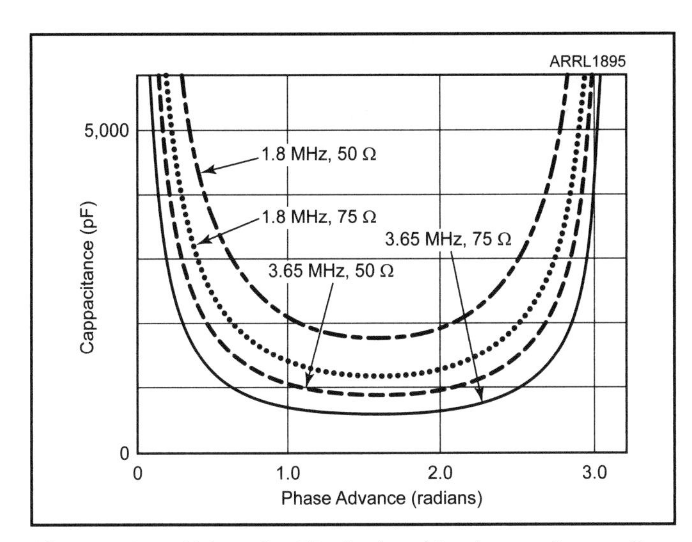


**Figure 5.18 — Values for C3 picofarad in phase advance Ω network.**

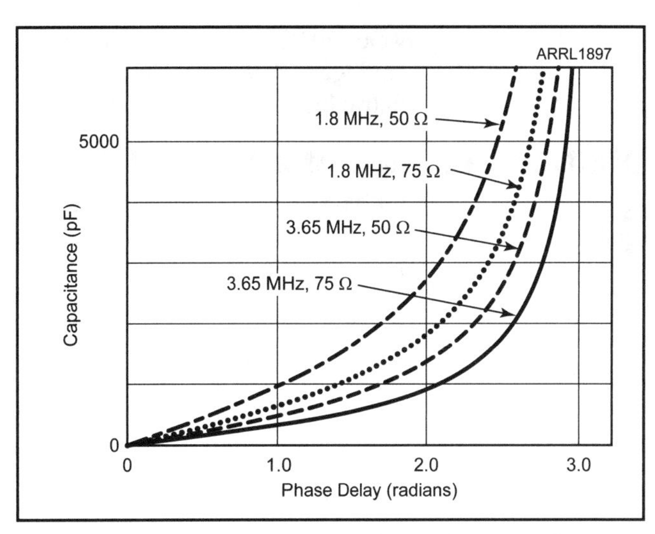


**Figure 5.20 — Capacitor value for C4 in picofarads for Ω network phase delay.**

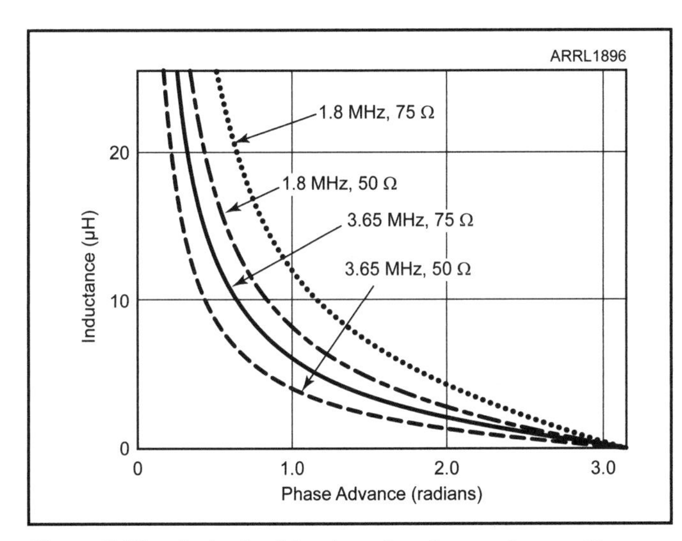


**Figure 5.19 — Inductor L3 values for phase advance Ω network.**

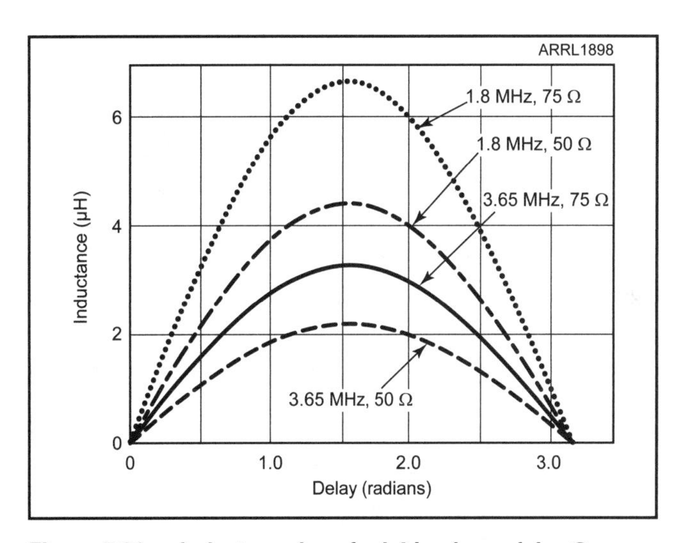


**Figure 5.21 — Inductor values for L4 in phase delay Ω network.**

to use multiple loops configured either as broadside and/or end-fire arrays. For a two-element broadside loop array, it is necessary to use the resistive termination-type loop antenna as utilized in the EWE and K9AY loops. However, for end fire configurations, the resistors may be eliminated and thus greatly increase the gain without compromise of front-to-back ratio.

The diagram in **Figure 5.22** shows two simple triangular loops. The directivity will be end-fire (either to the left of right of the diagram). The exact dimensions are not critical, but the loops should be as close to identical as possible. There are no ground connections or rods needed for this type of array, only two opposing relatively short supports (trees, poles, etc.) making it easy to pre-assemble and install for contest or DXpeditions.

The total horizontal length is 100 feet. A height for the bottoms of the loops of 8-10 feet is suggested for permanent installations to keep people and animals from contacting the loop wires. The entire array can be lowered if such concerns do not apply (i.e. temporary installations). Identical lengths of coax can be used with the same phasing techniques described above for simultaneous reception, in this case, from two directions.

Using a 135-degree lead on the lower upper (north) loop the pattern is to the south, and vice versa. (**Figure 5.23**) Since a small loop has an inherent bi-directional pattern (reducing side response) the pattern directivity is enhanced over an azimuthally omni-directional vertical.

A second identical orthogonal array could provide four switchable directions (the equivalent of a four-square). The –15 dBi gain shown here is considerably better than –26 dBi gain for a single K9AY loop and just a bit more than –11 dBi for a typical 1-wavelength Beverage. A wide range of varieties can be derived for many specific requirements.

In **Figure 5.24** we have an elevation gain *EZNEC* diagram for the twin-loop end-fire array.

**Figure 5.25** shows the same two element loop at 3.65 MHz. Notice the excellent gain. By adding a third loop in the middle of the original two on 80/75 meters, the directivity is significantly improved. In **Figure 5.26** we see the elevation pattern at 3.65 MHz with two loops.

## Digitized Version of a Phased Array

In the November 1987 issue of *Ham Radio* magazine, I published an article called "Tomorrow's Receivers: What Will the Next 20 Years Bring?"[3] In that article I included the first known block diagram of what today is known as a software defined radio, or SDR. I called it a "Digitized Radio of Tomorrow." The block diagram includes the antenna, preamp, A/D converter, DSP, and DAC to convert the audio to analog for human consumption.

Modern phased arrays, in effect, use SDR type techniques

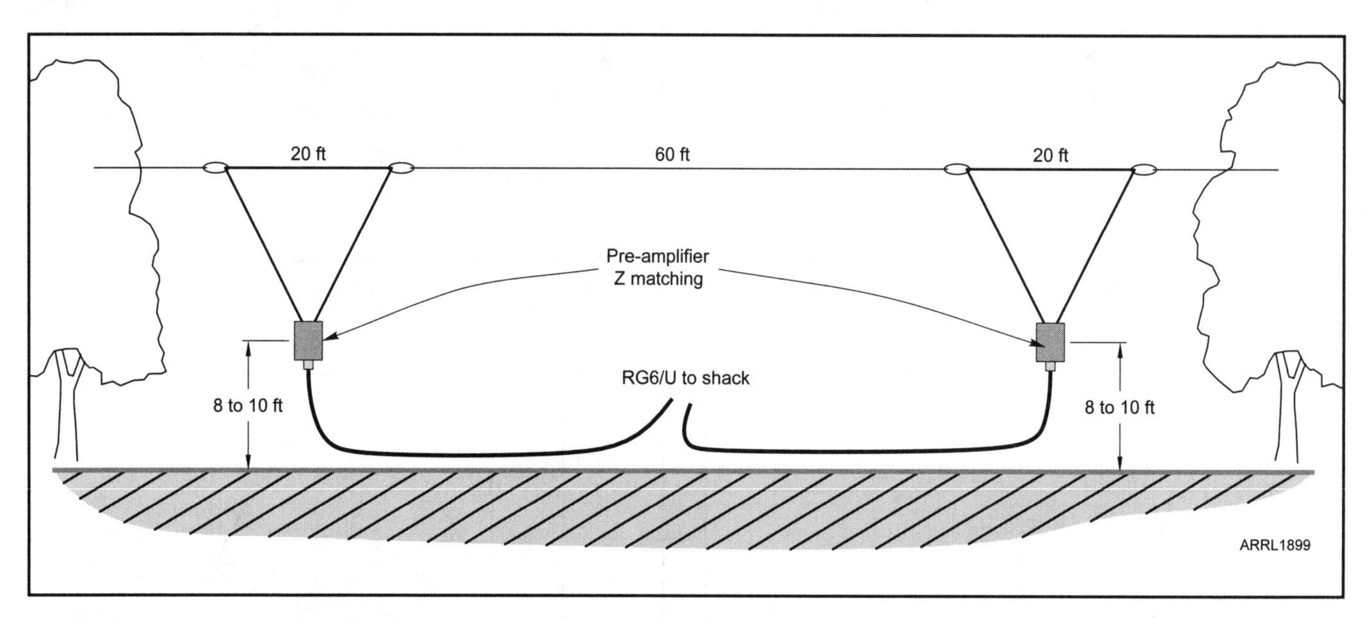


**Figure 5.22 — Two simple triangular loops. The directivity will be end-fire (either to the left of right or the diagram).**

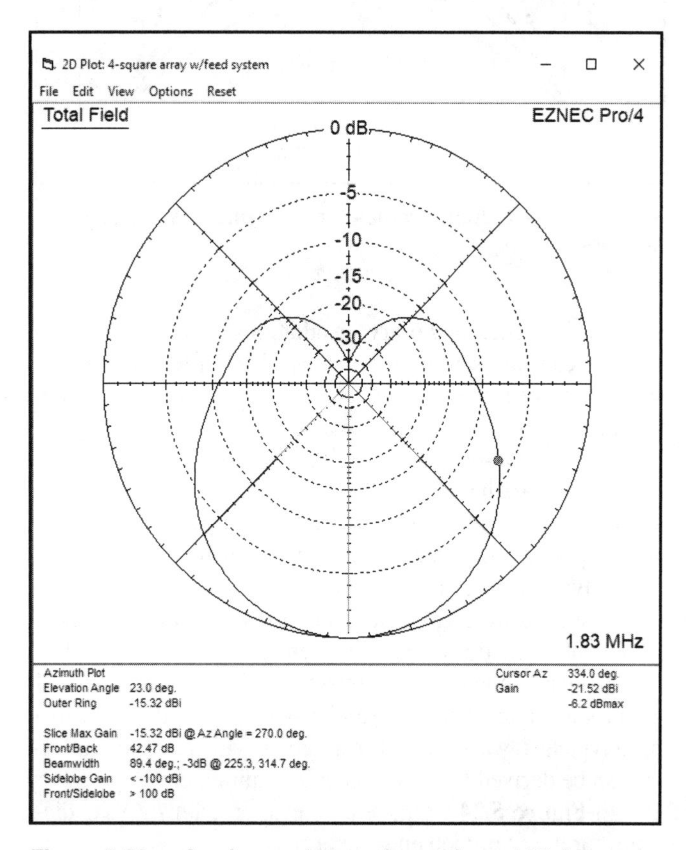


**Figure 5.23 — As shown above, when using a 135-degree lead on the lower upper (north) loop, the pattern is to the south, and vice versa.**

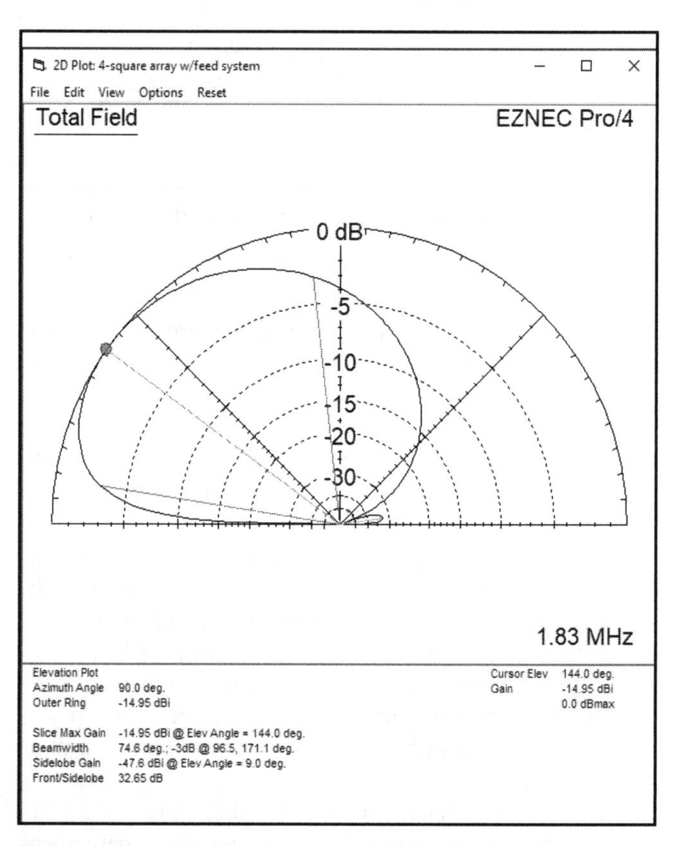


**Figure 5.24 — An elevation gain *EZNEC* diagram for the twin-loop end-fire array.**

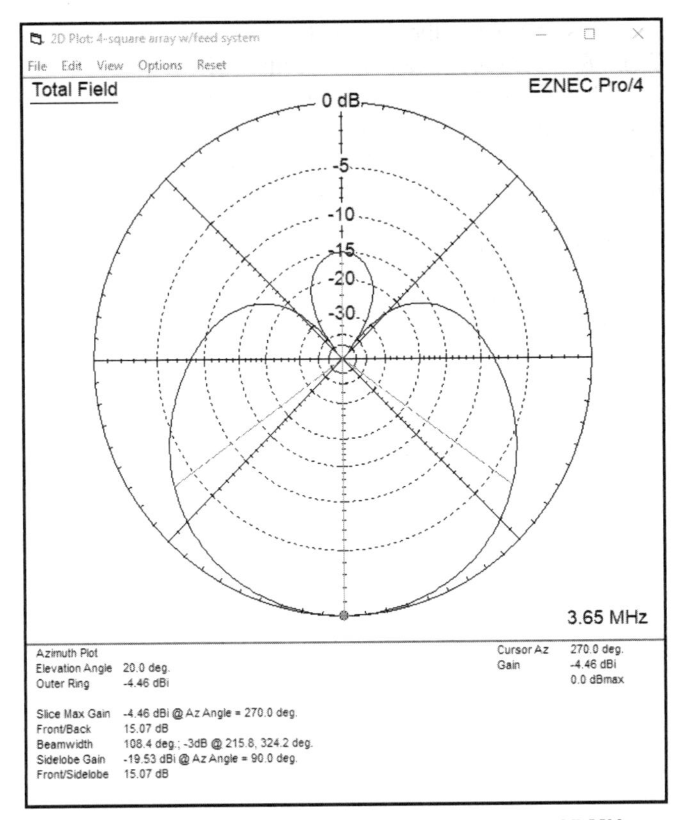


**Figure 5.25 — Modeling the two-element loop at 3.65 MHz.**

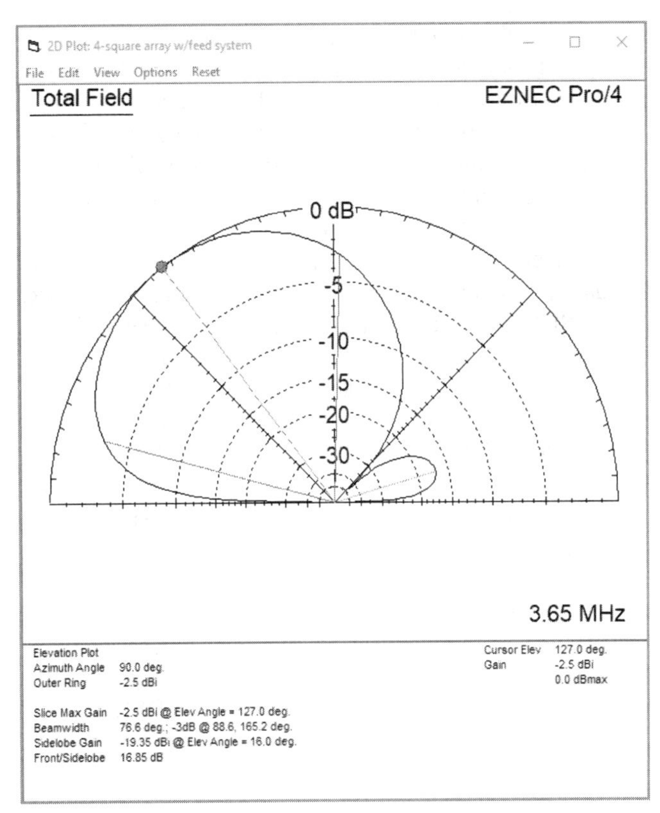


**Figure 5.26 — The elevation pattern at 3.65 MHz.**

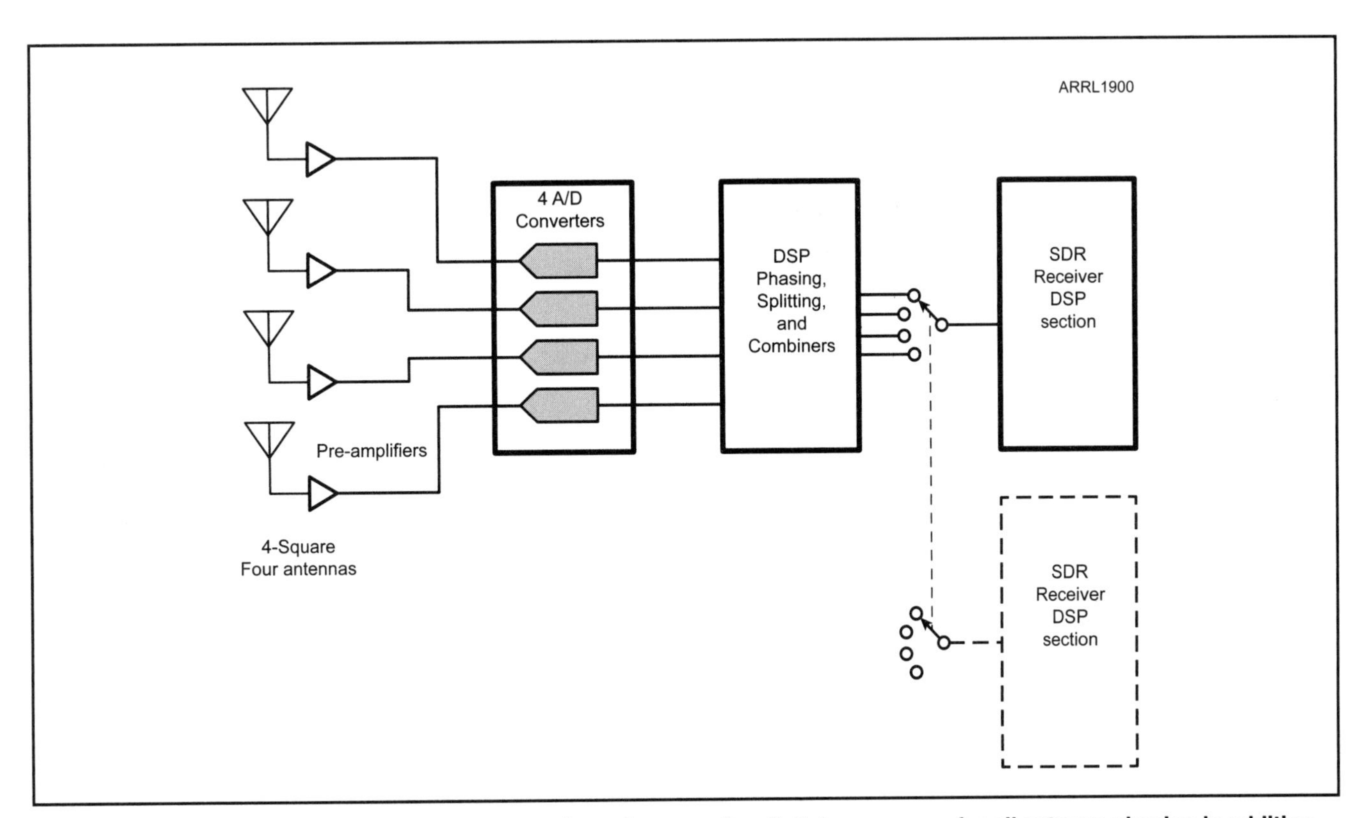


**Figure 5.27 — A digitized version of the 4-square phased array using digital processors for all antenna phasing in addition to one or more SDRs.**

combined with the antennas to create systems. To show an example, I include a block diagram of the 4-square array detailed above (in analog form) as a digitized phased array.

In such an array, the antenna outputs are amplified and then digitized through A/D converters. Then all phasing, splitting and combining are accomplished in the digital domain while simultaneously providing "N" number of separate receivers. Thus, the necessary signal processing for the 4-square is accomplished inside the multiple receiver SDR. Future SDR receivers might include programmable digital processors that could handle multiple antenna inputs that would be programmable by the user. This feature would afford phased array capability to an SDR receiver.

**Figure 5.27** illustrates a digitized version of the 4-square phased array using digital processors for all antenna phasing in addition to one or more SDRs. The antenna output signals are digitized after suitable amplification permitting maximum possible dynamic range in the A/D converters. The signals are then fed to a digital processor (DSP) for splitting, phasing, and combining. In effect, the signal processing functions are identical to what was described in the analog version presented above, but here it is all performed in DSP.

This completes the first section of this book. Chapters 7 and 8 of the second (and last) section will build upon this and all the previous chapters to apply these principles to understanding specific array forms. However, the first "applied" chapter will deal with the idiosyncrasies of antennas' inter-reactions with our own planet.

## References

[1]Layton, Jack, *Directional Broadcast Antennas: A Guide to Adjustment, Measurement & Testing*, Tab Books, 1974

[2]Terman, Frederick, *Radio Engineers' Handbook*, McGraw-Hill, 1943

[3]Zavrel, Robert J., "Tomorrow's Receivers, What Will the Next 20 Years Bring?," *Ham Radio* Nov. 1987.

# Part 2
# Applied Antenna Physics

# 6 Dielectric Effects Upon Radio Waves

In Part 1 we defined the most important physical properties of electromagnetic waves and how antennas radiate and receive them. These explanations assumed that the antennas were located in free space. We can assume the free space model is all we need if we are indeed using the antenna in free space, for example satellite-to-satellite communication (without the Earth in the way!). Obviously almost all antennas are located on or near the Earth's surface. The presence of the Earth has a profound effect upon the performance of an antenna located on or near its surface. We can divide these effects into two basic types: near field and far field effects.

In effect, *all* material (desired or undesired) within the near field of an antenna must be considered part of the antenna proper. The usual goal is to design and mount an antenna so as to minimize the effect support structures, vegetation, buildings, and other materials have on the antenna's intended performance. Sometimes the ground is intentionally made part of the antenna, such as with ground-mounted verticals. The effect of the ground proper has a very great impact on ground-mounted vertical antennas, and this will be covered in detail in the next chapter.

In the far field, the antenna's radiation is no longer affected by the adjacent Earth and other materials. Only the characteristics of the antenna described in free space will be affected. The far field effects are due to the *propagation* of waves — *not* due to the Earth becoming *part* of the antenna. In practice, there is a "grey" area between the far and near fields. Therefore, in many instances *both* effects are present, which leads to complications for modeling the antenna — not to mention forming an intuitive understanding of what is actually happening. Although there are numerous physical mechanisms for wave propagation due to the Earth, we will cover the two most influential on forming the final gain pattern.

## Reflection and Absorption

The bad news (for antenna calculations) is that the Earth's surface is composed of a very wide variety of materials both natural and man-made, thus complicating predictive modeling efforts. The distribution of these varied materials can be quite uniform over a large area, such as sea water. It can also be very non-uniform, such as when a line of exposed rock protrudes from deep soil. In addition to surface discontinuities, the solid ground is most often stratified — such as top soil over sub soil over bedrock. Radio waves will penetrate ground, and thus these lower stratified layers will affect both reflection and absorption. Furthermore, most propagation modeling tools also assume that the Earth is an "infinite" smooth plane, which, of course, it is not. The real topography of the Earth will distort smooth plane assumptions. The depth of ground penetration is a function of frequency and the characteristics of the dielectric ground. In general, lower frequencies will penetrate the ground to greater depths. Therefore very lossy bedrock under a "good" soil may become critical at medium or low frequencies, while at HF the bedrock may have no effect whatsoever.

Propagation effects can even take place in the *very far field.* For example, a multi-hop path to the other side of the world will use ionospheric propagation, but it will also use reflections from the Earth's surface. If one of those reflection areas is on sea water, the condition of the sea can determine if the path is "open" or not (**Figure 6.1**). If the waves become an appreciable fraction of a wavelength in height, the sea will represent a dispersive reflection surface and scatter much more power than a presumed smooth surface. Mountain-

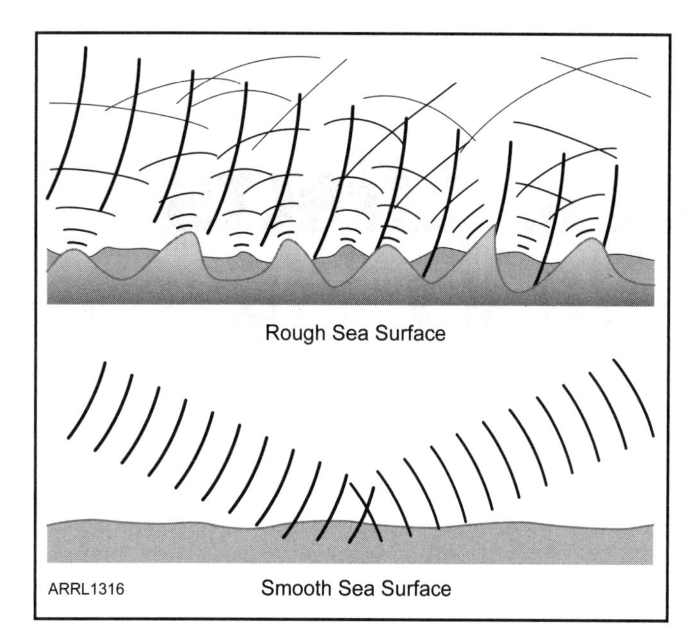


**Figure 6.1 — When considering a path, all factors of the reflective surfaces must be considered. Advanced propagation tools could include reflective coefficients for various land masses, as well as real-time sea swell heights.**

ous areas are known to be problematic for the same reason, whereas flat land is a better reflector. Although the mechanism is reflection, the *effect* is attenuation due to dispersion.

The good news is that these models often *do* provide fair to excellent approximations of what a *system* may provide. They are certainly better than using a free space model!

It is also fortunate that the actual geophysical terms relevant to radio wave interaction can usually be reduced to two terms: *conductivity and permittivity.* Therefore sufficiently accurate models can be derived by reducing the Earth to these two parameters. Therefore the subject of electromagnetic geophysics (at least for radio waves and thus antenna performance) is really the subject of how those waves interact at the boundary between free space and the lossy dielectric of the Earth as well as down to some depth.

## Conductors and Dielectrics

We recall from Maxwell's 4th equation

$$\oint \boldsymbol{B} \bullet dl = \mu_0 I + \varepsilon_0 \mu_0 \frac{d\Phi_E}{dt} \quad \text{(Equation 6.1)}$$

The term $\mu_0 I$ is the conduction current. However, if there is a resistance in the current path (1/G), we can rewrite this term as *GE.* Also the term

$$\varepsilon_0 \mu_0 \frac{d\Phi_E}{dt}$$

represents the time-dependent displacement current (current causing time-dependent charging and discharging of a capacitor). If we represent this term by a sinusoidal *E* field variation, it becomes (in complex notation) $j\omega k_d E$. Thus we have

$$\oint \boldsymbol{B} \bullet dl = GE + j\omega k_d \boldsymbol{E} \quad \text{(Equation 6.2)}$$

In this chapter we will assume the μ term in Equation 6.1 will be $\mu_0$ because we will only consider materials that have μ values identical to free space. Under most circumstances, μ of the Earth, from the surface down to about 10 meters, can be assumed to be the same as free space or very close to it. Therefore we can now write a ratio of the two terms as

$$DF = \frac{G}{\omega k_d} \quad \text{(Equation 6.3)}$$

where DF is the dielectric dissipation factor, *G* is the conductivity (frequency independent), and $\omega k_d$ is the *frequency dependent* dielectric constant of the material. *Conductors* will have very high conductivity, thus DF is >>1. Good dielectrics (insulators) will have very poor conductivity, thus DF becomes <<1. Some readers may correctly see a correlation with the *Q* of a capacitor. Indeed,

$$\frac{G}{\omega k_d} \approx \frac{1}{Q} \quad \text{(Equation 6.4)}$$

Therefore we can determine the theoretical maximum Q of a capacitor if we know the DF of the dielectric. A "perfect" dielectric will exhibit a $k_d$ that is constant for all frequencies. There are no materials that have this property. Only free space has such a frequency-independent characteristic. The DF of soil and rock is *very* complex and variable with temperature and moisture level.

## Wave Propagation in Dielectrics

Let us assume first, a perfect dielectric with a dielectric constant $k_d$. The speed of light, and therefore of a radio wave will be

$$c_d = \frac{1}{\sqrt{\varepsilon_d \mu_0}} \quad \text{(Equation 6.5)}$$

where $\varepsilon_d$ is the permittivity of the dielectric. We can simplify the relationship by multiplying the free space speed of light (*c*) by a simple coefficient to derive the speed of light inside a dielectric material

$$c_d = c\sqrt{\frac{1}{k_d}} \quad \text{(Equation 6.6)}$$

This is the most important effect upon an electromagnetic wave *inside* a perfect dielectric (zero conductivity). The frequency remains constant, but the speed slows and therefore the wavelength decreases.

## Snell's Law

Let us assume that a free space plane wave strikes a dielectric surface with a $k_d = 4$. The angle of incidence is $\theta_i$. The reflected wave will be the "mirror image" of the incident wave. However, the transmitted wave, propagating

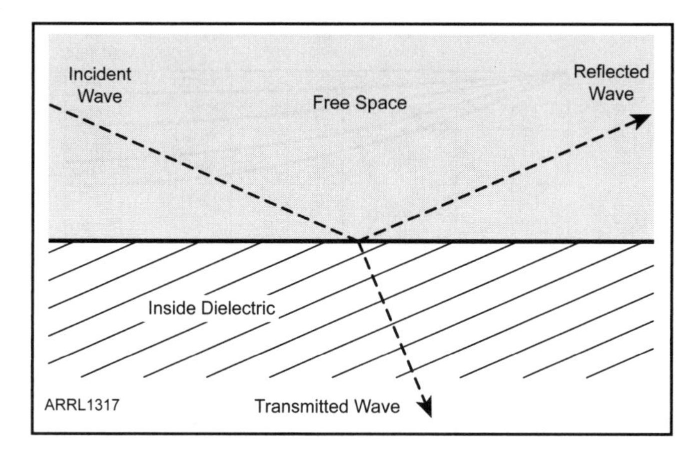


**Figure 6.2 — This figure shows the relationships among the three waves resulting from the boundary of two dissimilar dielectric mediums, in this case free space and a perfect dielectric with k = 4.**

into the dielectric, will be traveling at ½ *c* by Equation 6.6. **Figure 6.2** shows the effect upon the angle $\theta_i$ *through* the dielectric. Since the wave velocity is now ½ *c*, the wavelength inside the dielectric is now half the wavelength in free space. This has the effect of *changing* the angle of the transmitted wave. If you imagine the incident wave striking the boundary in slow motion, and the wave slows down inside the dielectric, it is easy to envision the cause of this transformation of θ where:

$$\frac{\sin\theta_i}{\sin\theta_t} = \sqrt{\frac{k_t}{k_i}} \qquad \text{(Equation 6.7)}$$

In other words, the ratio of the *sin* values of the angles is equal to the inverse-square ratios of the dielectric constants of the two medium. In this case the two medium are free space and a material with dielectric constant of 4.

Notice that when the incident wave is striking the boundary from a perpendicular direction, in this case, from straight up $\theta_i$ approaches 0 degrees, but $\theta_i$ also approaches 0 degrees. How can this be, when

$$\sqrt{\frac{k_t}{k_i}} = 2$$

This is called a *limit* in mathematics. In this case, where the *sin* values become infinitesimally small, the factor of 2 becomes less and less significant. So, for practical purposes the angles become equal.

The reflected wave always will "mirror" the incident wave, while the transmitted wave's direction will change at the boundary. Again, if the dielectric material is "perfect," i.e. zero conductivity, there is no dissipation of power. The incident wave's power is simply divided into the transmitted wave and the reflected wave, where

$$P_i = P_t + P_r \qquad \text{(Equation 6.8)}$$

The ratio of $P_t$ and $P_r$ is a function of the ratio of the dielectric constants and the values of θ.

Inside an imperfect dielectric (such as soil), the situation becomes more complicated. A finite value of conductance implies some finite value of resistance, which in turn will cause attenuation of a wave inside the material. This will be true for directed as well as radiated waves.

## The Earth as a Dielectric

Any material comprising the "ground" will have a higher value of ε and some non-zero value for conductivity. The boundary between these two dielectrics forms a discontinuity that has profound effects upon electromagnetic fields.

Again, the wave equation for free space is:

$$\frac{\partial^2 \boldsymbol{E}}{\partial x^2} = \varepsilon_0 \mu_0 \frac{\partial^2 \boldsymbol{E}}{\partial t^2} \qquad \text{(Equation 6.9)}$$

In a dielectric, $\varepsilon_0$ has changed to $\varepsilon_d$, where $\varepsilon_d$ is always $> \varepsilon_0$. Inside a non-free space dielectric, a wave is still a wave, but with two profound modifications: the speed of light has slowed down and the intrinsic impedance has decreased where

$$c_d = \sqrt{\frac{1}{\mu_0 \varepsilon_d}} \qquad \text{(Equation 6.10)}$$

and

$$Z_d = \sqrt{\frac{\mu_0}{\varepsilon_d}} \qquad \text{(Equation 6.11)}$$

Also from Chapter 4

$$k_e = \frac{\varepsilon_0}{\varepsilon_r} \qquad \text{(Equation 6.12)}$$

and

$$V_{dielectric} = \frac{c}{\sqrt{k_e}} \qquad \text{(Equation 6.13)}$$

where *V* is velocity.

## Effect on Wave Polarization

We have thus far only considered "a wave" incident upon the dielectric. Vertical and horizontal waves interact with the dielectric differently. This is particularly true of the relationship between the phase of the incident and reflected waves.

## Horizontal Polarized Wave Reflection

When a wave's *E* field is parallel to the surface of the reflecting Earth dielectric, the wave is *horizontal* (**Figure 6.3**) According to Snell's Law, the reflected wave angle will mirror the incident wave. However, some of the power will be transmitted into the dielectric and thus lost as heat in the dielectric's resistance. **Figure 6.4** shows how the direct wave

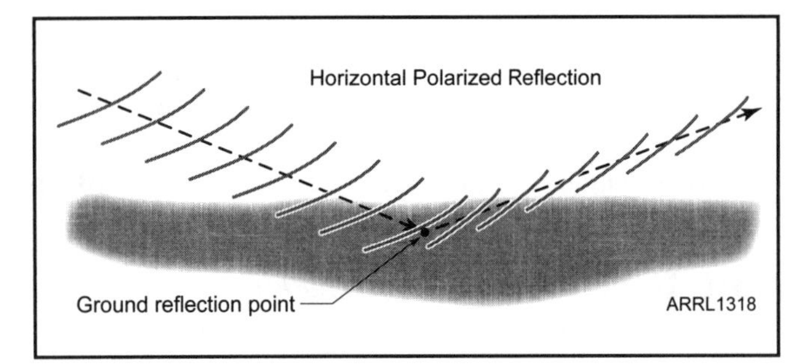


**Figure 6.3 — Reflection of a horizontal wave.**

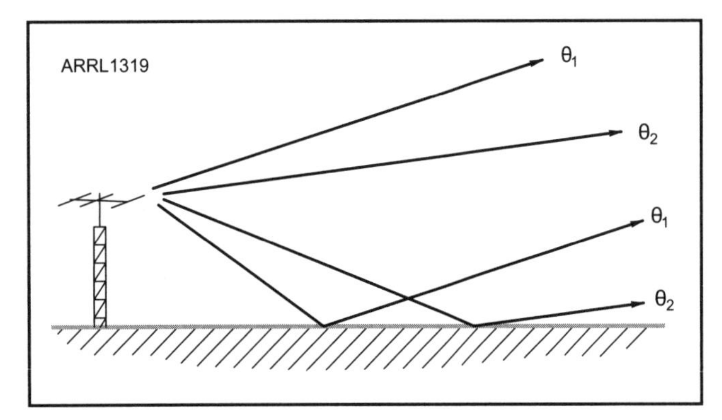


**Figure 6.4 —This figure shows how the direct wave and reflected wave at two take-off angles will differ in the total phase lag (distance traveled). Since the reflected wave reverses phase about 180 degrees when reflected, it will tend to null the direct wave when the total distances are in phase and reinforce when they are out-of phase (two 180 degree phase shifts).**

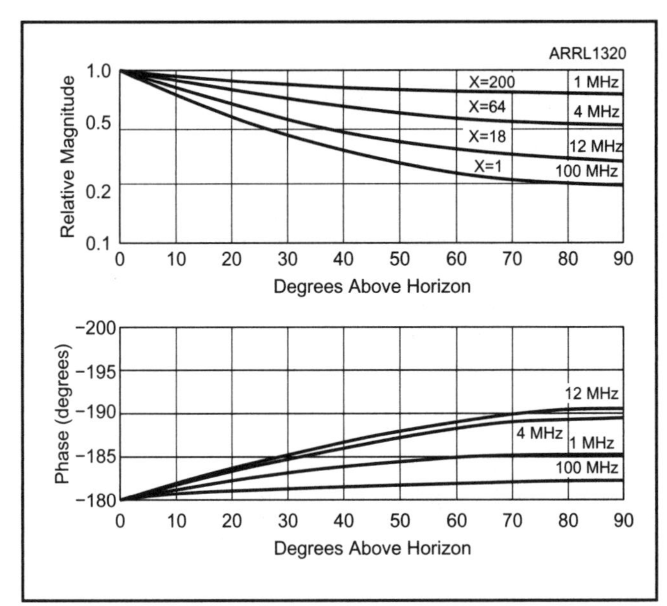


**Figure 6.5 — Magnitude and phase of the plane wave reflection coefficient for horizontal polarization over average ground.**

and the reflected wave will differ. **Figure 6.5A** shows the relative magnitude of the reflected wave and Figure 6.5B shows the relative phase of the reflected wave. For horizontal waves, the phase reversal is nearly 180 degrees for most incident angles. **Figure 6.6** shows a simulation of a ½ λ horizontal dipole at 2 λ over real ground.

## Vertical Polarized Wave Reflection

A vertical polarized wave's *E*-field is *perpendicular* to the Earth's dielectric surface, or "vertical." But upon careful review there can be no "true" vertical polarization that is being reflected. **Figure 6.7** shows why this is true. The reflection angle resulting in a 90 degree phase shift corresponds to the minimum magnitude of the reflected wave. This is defined as the Brewster or pseudo-Brewster angle. **Figure 6.8** shows the relative magnitude and phase of the reflected wave for vertical polarization.

In the case of a ¼ λ ground-mounted vertical, nulling due to reflections is restricted to very low (grazing) angles. The point source of radiation is simply too low to provide

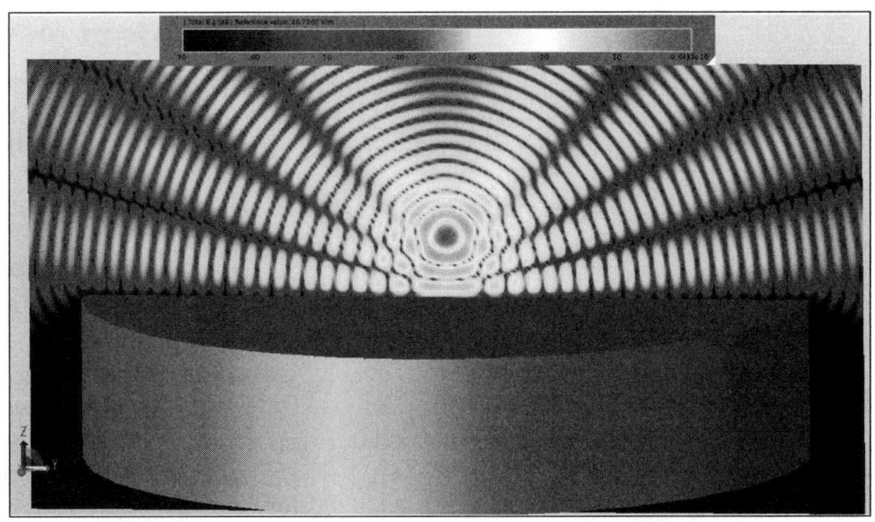

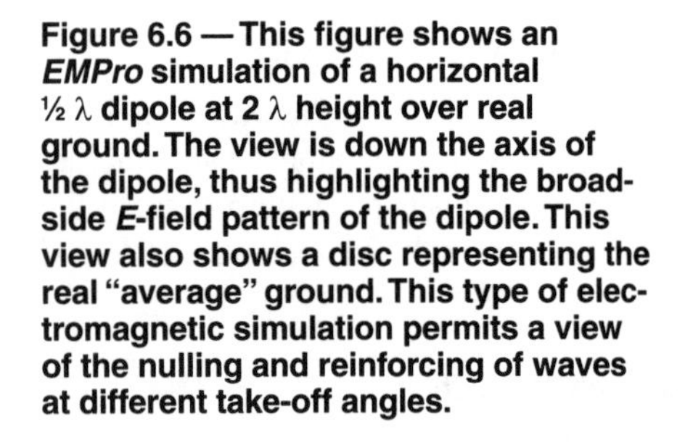

**Figure 6.6 —This figure shows an *EMPro* simulation of a horizontal ½ λ dipole at 2 λ height over real ground. The view is down the axis of the dipole, thus highlighting the broadside *E*-field pattern of the dipole. This view also shows a disc representing the real "average" ground. This type of electromagnetic simulation permits a view of the nulling and reinforcing of waves at different take-off angles.**

Figure 6.7 — This figure shows three "vertical" polarized waves propagating toward the ground. The only true vertical polarization occurs when the wave is at 0 degrees elevation angle (and thus no reflection). As the angle increases, the wave begins to combine a horizontal and a vertical component to the polarization. Finally, a wave incident on the ground from the zenith (straight up) cannot have any vertical component (unless it is not a true TEM wave, such as in a waveguide). Thus, the computation of the reflection characteristics of "vertical" polarization becomes quite complicated. It is necessary to combine the vertical and horizontal components to derive an accurate representation.

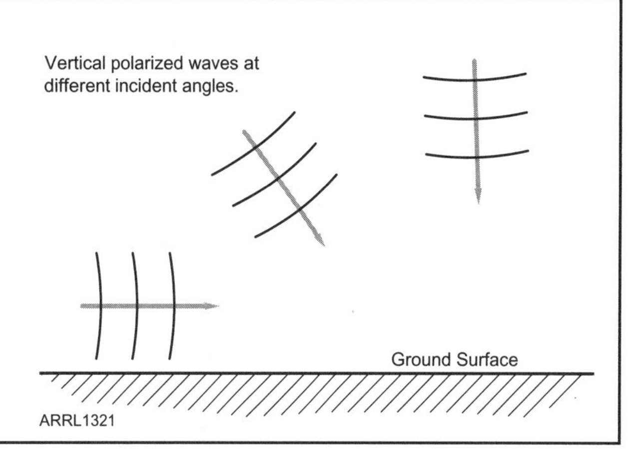


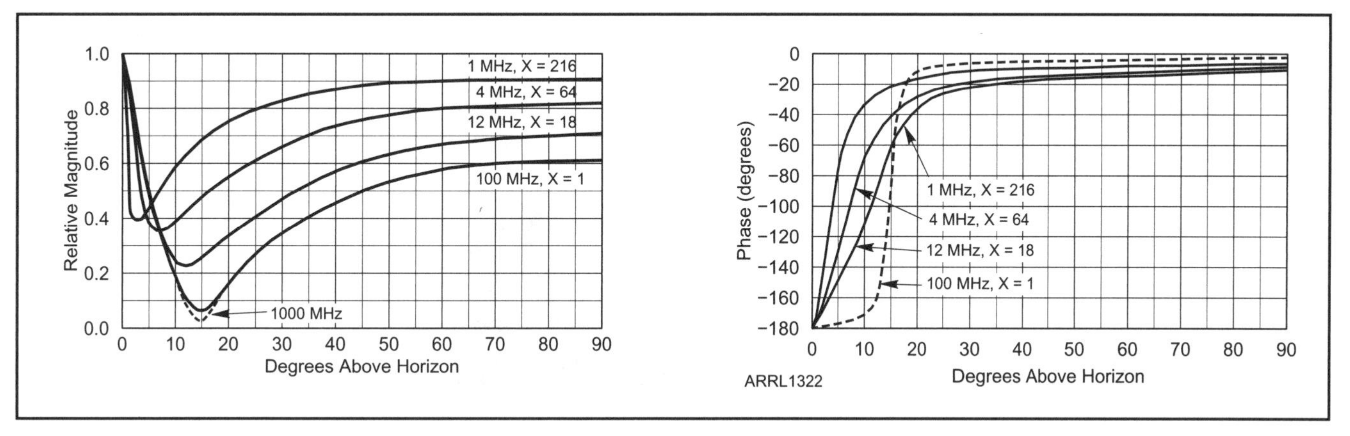


Figure 6.8 — In this figure we show the magnitude and phase difference over average ground of the vertical component of an incident wave. Notice the striking difference between horizontal polarization shown in Figure 6.5 and vertical polarization shown here.

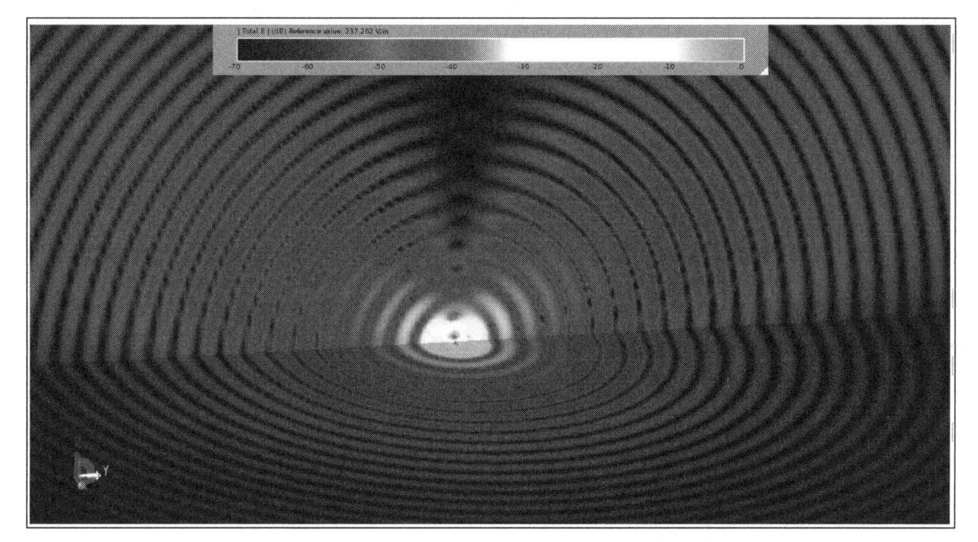

Figure 6.9 — This figure shows free space *E*-fields and ground surface *E*-fields of a ¼ λ vertical using four ¼ λ ground radials over average ground. This figure shows that the field attenuation rate of the free-space wave is the same as the field attenuation rate of the surface ground wave. Therefore the attenuation of lower radiation angles is affected almost completely by ground attenuation. Cancellation of waves by reflection is nearly nonexistent due to the very low elevation of the antenna's current maximum. In the case of an infinitesimal height, ground reflection is nonexistent.

ground-reflected waves at higher angles, so wave attenuation is primarily due to ground attenuation and the requirement for the values of the *E*-fields to remain constant along the boundary of the dielectric and free space. Therefore, we do not see ground reflections causing nulls and reinforcement in the resulting pattern in **Figure 6.9**. **Figure 6.10** illustrates vertical polarized wave reflection.

Unlike a ground-mounted vertical, with the elevated ½ λ vertical dipole, the center of radiation is located at a sufficient height to provide very significant reflected waves,

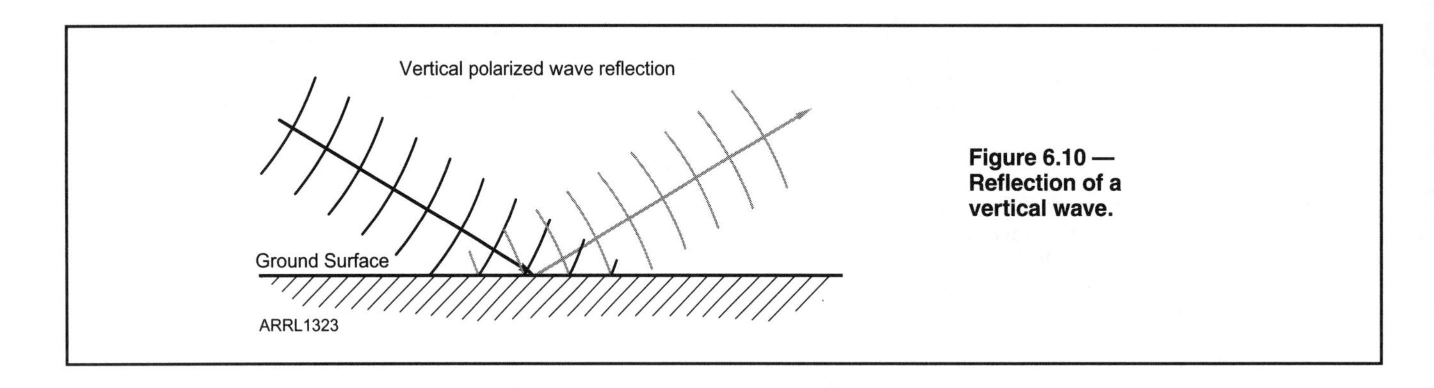


**Figure 6.10 — Reflection of a vertical wave.**

**Figure 6.11 — This is a plot of an elevated ½ λ vertical dipole (two wavelengths high) over a large slab of average ground. In this case we can clearly see the effects of reflected waves, with interfering and reinforcing fields creating the final pattern.**

even at high angles. **Figure 6.11** shows a simulation of a ½ λ vertical dipole at 2 λ over real ground.

## Assumption of an Infinite Plane

Antenna modeling software usually assumes that the antenna is above an infinite plane representing the Earth. Of course the Earth is not a plane — it's a sphere (almost). For many practical solutions, the infinite plane assumption is acceptable, but for some modeling problems it is not.

Even at hundreds of wavelengths away from the antenna,

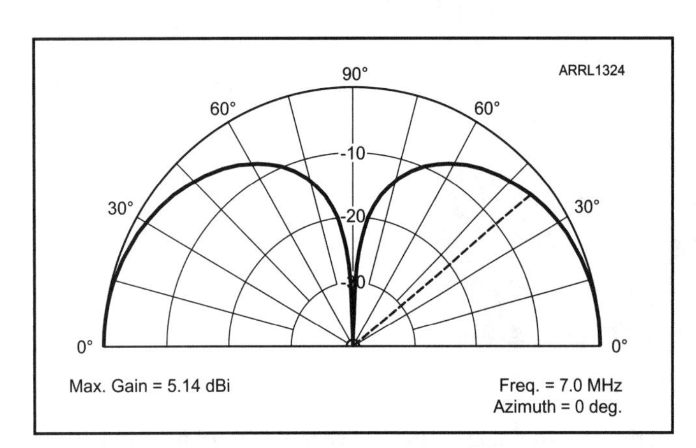


**Figure 6.12 — In this figure, we see the maximum gain from a ¼ λ ground-mounted vertical over a perfect conducting infinite plane to be at 0 degrees elevation angle. This approximation can never be realized with practical antennas, especially at lower frequencies.**

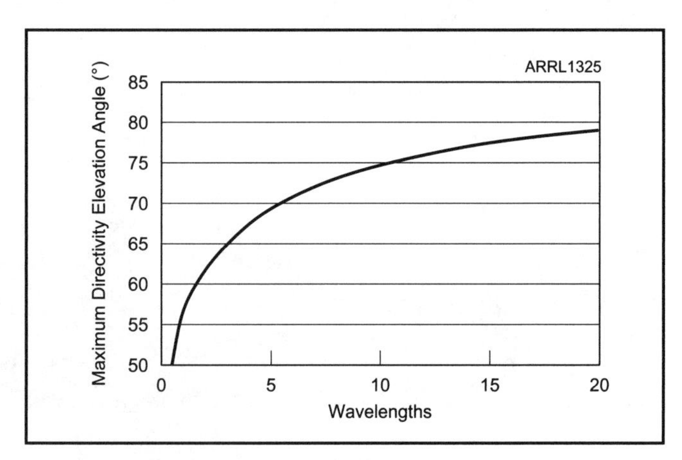


**Figure 6.13 — This figure plots the results of an empirical study from Croatia (see the references in Appendix C). As can be seen, as the perfectly conducting ground plane is made larger relative to the wavelength, the actual maximum radiation angle *never* reaches 90 degrees (on the horizon). Even at 20 λ the maximum gain angle is only 11 degrees above the horizon.**

the maximum gain does not approach the 0 degree angle, since the Earth's curvature will not permit the convergence (**Figure 6.12**).

Extrapolating this curve and using MF (medium frequency) wavelengths, we can see the Earth's curvature will serve as a fundamental limit to how low the maximum gain elevation angle can be (**Figure 6.13**). Figure 6.9 implies even more low-angle attenuation over "average ground". We will explore how these terms relate to vertical antennas in the next chapter.

# Vertical Antennas

In Chapter 6 we discussed the physics at a boundary between free space and a dielectric for radio waves. We also discussed the differences between horizontal and vertical polarized waves at these boundaries. In this chapter we will focus upon the characteristics of vertical antennas, return to complete the discussion of their relationships to the Earth, and then conclude with comparisons among various vertically polarized configurations.

Vertical antennas may be placed above the surface of the Earth or mounted on the Earth proper. Ground-mounted verticals actually use the Earth as part of the antenna, or as the "image" of the antenna. Verticals that are mounted above the Earth may also include the Earth as part of the antenna, depending upon how far above the Earth the "lowest" part of the antenna is placed. For example, a ½ λ vertical dipole with one end just above the Earth actually may be considered a ground-mounted vertical, although a direct connection to the Earth will significantly change the antenna's characteristics.

An antenna connected directly to the ground is a "driven ground" array. An antenna located near the ground, but not connected to the ground proper, uses the ground (for better or worse) as a parasitic surface, but with very complex properties. As will be shown, the gain response of driven and parasitic ground antennas mirrors the difference in gain between driven and parasitic arrays, as in Chapter 5.

## Ground-Mounted Vertical Antennas

Since the ground proper is actually part of a ground-mounted vertical antenna, its relationship to the ground is of particular importance. Of even greater concern is when compromised *ground systems* are employed. In Amateur Radio publications, there has been intense interest in how to optimize ground systems for better overall antenna efficiency. For professional AM broadcasters, very stringent ground requirements are mandated so that minimum ground losses can be assumed. In effect, the reality of compromised ground systems and complete freedom of antenna design has resulted in much more research in amateur than professional circles. In AM broadcasting, antenna parameters have been established for many decades and are highly accurate in practice. In AM broadcast engineering, lossless ground is assumed and only ground-coupled vertical elements are mandated. That makes life much easier!

The general equation for radiation resistance, $R_r$, from Chapter 3 is in an awkward form for easy use with vertical antennas. Special attention must be given to $R_r$ in ground-mounted verticals, especially for amateur applications where compromised ground systems are often used and/or the vertical element is shorter than ¼ λ.

## Radiation Resistance of a Ground-Mounted Base-Fed Vertical Antenna

The height (in wavelengths) of the vertical section can be used to provide a good approximation of the radiation resistance, $R_r$, and the idealized pattern of the antenna. Since ground-mounted verticals radiate primarily "above ground," we characterize them over a hemisphere, not a sphere as we previously assumed for free space antennas. We will begin to quantify the performance characteristics of a ground-mounted vertical by deriving its $R_r$. In Chapter 3 we defined the following relationship:

$$R_r = \frac{S(\theta,\phi)_{max} r^2 \Omega_A}{I^2} \approx \frac{P_r}{I^2} \quad \text{(Equation 7.1)}$$

There is an alternative and more useful method to determine the equivalent of the power term in the general equations for $R_r$ in ground-mounted vertical antennas.

## Effective Height

A critical term in our to-be-derived equation for $R_r$ is $h_e$ which is the *effective height* of the antenna. $h_e$ is another potentially confusing term in that it is a *function* of the physical height of a vertical antenna (including the ground image), but should never be equated to it. It should also not be confused with "electrical height." Only in very specific circumstances does $h_e$ approach the actual physical height, $h_p$. $h_e$ is always less than $h_p$. The two terms can be equated using only the physical height (without the image) in some specific case. The general definition of $h_e$ is

$$h_e = \frac{V}{E} \quad \text{(Equation 7.2)}$$

or

$$V = h_e E \quad \text{(Equation 7.3)}$$

The concept is quite simple. The voltage induced into an antenna (of the same polarization) is proportional to the incident *E* field and the "effective height." The computation of $h_e$ is a bit non-intuitive for this simple definition. The relationship between $h_e$ and the physical height $h_p$ is simply multiplying the *average current* along the antenna by $h_p$ and then dividing by the maximum current along the same antenna. For a linear vertical antenna whose $h_e$ is greater than about ⅛ λ, the current distribution will be sinusoidal. Therefore, we need to calculate the average of a sine wave (or a portion thereof) by integrating current over the physical height. This integral simply computes the *average* value of a *portion* of a sine wave. The portion of the sine wave to be averaged is defined by two phase points along the sine wave.

$$h_e = \frac{1}{I_0}\int_{\alpha}^{\beta} I(z)\,dz = \frac{I_{av}}{I_0} h_p \quad \text{(Equation 7.4)}$$

In this equation we use two phase points α and β rather than an arbitrary physical height. This is because the average current can be different over the same physical length depending upon which portion of the sine wave is present along the length. We will show an example a bit later.

For computations of $h_e$ it is convenient to take the maximum current ($I_0$) to be 1 A and the λ to be 1 meter to simplify the equations. We then represent the physical height $h_p$ as the appropriate fraction of λ, which we assume is 1. Therefore, the term $I_{av}/I_0$ is a dimensionless number which is less than 1. $h_e$ and $R_r$ are independent of the power fed to the antenna, and thus also independent of the actual current value or the actual wavelength. We are only interested in the *average* current referenced to the *maximum* current. This ratio is independent of the amount of power applied to the antenna. Also $h_e$ and $R_r$ are independent of λ as long as we scale the dimensions of the antenna and thus the two phase values. For example, a ¼ λ vertical will show the same results no matter what the wavelength, as long as the *actual* height is ¼ λ at the *actual* wavelength involved.

Using radian measure and integrating radian values for the average current also permits us to use radian values as $h_e$ For example, a ¼ λ vertical would be π/2 radians high.

A ground-mounted vertical antenna's *image* is critical for grasping a conceptual understanding of ground-mounted verticals as well as setting up equations that define their parameters. The image is considered to be an identical vertical antenna that extends vertically *underground*, mirroring the actual vertical element above ground. In the calculation of $h_e$ the image's $h_p$ must be added to the $h_p$ of the vertical element, or $h_e$ of a ground-mounted vertical is *twice* the value of the computed value for $h_e$ of the vertical element taken alone.

There are some exceptions to this rule. In the case of verticals higher than ¼ λ, the image may actually be shorter than the physical length. Also, for very lossy dielectrics (very low conductivity and dielectric constant) that approach the characteristics of free space, the image disappears and significant radiation occurs into the dielectric. This radiation must be taken into account when calculating $R_r$. The disappearing image is due to the gradual decrease in ground conductivity, which lowers the average ground current — eventually rendering the image insignificant. $R_l$ only accounts for heat loss of RF current inside the dielectric, not radiated power propagating inside the ground dielectric.

The power delivered to the load that is matched to a receive antenna is:

$$P = \frac{V^2}{4R_r} \quad \text{(Equation 7.5)}$$

From Chapter 2 we calculated that half of the power received by an antenna is reradiated back into space and that only $A_e$ is significant for the transmit application. Now from Equations 7.3 and 7.5 we derive:

$$P = \frac{h_e^2 E^2}{4R_r} \quad \text{(Equation 7.6)}$$

And from Equation 2.16

$$P = SA_e = \frac{E^2 A_e}{Z_0} \quad \text{(Equation 7.7)}$$

where

$$S = \frac{E^2}{Z_0}$$

So,

$$h_e = 2\sqrt{\frac{R_r A_e}{Z_0}} \quad \text{(Equation 7.8)}$$

Therefore

$$R_r = \frac{h_e^2 Z_0}{4A_e} \quad \text{(Equation 7.9)}$$

To confirm that this equation results in a resistance, we

can look at the dimensions within the equation, ignoring the coefficients. Recalling Equation 7.4

$$h_e = \frac{I_{av} h_p}{I_0}$$

therefore

$$R = \frac{m^2 I^2 R}{m^2 I^2} \quad \text{(Equation 7.10)}$$

The $m^2$ and $I^2$ terms cancel, leaving R. Therefore, this special form for the general equation for $R_r$ provides a pure resistance. It is critical to understand that in a ground-mounted vertical, the physical height ($h_p$) *also includes* the image. Notice also that the current squared term in the denominator is the maximum current term, identical to the other two equations thus presented for $R_r$. All three equations are equal or very close to it.

## Example of the Effects of $h_e$ on $R_r$

In this example we use two identical ¼ λ ground-mounted vertical antennas, except one uses top loading to place the current maximum halfway up the vertical element. The other is a simple vertical with no top loading, and thus the current maximum appears at the base. The physical height $h_p$ is the same in both cases, ¼ λ.

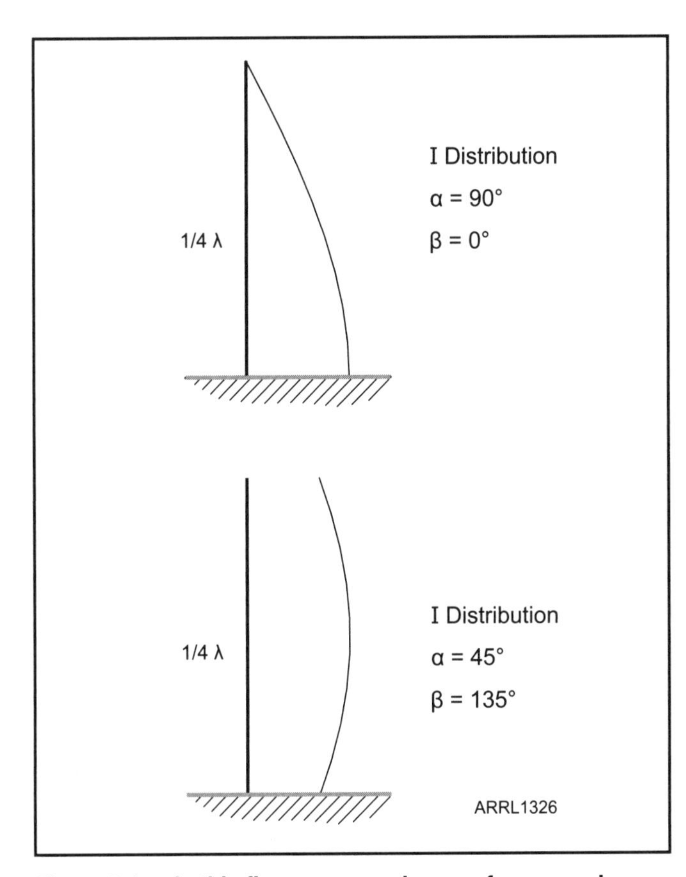


**Figure 7.1 — In this figure we use degrees for convenience. However, in the actual calculation of the definite integral, radians are used.**

**Figure 7.1** shows two possible current distributions along a ¼ λ vertical. The upper shows the current at zero at the top of the vertical while the lower shows the current at the top to be 0.707 compared to the maximum value, at ⅛ λ height, which is halfway up the vertical element. The usual technique involves top loading to move the current maximum "up" the vertical element. The average value of the top configuration is about 0.64, while the average current in the lower figure is about 0.92. Since the ground is actually providing the other ¼ λ section of the vertical, the *total* $h_p$ of the antenna is 0.5 λ. Therefore, $h_e$ for the unloaded vertical is 0.32 λ, and it is 0.46 λ for the optimized loaded vertical. On the other hand, raising the current maximum also increases the gain, in this case very slightly, and thus the equivalent $A_e$.

We can now use Equation 7.9 to actually calculate the values of $R_r$ for these two cases.

Simple ¼ λ vertical: 36 Ω
Top loaded ¼ λ vertical: 74 Ω

It is apparent that careful placement of the current distribution along a vertical's element can provide appreciable efficiency gain for a fixed value of $h_p$. After comparing the two ¼ λ examples, it is also apparent that maximum $R_r$ — and thus maximum efficiency — will occur when the current maximum is halfway up the vertical element, assuming that the ground loss is the same.

## Directivity of a Ground-Mounted Vertical Antenna

From Equation 7.9, the other critical term for calculating $R_r$ is $A_e$. In Chapter 3 we showed the directivity of a ½ λ dipole in free space. The ¼ λ vertical is often described as a dipole, with the ground (the image) representing the other half of the antenna. The ground also divides free space into two hemispheres with the radiation "forced" into the upper hemisphere. Therefore, we might expect that the gain will be higher from a ground-mounted vertical than a free space dipole since the power in the lower hemisphere (in the free space dipole) is not "added" to the upper hemisphere. Intuition suggests that the gain should be doubled. The maximum gain of the dipole is 2.15 dBi, while the maximum gain from the ¼ λ vertical is 5.15 dBi, or exactly 3 dB higher, or twice the power gain of the dipole. Also, the directivity of the vertical is simply the same pattern as the dipole, except that the dipole's pattern is cut into two (**Figure 7.2**). The same comparison can be made between an extended double Zepp and a ⅝ λ vertical, as in **Figure 7.3**. Of course, this assumption is for the idealized case of an infinite perfectly conducting ground. For conditions of "average" or "real" ground, the mechanisms were shown in Chapter 6.

Unfortunately, these very encouraging graphs become somewhat disappointing when real ground is substituted for perfect ground, as we shall see.

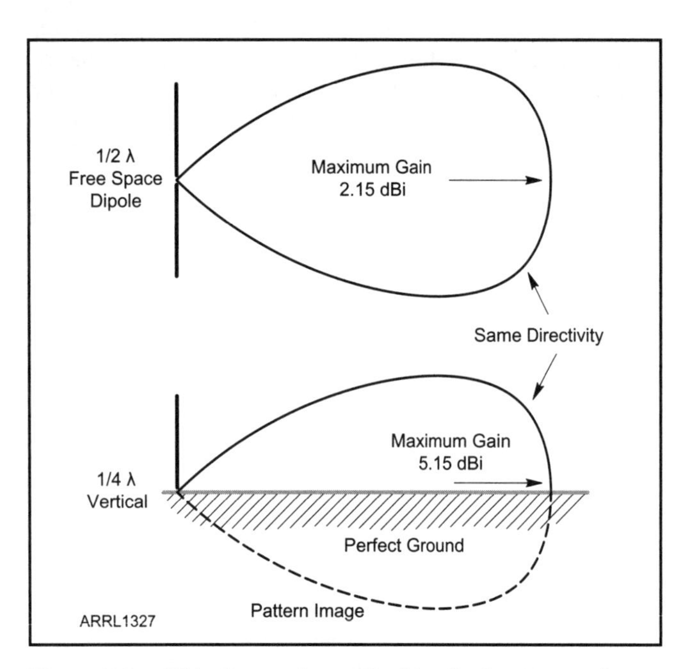


**Figure 7.2 — This figure shows the identical patterns of a ½ λ dipole in free space and a ¼ λ ground-mounted vertical. Half the power from the dipole is radiated into the lower hemisphere of free space, while all the power from the vertical is concentrated into the upper free space hemisphere. The 3 dB increase assumes a perfect conducting infinite ground.**

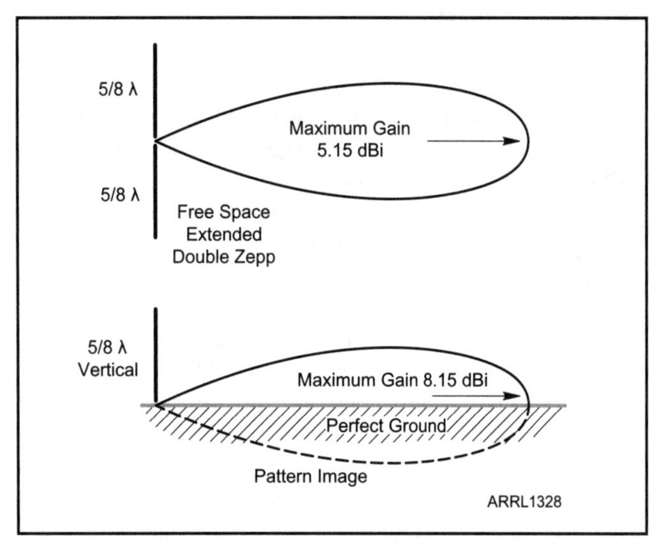


**Figure 7.3 — Exactly the same comparison can be made of an extended double Zepp in free space and a ⅝ λ vertical. The gain of an extended double Zepp in free space is about 5.15 dBi, or about 3 dB more than a ½ λ dipole in free space. As can be seen in the graph, the gain of a ⅝ λ vertical is also about 3 dB over a ¼ wave vertical. As the vertical height is extended higher than ⅝ λ the low-elevation gain begins to decrease, being replaced by a higher and higher angle pattern. Again, a perfect conducting infinite ground is assumed.**

## Ground Influences on Ground-Mounted Verticals

In professional applications, such as AM radio broadcasting, extensive ground radials systems are mandated by regulation. The use of 120 buried ¼ λ radials can usually be assumed to reduce antenna ground losses ($R_l$) to a very small value, say 1 or 2 Ω (in calculating power they are assumed to be 0). In such professional applications, $R_l$ is taken to be zero and thus

$$I^2 = \frac{P_r}{R_{rp}} \qquad \text{(Equation 7.11)}$$

where $P_r$ is the licensed power output. $R_{rp}$ is used since the actual feed point may be a bit below the actual current maximum. Often the broadcast tower is specified as physically ¼ λ high ($h_p$), which is about 5% higher than a resonant vertical. If no losses can be assumed, however, the radiated power can be calculated using the $R_{rp}$ at any point and feeding the appropriate current.

In contrast, for the average amateur installation, both $h_p$ and the ground system will be limited by cost and/or the availability of suitable space. Extensive empirical testing has provided insightful approximations, especially when solidly based upon physical principles. The usual intent of all this work centers on the improvement of antenna efficiency. Directivity can usually be assumed by standard models as there is generally insignificant modification of directivity due to *antenna* ground losses in the near field. However, gain reduction directly results from these antenna ground losses. In contrast, ground losses in the far field can have a dramatic effect upon the final directivity and thus gain of a vertical. These losses are a *propagation* effect in that they are not part of the antenna proper. Therefore the directivity of the *system* is determined in the far field, and thus little if anything can be done to affect it, except move the antenna to a better location.

Thus far we have derived the specific equation for $R_r$ of a ground-mounted vertical antenna. This value is of primary importance for understanding and designing vertical antennas. Since such an antenna uses the Earth as part of the antenna, the characteristics of the Earth in the vicinity of the vertical have a profound effect upon the antenna's performance. Of particular interest is the ground's contribution to the efficiency of the vertical. $R_r$ is usually taken to be an independent term in the calculation of antenna parameters. The general equation for determining the efficiency of a vertical antenna is

$$Efficiency = \frac{R_r}{R_r + R_l} \qquad \text{(Equation 7.12)}$$

and

$$Gain = Directivity \times Efficiency \qquad \text{(Equation 7.13)}$$

In Equation 7.12, $R_l$ is the loss resistance of both the vertical element and losses due to ground resistance (**Figure 7.4**).

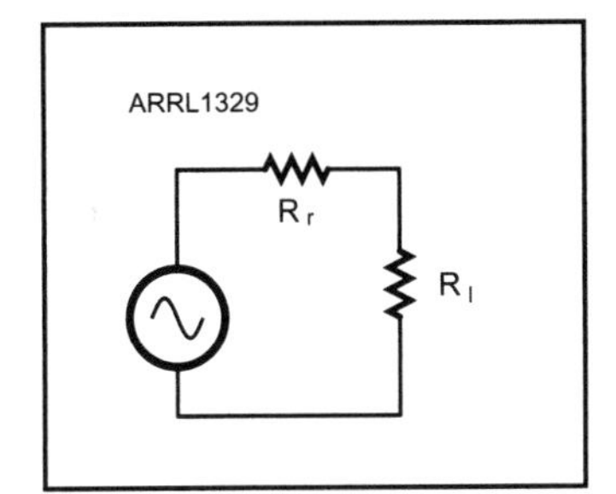


**Figure 7.4 — This figure shows the simple equivalent circuit of an antenna's $R_r$ and $R_l$. Power dissipated by $R_r$ is radiated, and power dissipated by $R_l$ is lost to heat.**

Some authors also include $R_l$ in matching losses. Although these matching losses are not part of the antenna *per se*, they *do* become important for system-level efficiency calculations.

This efficiency equation is deceptively simple. As we have shown, the $R_r$ term requires some rigorous computations for an accurate derivation. The $R_l$ term is exceedingly difficult to predict in compromised ground systems. However, there *are* only two terms — $R_r$ and $R_l$. Therefore, at a first approximation, we can increase efficiency by maximizing $R_r$ or minimizing $R_l$ or a combination of both. Other terms come into play in the overall efficiency, but this basic equation is of prime importance.

As will be shown later, it is also critical to note that $R_l$ represents *ground losses* and not a fixed resistance. Somewhat paradoxically, this "distributed" ground resistance actually converges upon an equivalent lumped resistance shown as part of the feed point impedance. This equivalent resistance must be normalized to $R_r$ to satisfy Equation 7.12. Also the resistance is distributed over and under the ground. Simply stated, resistive losses depend upon the current applied through them. With a distributed resistance, the current distribution over that distributed resistance will determine the conducted dielectric losses. Therefore the solution demands a partial differential that will be developed later. Fortunately, however, all contributions to the distributed ground loss can be simplified to a single resistor and applied to Equation 7.12.

The $R_l$ of the vertical element proper can usually be ignored except when the $h_p$ of the element is much less than ¼ λ. In such cases, $R_r$ can fall to less than 1 Ω or so, and thus $R_l$ can become the dominant part of the power dissipation, and thus efficiency will fall to very low values. Ground losses are a concern in all cases, but it is important to consider *both* $R_l$ *and* $R_r$ in the all-important efficiency equation. We will treat the special case of short vertical antennas in later sections.

The directivity of the vertical will be mostly influenced by the vertical's height and the current distribution along that physical height, as with any antenna. Lowering ground losses increases the efficiency, and thus increases the gain of the vertical. Lowering ground losses in the near-field ground plane has an insignificant effect upon the directivity of the antenna, but *does* increase the efficiency and thus the gain by Equation 7.13.

The ground-mounted vertical can be broken into three fundamental basic components that directly affect the directivity, efficiency, and gain of the overall system:

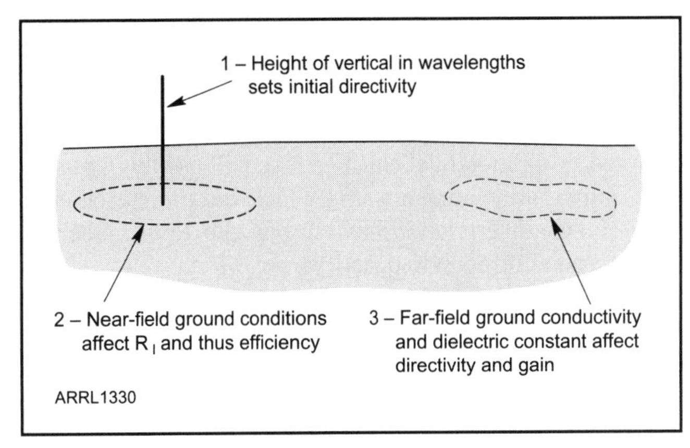


**Figure 7.5 — This figure shows the logical progression of the factors that determine a ground-mounted vertical's overall performance. 1) We begin with the height of the vertical, assuming a perfect ground, which determines the $R_r$, initial directivity, and thus maximum possible efficiency and gain. 2) We subtract from the ideal efficiency and gain using $R_l$ of the near-field ground. The directivity is little affected by the near-field losses, but the efficiency and thus gain are directly affected. 3) The far-field ground contributes additional losses which can greatly modify the directivity (particularly at low elevation angels) and overall system efficiency and gain at all elevation angles.**

1) The vertical radiator
2) The near-field ground
3) The far-field ground

The height of the vertical radiator and the distribution of current along that height largely determine the maximum gain (assuming no ground losses), $h_e$, and thus also the $R_r$ of the vertical. The near-field ground plane is the main contributor to $R_l$ and thus, along with $R_r$, determines the *antenna* efficiency by the simple Equation 7.13. The limited conductivity of the far-field ground will attenuate the power in the far field particularly at low elevation angles and thus further reduces the *system* efficiency. The usual method of reducing near-field losses is the use of ground radials in an attempt to provide a low resistance return path for the antenna ground current.

The electrical characteristics of the soil, rock, water, organic material, salts, and so on, have a profound effect upon the operation of a ground-mounted vertical antenna. Since the ground is part of such an antenna, it contributes directly to the efficiency, directivity, and gain of the antenna.

There is also a secondary effect that becomes somewhat ambiguous to the performance of the antenna proper. In the far field, ground attenuation is usually significant and thus shapes the final transmit (and receive) directivity of the antenna (**Figure 7.5**). Because this is in the far field, we usually consider these effects to be *propagation* effects rather than effects due to the antenna proper.

The attenuation mechanism of "real" ground can be simplified into the one single term $R_l$ assuming that the vertical section's resistance << $R_l$. If $R_l$ could be determined easily, then the basic characteristics of the antenna would be easily modeled. Unfortunately, determining accurate ground characteristics and how they *actually* combine with a radial system to create $R_l$ depends upon multiple variables that are difficult to quantify.

Quantitatively, the difficulty of the problem can be summarized by a differential equation. We will use this equation as a step to formulating a matrix which will more accurately quantify the interrelationships among the terms. Each of these terms is difficult to quantify:

$$R_l = \frac{\partial R_l}{\partial G} + \frac{\partial R_l}{\partial rad_\#} + \frac{\partial R_l}{\partial rad_l} + \frac{\partial R_l}{\partial Z_{feed}} \quad \text{(Equation 7.14)}$$

If these terms were mutually independent, we would have this simple equation to define $R_l$. Unfortunately, these terms *are* mutually dependent, thus Equation 7.14 can only be used as a conceptual step.

1) $\frac{\partial R_l}{\partial G}$

This term defines the soil conductivity's contribution to the overall $R_l$ and represents the most important term. If the "soil" is a perfect infinite plane conductor, there would be no need for radials, and radiation resistance would not matter since $R_l$ is zero.

2) $\frac{\partial R_l}{\partial rad_\#}$

This term defines the reduction in $R_l$ as a function of the number of ground radials used. $R_l$ will decrease as the number of radials increases. However, there is yet another complexity in that the *number* of radials often forms a trade-off with the *length* of the radials.

3) $\frac{\partial R_l}{\partial rad_l}$

This term defines the reduction in $R_l$ as a function of the length of the radials with the above caveat. As we will imply later, *many* short radials are often more advantageous to fewer long radials.

4) $\frac{\partial R_l}{\partial Z_{feed}}$

This term defines the *effect* the feed point impedance ($Z_{feed}$) has upon the overall $R_l$. $R_l$ will not vary by changing $Z_{feed}$. However, feeding a radial system with a higher voltage will result in lower *power* losses, especially losses close-in to the vertical element. An example is the use of a ½ λ vertical which significantly reduces ground losses when compared to a ¼ λ vertical over the same ground system. Power loss is $P_{loss} = I^2R$ and current will be lower for a higher feed point impedance.

Equation 7.14 is provided as a first approximation to the actual quantification of the problem. The solution for $R_l$ is actually more complex than solving a simple differential equation. All the variables listed above are mutually dependent upon each other. When each term is *partially* dependent upon all the other terms, that is, they are all mutually interdependent, we can use a matrix to quantify this particular type of problem. For example, if we want to know the interrelationship between the number of radials and the length of the radials as it affects $R_l$, we look at the terms

$\frac{\partial R_l}{\partial 32}$ and $\frac{\partial R_l}{\partial 23}$

Now we will redefine Equation 7.14 into a more accurate and formal relationship. If we define $R_l$ as a scalar dependent upon a system of mutually interdependent partial differentials, we can set up a matrix that, in principle, will capture the most important variables that define a vertical's $R_l$.

Matricies are very useful tools in an area of mathematics called linear algebra. A matrix can be used to simplify simultaneous equations. Let's say we want to solve to $R$, and $R$ is a function of two other variables (X and Y), and the system of equations is:

$$3R + 2X - Y = 1$$

$$2R - 2X + 4Y = -2$$

$$-R + 1/2X - Y = 0$$

Notice that all values of the variables are mutually dependent upon each other's values. The solution is $R = 1$ and $X = -2$ and $Y = -2$. "1" could easily be an equivalent resistive loss. I will not present the method of solution. When a system of equations becomes very complicated, suffice it to say that use of a matrix becomes a very convenient "shortcut" to a solution. If we use partial differentials as the elements, the form is as follows:

$$\frac{\partial R_l}{\partial \boldsymbol{X}} = \begin{bmatrix} \frac{\partial R_l}{\partial x_{11}} & \cdots & \frac{\partial R_l}{\partial x_{14}} \\ \vdots & \ddots & \vdots \\ \frac{\partial R_l}{\partial x_{41}} & \cdots & \frac{\partial R_l}{\partial x_{44}} \end{bmatrix} \quad \text{(Equation 7.15)}$$

Here we use $\boldsymbol{X}$ to identify the matrix. We use four variables which we defined above as $x_{1-4}$ where

$x_1$ is the ground conductivity ($g$)
$x_2$ is the number of radials ($radial_\#$)
$x_3$ is the length of the radials ($radial_l$)
$x_4$ is the base feed point impedance ($R_{feed}$)

The 16 partial differential terms that comprise this matrix represent all the important interrelationships that define losses associated with a ground-mounted, base-fed vertical. For example, in free space a ground plane antenna will have infinite "ground" resistance, but

$$\frac{\partial R_{radial}}{\partial R_g}$$

is zero since there is no effect upon the radial's contribution by the ground, so $R_l$ approaches zero. As the same ground plane antenna is lowered onto a ground dielectric,

$$\frac{\partial R_{radial}}{\partial R_g}$$

becomes significant and $R_l$ becomes higher. Finally, when the radials come in contact with the dielectric, $R_l$ is maximum for the given antenna. Ground plane antennas whose radial systems are very close to the Earth($\ll$ ¼ λ) have been a subject of intense study in the amateur community, mostly to empirically determine these differential terms.

Thus far there are two key variables that have not been treated in the ground-mounted vertical discussion: the frequency of operation and the height of the vertical above ground. The main difference realized when the frequency is changed is the characteristics of the ground and the ground's interaction with the other variables. Therefore, the ground coefficients of the matrix will differ for different frequencies. The only situation where the outcome is independent of the ground characteristics is when the ground is assumed to be an infinite perfect-conducting plane. The height of the verticals in the above discussion have been zero, since they are all ground-mounted. A following section will deal with "raised" vertical antennas, or *ground plane antennas.*

## Experimental Method to Derive $R_l$

Equation 7.15 is formidable. It is offered as an approach to quantify the variables and interdependencies of the problem. Empirical tests remain the best method to *approach* some maximum efficiency for a given set of constraints. The extensive empirical work that has been done has reflected one or more of the differential terms expressed in Equation 7.15. The "holy grail" would involve combining the results of this work into a unifying equation like 7.15.

Having quantified the overall problem, let's return to the other extreme at problem-solving: simple empirical measurements. There is a non-ambiguous method to determine $R_l$ for a given location for a vertical ¼ λ or shorter. Install a ¼ λ resonant vertical over the ground plane. The base feed point impedance should show as little reactive value as possible (a pure resistance, but some small value is acceptable). We are only interested in the real portion of the impedance. The value for $R_r$ will be very close to 36 Ω. Therefore the efficiency will simply be

$$R_l = R_{feed} - 36\ \Omega \qquad \text{(Equation 7.16)}$$

As you add radials, change their lengths, and so on, the feed point should change accordingly. There are many empirical studies in print. These can form the basis of an approximation for a general case. However, the multiple differentials involved will be different for any given location. There is no substitute for empirical testing at a given location. As you experiment simply use Equations 7.16 and 7.13 to maximize efficiency. If you attempt to use a vertical shorter than ¼ λ, then you must calculate the $R_r$ and adjust Equation 7.16 accordingly. For verticals higher than ¼ λ, Equation 7.16 will not provide an accurate answer but can provide relative data.

The results of this exercise will provide a very good approximation for efficiency of a ¼ λ vertical at a given location. It will also provide a first approximation for efficiency of verticals of different heights and loading. Equation 7.15 is a method for quantifying *all* the important variables involved.

There is also a possibility of another term that implies $R_r$ may itself be influenced by $R_l$. This is certainly the case for very low conductive dielectrics. For example, if the "ground" under a vertical/radial system is gradually made more lossy, it will begin to approach the characteristics of free space. Thus the "ground-mounted" vertical has morphed into a ground plane with significant radiation *into* the ground. This will clearly affect both $h_e$ and $A_e$ and thus modify $R_r$ (see Appendix B).

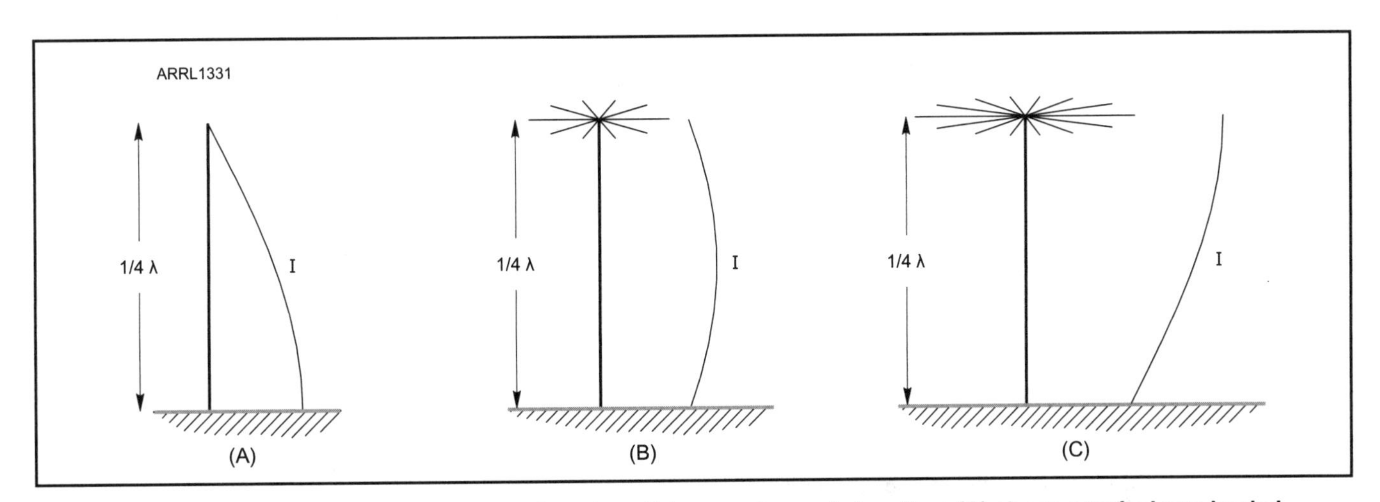


**Figure 7.6 — This figure shows three configurations for a ¼ λ ground-mounted vertical. (A) shows a typical non-loaded array and its current distribution. (B) shows the current distribution optimized for maximum $R_r$. (C) shows the current distribution for maximum $R_{feed}$. The directivity of these three configurations is, for practical purposes, nearly identical.**

## Feed point R vs Maximum $R_r$

We have assumed that maximizing $R_r$ will, by definition maximize efficiency. This remains a correct assumption *provided that* the term

$$\frac{\partial R_l}{\partial Z_{feed}}$$

does not become more significant in reducing $R_l$. We have quantified and shown a simple method to maximize efficiency by maximizing $R_r$. We have also found that the effective losses can be minimized by placing the current *minimum* at the feed point impedance (also called voltage fed), again through top loading. **Figure 7.6** shows three configurations for a ¼ λ ground-mounted vertical.

For top loading there is yet another important term, R losses in the capacitive hat. Loading inductors are often placed in series with the capacitive element(s) to increase their effect. These inductors can cause losses due to resistance in the windings. Quantitatively, the effect of these loss resistances can simply be added to $R_l$. Therefore, in an attempt to keep raising the current maximum farther up the vertical, the top loading structure eventually reaches a point of diminishing returns, in that the increased top losses begin to approach the diminishing ground losses. Another problem with large top loading structures is unwanted radiation. As the loading conductors increase in length they begin to radiate. This radiation is usually highly undesirable since it will be horizontally polarized and at a *high* elevation angle maximum. The usual desired directivity from a vertical antenna is a *low* elevation angle maximum.

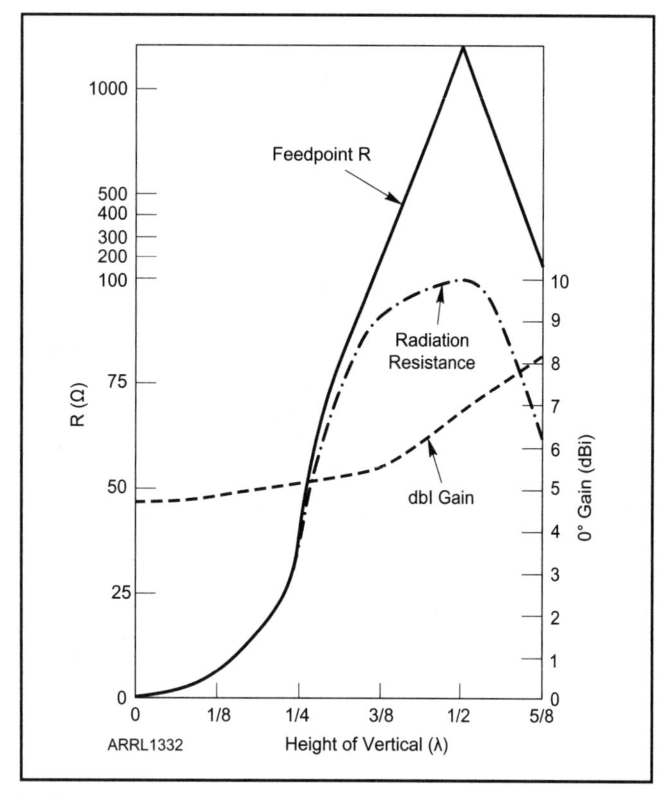


**Figure 7.7 — This graph shows the approximate values for the real portion of the feed point impedance and the radiation resistance as a function of a vertical's physical height. This data is for an unloaded linear vertical element over a perfect ground, no loss resistance and where the diameter of the element << λ. It also plots the gain at the 0 degree take-off angle (on the horizon) in dBi. The gain begins to decrease rapidly for heights greater than ⅝ λ.**

## Verticals Higher than ¼ λ

We have thus far focused on the ¼ λ vertical element. We have also shown how placing top loading on this antenna can significantly increase efficiency by redistributing the current and increasing the effective height of the vertical. Over a perfect ground, low angle directivity increases as the vertical is made higher. However, when a linear vertical antenna $h_p$ is raised above ⅝ λ, the low-angle directivity begins to decrease. With further increase in height the directivity eventually converges in a very high angle, or "straight up." For most practical purposes, there is no reason to build a vertical higher than ⅝ λ $h_p$. **Figure 7.7** shows a comparison of vertical antennas' fundamental relationships up to ⅝ λ $h_p$.

## Raised Vertical Antennas

As we raise the vertical antenna together with its radial system above ground, we *gradually* leave the characteristics of a ground-mounted vertical and begin to see characteristics of a *ground plane antenna* (**Figure 7.8**). Since we have explored the ground-mounted vertical with some depth, we will now consider the opposite extreme, a ground plane antenna in free space. The ground plane antenna in its simplest form is a ¼ λ vertical section fed against two or more radials. The usual configuration is three or four radials.

A ground-mounted vertical antenna can be assumed to have a complete image that mirrors the vertical element below ground. Since there is no "ground" under a free space ground plane, it is interesting to investigate what effect the radials have upon substituting for "ground."

We have already quantified $h_e$, $R_r$, and directivity for

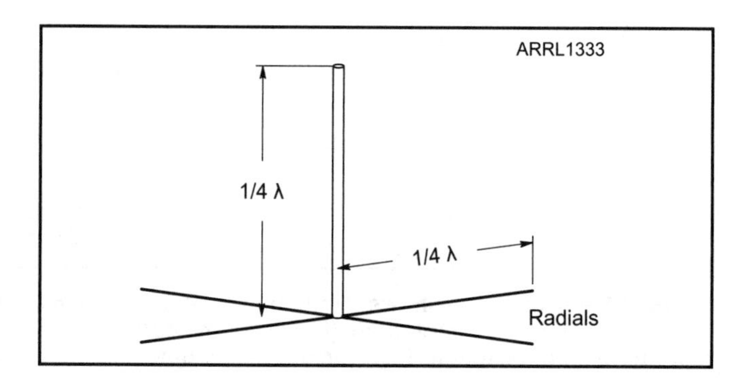


**Figure 7.8 — This figure shows a typical ¼ λ ground plane antenna. The ground plane can be considered identical to a ¼ λ ground-mounted vertical with four radials, but with no ground under it.**

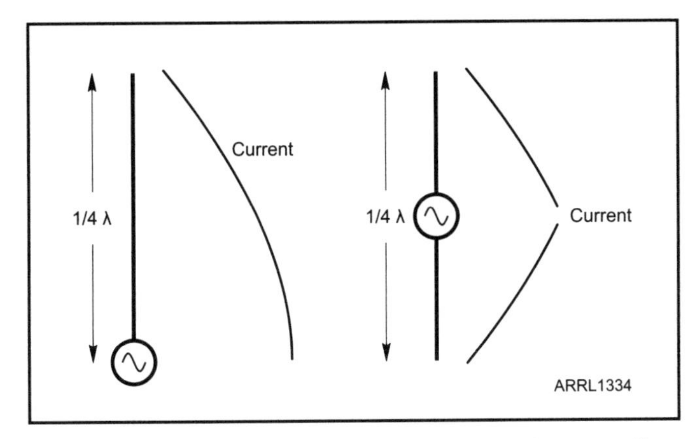


**Figure 7.9 — In this figure we see the effects of current distribution along a ¼ λ vertical antenna, one with end feeding and the other center fed.**

ground-mounted verticals with an assumed "full" image. Since the image directly effects $h_e$ and thus $R_r$, we are in a position to explore the free space ground plane.

The word "counterpoise" can reflect several definitions. In this text we will define it as a raised ground system for a vertical antenna. A counterpoise may consist of a single wire, multiple wires, a flat conductive plane, or other configurations. Unlike a large dielectric (such as the Earth) under the vertical, a counterpoise does not create an image under the vertical unless the size of the counterpoise becomes very large compared the wavelength.

**Step 1**: ¼ λ vertical with no radials in free space, simulating *no* counterpoise (**Figure 7.9**). In both cases, the directivity is about 1.82 dBi but $R_r$ is about 25 Ω for the end-fed and about 12 Ω for the center fed. This result is very close to the expected ratio using 0.64 $h_e$ for the end fed and about 0.5 for the center fed, everything else being about equal. Also notice that the directivity is just a bit better than the infinitesimal dipole from Chapter 3, which was 1.76 dBi. The directivity pattern, as might be expected is a doughnut.

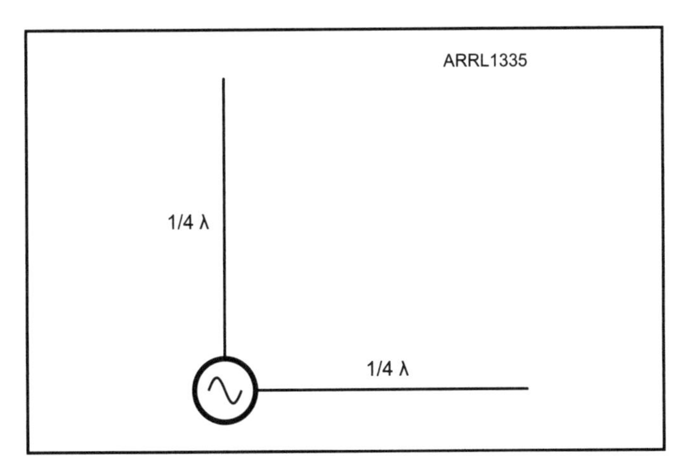


**Figure 7.10 — $R_r$ is now about 42 Ω, but the directivity actually slightly decreases from the Step 1 configuration to about 1.66 dBi. Also, as expected, there is equal gain in both vertical and horizontal polarization.**

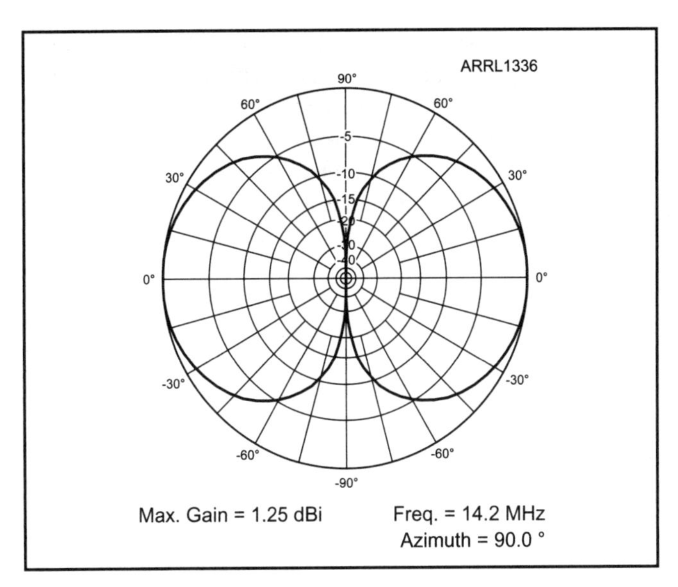


**Figure 7.11 — This is the free-space directivity of the ground plane antenna shown in Figure 7.8. The vertical element is on the Y-axis, while the ground radial plane is normal to the X-axis. Notice that the ground plane radial wires have no effect upon forcing all radiation into the upper hemisphere as with the ground-mounted vertical.**

**Step 2**: ¼ λ vertical with 1 radial, often referred to as an inverted-V configuration, also in free space (**Figure 7.10**). The Step 2 example is identical to the development of a ¼ λ transmission line from a dipole in Chapter 4. There is now a counterpoise relative to the vertical element. As we know, if this were a vertical ½ λ dipole, the gain would be higher, the $h_e$ would reflect an $h_p$ of ½ λ and $R_r$ would 73 Ω.

The feed point has risen from 22 Ω to about 42 Ω. In the case of a single wire counterpoise, the counterpoise becomes part of the antenna. With a multiple wire radial counterpoise, the fields cancel and do not contribute to $h_e$, and therefore gain.

**Step 3**: ¼ λ vertical with a four-radial configuration, also in free space. Now we return to the typical ground radial shown in Figure 7.8. The results (**Figure 7.11**) are counter-intuitive: the directivity has dropped again slightly to 1.25 dBi, the directivity pattern is nearly identical to the doughnut pattern in Step 1 (symmetrical in the upper and lower hemispheres), and $R_r$ is also very close to Step 1, about 22 Ω.

The only advantage the radials provide with a ¼ λ ground plane is to create a counterpoise for the vertical element. In a practical sense, complications of end-feeding a ¼ λ element are greatly reduced by including a counterpoise. This is a very good reason to use radials. The radials accomplish little if anything else; again, from a practical sense, there is no effect on gain or $R_r$

## Raised Vertical Antennas Over Earth

**Step 4:** The presence of the Earth under a ground plane has a very profound effect upon its final directivity. At elevations down to about ½ λ, $R_r$ remains relatively constant at about the free space 22 Ω value. Below this height some anomalies appear until $R_r$ returns to about 36 Ω when the vertical again is ground-mounted. At the higher elevations, *propagation* effects from the ground dominate the formation of the final directivity pattern, but the antenna proper is largely unaffected. But, as the vertical is lowered to ground level, the image gradually returns — thus increasing *antenna* gain, $h_e$ and $R_r$. In effect, as the vertical is lowered closer to the ground, the ground itself gradually becomes part of the antenna.

Much empirical work has been performed, mainly by amateurs, to determine the trade-offs between ground-mounted verticals and *slightly* elevated radials. In effect, this work attempts to resolve several more partial differentials involving the scalar $R_l$. These include the same terms that appear in the $R_l$ matrix above, but the addition of radial height as an additional partial variable. The trade-off is the decreased image in the ground vs. an effective decrease in ground losses as described in the above paragraph.

**Step 5**: ½ λ dipole lowered from a high elevation to ground-level. In free space the vertical dipole will have an *antenna* gain of 2.15 dBi (power gain of about 1.64) and $R_r$ of about 73 Ω. With one end nearly touching the ground, the idealized gain becomes 6.94 dBi (power gain of 4.94). In effect, the dipole operating just above or connected to the ground behaves like a 2-element collinear antenna with the image providing the other ½ λ. When lowered close to the Earth, there appears an image, also ½ λ. Therefore, $h_e$ doubles and $h_e^2$ quadruples. Aperture increases by 4.94/2.64 = 2.97, and therefore $R_r$ increases to 4/2.97 (73 Ω) or about 97 Ω.

## Short Vertical Antennas

For very short vertical antennas ($h_p$ < ⅛ λ), the small portion of a sine curve can be approximated by a straight line. Therefore, the average current will be close to ½ the maximum current value. See **Figure 7.12**.

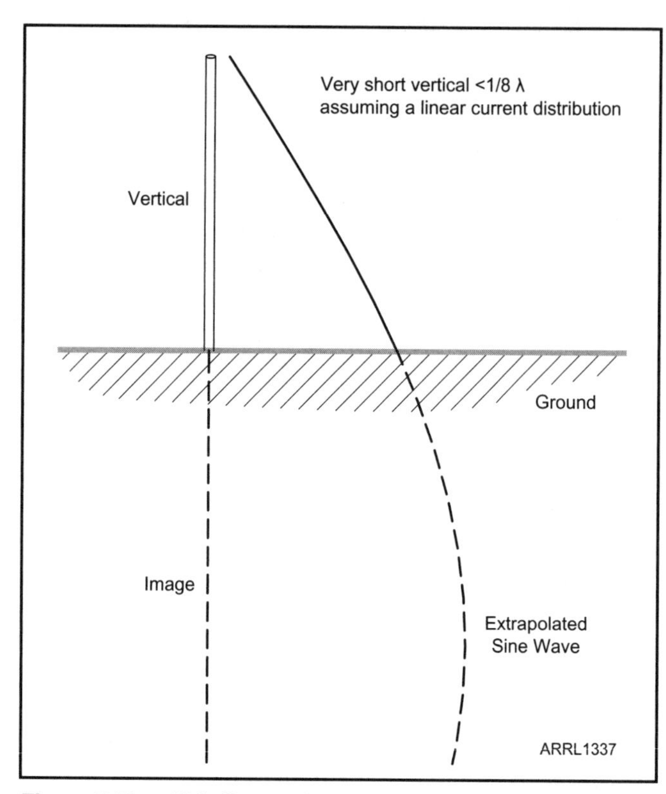


**Figure 7.12 — This figure shows how a very short vertical's current distribution can be assumed to be "linear." It is actually a small portion of a sine curve.**

## Data for Various Ground-Mounted Unloaded Vertical Physical Heights

**Table 7.1** shows calculated data for various vertical physical heights with no top loading. For the three shortest lengths, the current distribution is assumed to be linear, therefore the average current is simply 50% of maximum (we don't need to solve the integral in Equation 7.4). For about ⅛ λ and higher the sinusoid distribution is assumed and Equation 7.4 must be used.

In Table 7.1 are the physical heights in wavelengths of various vertical elements. $h_e\lambda$ is the effective height stated as a fraction of the wavelength; $G_{iso}$ is simply the numeric gain value; and $A_e\lambda$ is $G_{iso}l^2/4\pi$, where λ is normalized to 1. And finally, $R_r$ is from Equation 7.9.

As also shown in Figure 7.7, notice that the apertures (gain) for very short verticals (less than about ⅛ λ) remain nearly constant as we change $h_p$. For very short verticals (from 0 to about 1/50 λ), the dBi value derived from *EZNEC* modeling was exactly 4.76 dBi (a power gain of exactly 3). Also, since the gains and thus apertures do not change with antenna height for very short heights, we can rearrange terms by substituting 3λ/4π for $A_e$ in Equation 7.17, and simply multiply the isotropic aperture by 3:

$$R_r = 94.25\frac{h_e^2}{A_e} = 94.25\frac{h_e^2}{3\lambda^2/4\pi} = 395\left(\frac{h_e}{\lambda}\right)^2 \quad \text{(Equation 7.17)}$$

This is the simplified equation commonly published for short vertical antennas. With no top loading, the average current is about 50%. If appropriate top loading is installed, then the current average approaches 100% and $h_e$ is now nearly twice $h_p$. For an optimally top loaded short vertical, the average current is 100%, thus $R_r$ *quadruples.* This equation is simpler than a first glance implies. For short verticals with no

**Table 7.1**

**$R_r$ for Various Ground-Mounted Vertical Physical Heights With No Loading**

Notice that the maximum possible gain for very short verticals converges on the value for an infinitesimal high vertical, or twice the gain of an infinitesimal dipole.

| $h_p\lambda$ | $h_e\lambda$ | dBi | $G_{iso}$ | $A_e\lambda$ | $R_r\Omega$ |
|---|---|---|---|---|---|
| 1/32 | 0.031 | 4.76 | 3.00 | 0.239 | 0.4 |
| 1/16 | 0.063 | 4.77 | 3.00 | 0.239 | 1.6 |
| ⅛ | 0.125 | 4.86 | 3.06 | 0.244 | 6.0 |
| ¼ | 0.32 | 5.14 | 3.27 | 0.260 | 36.8 |
| ⅜ | 0.54 | 5.79 | 3.79 | 0.302 | 92.1 |
| ½ | 0.64 | 6.88 | 4.88 | 0.388 | 99.5 |
| ⅝ | 0.587 | 8.14 | 6.52 | 0.519 | 62.9 |

loading (a simple vertical whip antenna), $h_e$ actually equals $h_p$ since the average current is 0.5, and we double this including the image, which results in a coefficient of 1 between the two terms. Thus you can simply substitute the actual physical length (without the image) of the whip for $h_e$. A specific example of very short verticals is presented at the end of this chapter.

Table 7.1 summarizes the three critical issues regarding very short vertical antennas:

1) The directivity, and therefore potential gain, does not significantly decrease with shrinking height.

2) The major problem is a rapid decrease in $R_r$, which requires very careful design of the antenna and matching unit to maximize efficiency. For example, a 1⁄32 λ vertical would require only 0.4 Ω of loss in order to achieve only 50% efficiency!

3) With optimum top loading, $R_r$ quadruples, thus offering significant efficiency improvement.

## Ground Proximity to Current Maximums

We have already discussed the terms critical for maximizing a ground-mounted vertical antenna's efficiency. In addition to maximizing $R_r$, the real part of the feed point impedance is the other most critical term. For example, a base-fed ½ λ vertical's feed point impedance is very high, and since it is base-fed, the current minimum also appears at ground level. Since the actual value of ground resistance (at a specific frequency) remains constant, power loss through ground current will be less at high voltage points along the ground. This is similar to why high voltages are used for long-distance transmission lines: for a fixed R value of the line's wires, higher voltage will exhibit lower loss. Furthermore, as the distance increases from the base of the vertical, the *total* current increases but is distributed over a larger area of the ground surface. (The current is not confined to the surface, but does penetrate into the ground. For the sake of this argument, the ground is assumed to be a two-dimensional plane.) In the case of a ¼ λ vertical, the ground current is maximum at and very near to the antenna's base, and therefore all the ground current is forced to traverse through a small portion of ground.

The ground can be considered to be a two-dimensional "square resistance." However, the feed point becomes a *point source* for current on the square resistance. We can approximate this idea with a simple equivalent schematic diagram.

**Figure 7.13** shows how the current values decrease through each equivalent resistor as we move away from the vertical's base. Assuming a constant total current along a radial angle, we can calculate the total power dissipated as a function of distance from the feed point. Let us assume the value of all resistors is 1 Ω. Therefore, the power dissipated by the first ground ring (closest to the feed point) is 1 W ($1^2 R$). The second ring is 0.5 W ($2(0.5^2 R)$), and the third ring is 0.25 W ($4(0.025^2 R)$).

Of course the current is not constant as a function of distance from the feed point, but this simple model does serve to illustrate the point: for any current distribution ground

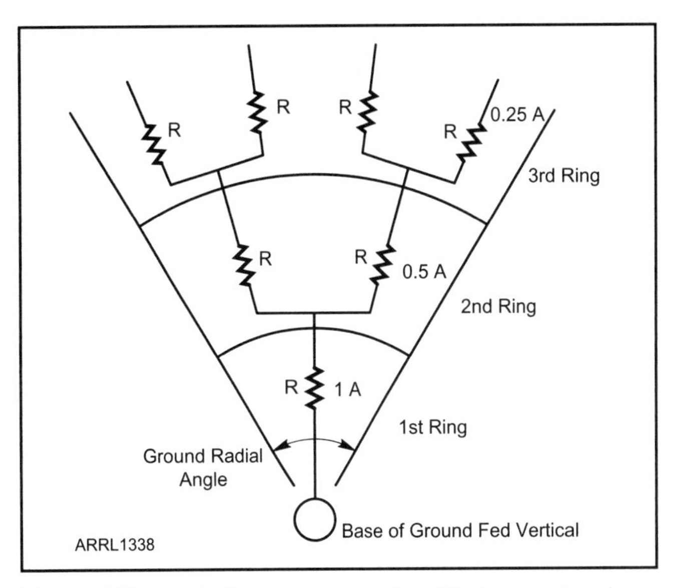


**Figure 7.13 — This figure shows a simplified ground resistance model. In this case, all R values are constant, but the current is distributed through more equivalent parallel but equal resistances as the distance increases.**

attenuation decreases as the distance from the feed point increases. When the voltage maximum is at the feed point, it is easy to see why ½ λ base-fed verticals have significantly lower ground loss than ¼ λ verticals for similar ground systems.

The actual current distribution as a function of distance results from yet another partial differential equation involving the sinusoidal function, dissipation due to ground attenuation, and dielectric effects upon the wave velocity in the ground and on the ground's surface.

This simple model also shows why wire radial ground systems are most effective. They inherently provide for the highest density of conductivity where it is most needed: closest to the feed point. They also provide this conductivity in the most optimum direction(s): normal to the feed point. It also points to the importance of the

$$\frac{\partial R_l}{\partial rad_\#}$$

term (the number of radials being used).

It should now be clear why avoiding maximum current at ground level inherently increases efficiency in a compromised ground system. The ½ λ vertical is one example, but other vertically polarized configurations can accomplish this same goal. We will show later that efficiency is only part of the system performance. We will now explore particular vertical polarized configurations and then compare their performance.

## The Near-Ground Inverted-V Antenna

The near-ground inverted-V in its simplest form is a ½ λ dipole supported by a single high support, and the two legs form a 90 degree angle with their ends near ground level

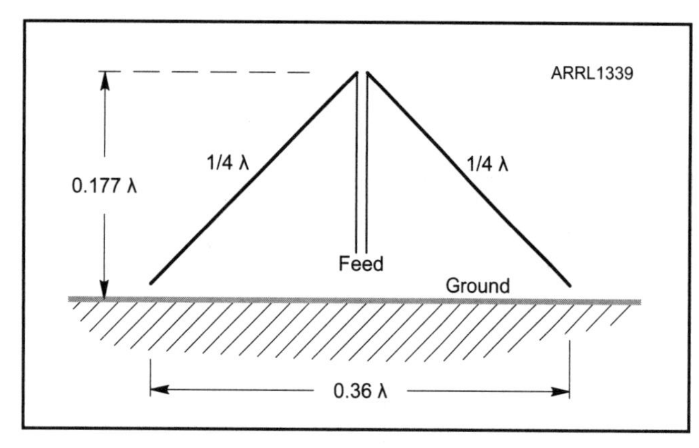


**Figure 7.14 — This figure shows a near-ground inverted-V antenna.**

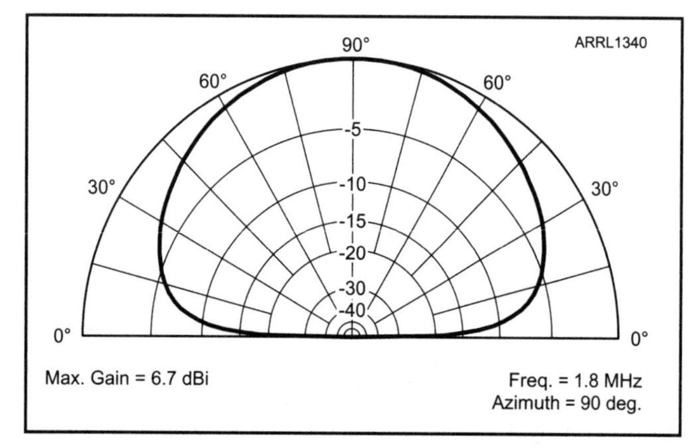


**Figure 7.15 — Near-ground inverted-V total field with the plane of the wires normal to the plane of the graph.**

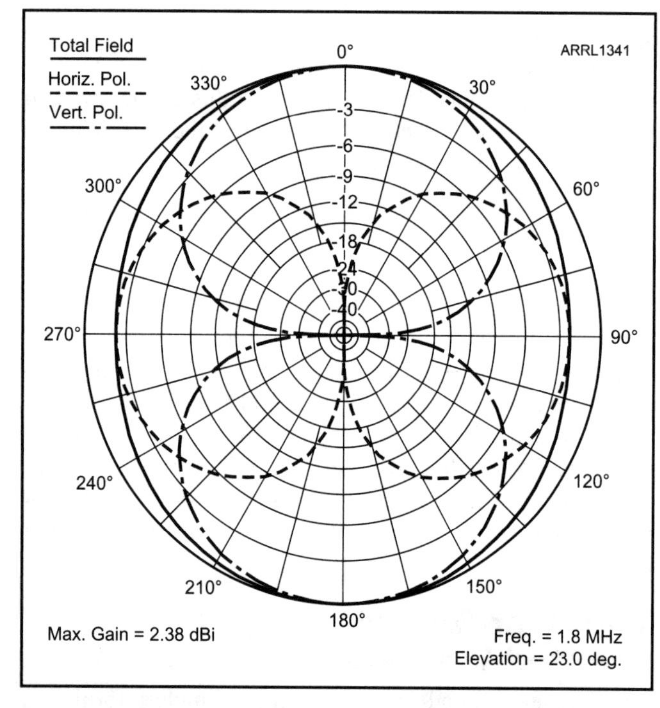


**Figure 7.16 — Near-ground inverted-V showing horizontal and vertical polarized fields at 23 degree elevation angle. The line of the wires is coincident with the Y-axis.**

as shown in **Figure 7.14**. Although seldom considered as a "vertical" antenna, it should be considered as such.

Since the inverted-V legs form 45 degree angles with the ground's surface, this antenna has both a horizontal and vertical polarization response. Broadside to the antenna (into and out of the page in Figure 7.14), the response is predominantly horizontal, while the end-fire response (to the right and left of the figure) is predominantly vertically polarized. If we compare the azimuth patterns of a single vertical element and that of a near-ground inverted-V, we can see that the low-angle gains are nearly identical (**Figures 7.15** and **7.16**).

The inverted-V, like its dipole cousin, has its current minimum at its ends. The ground, in effect, *does* form part of the antenna, as with the ground-mounted vertical. In addition, high-voltage coupling to the ground is by means of capacitive coupling. The capacitive values are very low, although the voltages are very high. Again, for *total* losses in such cases, a partial differential equation is required to solve the problem. The ground losses due to capacitive coupling from the high voltage ends is usually far less than the advantage of getting the current maximum(s) as far away from the ground as possible.

## Near-Ground Phased Inverted-Vs

We can take additional advantage of the vertical polarized end-fire low-angle directivity of the inverted-V by phasing two together. See **Figures 7.17** through **7.19**. As a practical matter, the feed point impedances of multielement, near-ground inverted V arrays drop to a few ohms. A simple technique to raise the feed point impedances is to use a folded V, similar to the folded dipole in Figure 3.12.

## Broadside Bobtail Curtain

The Bobtail Curtain (**Figure 7.20**) is an example of a broadside, voltage-fed vertical array that automatically

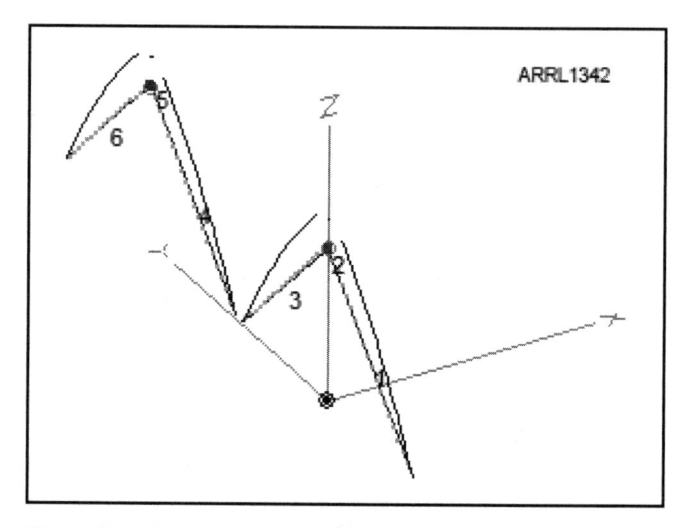


**Figure 7.17 — Two near-ground inverted-V antennas placed end-to-end and fed 180 degrees out-of phase results in substantial low-elevation angle gain.**

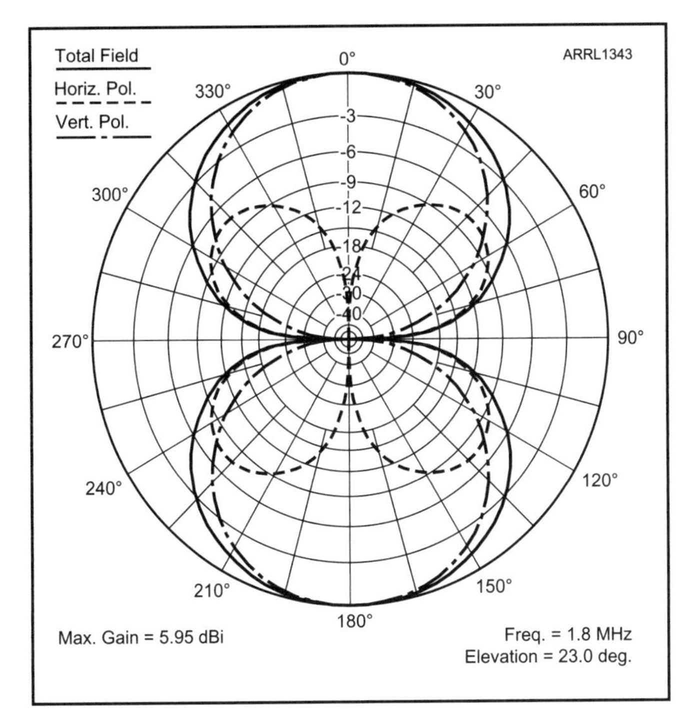


**Figure 7.18 — The bidirectional end-fire pattern of the two phased inverted-Vs over average ground. The line of the inverted-V antennas is along the Y-axis.**

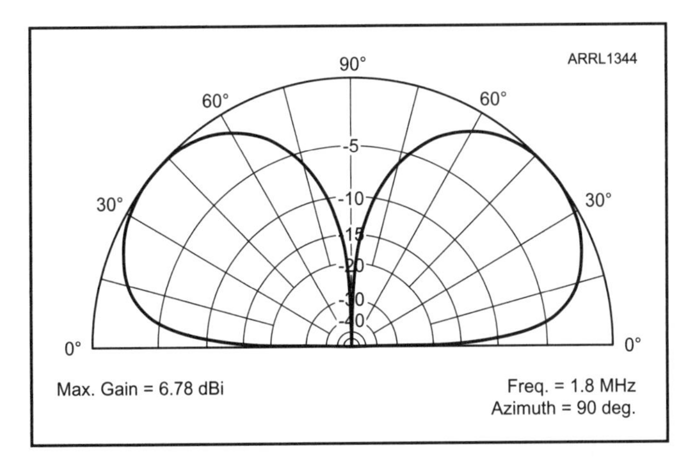


**Figure 7.19 — This shows the elevation angle plot of the end-fire phased inverted-Vs.**

provides for voltage maximums at the base of all three elements. The array is base-fed at the center vertical element. A minimum of radials are required only at the center-fed element, in practice to help stabilize the high feed point impedance. The three vertical elements are ¼ λ and the two horizontal wires are ½ λ making the array 1 λ long. These are the dimensions for optimum gain. However, "mini Bobtails" can also provide excellent results using shorter vertical elements and/or narrower width as long as the phases and amplitudes along the three elements are re-optimized for best results. **Figures 7.21** and **7.22** show the patterns.

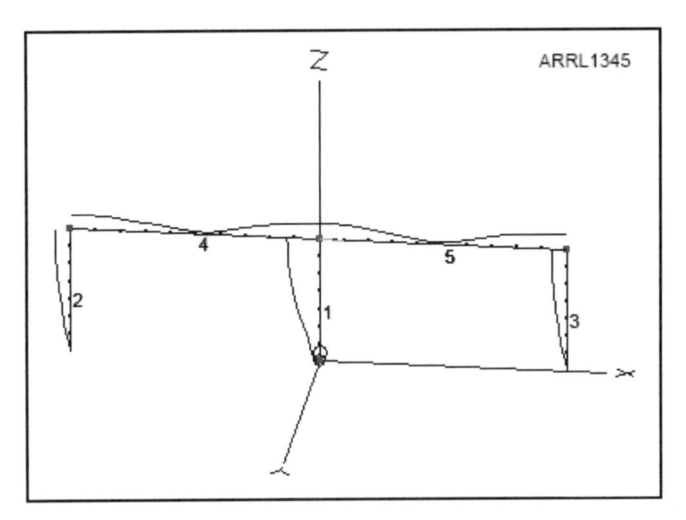


**Figure 7.20 — The Bobtail curtain.**

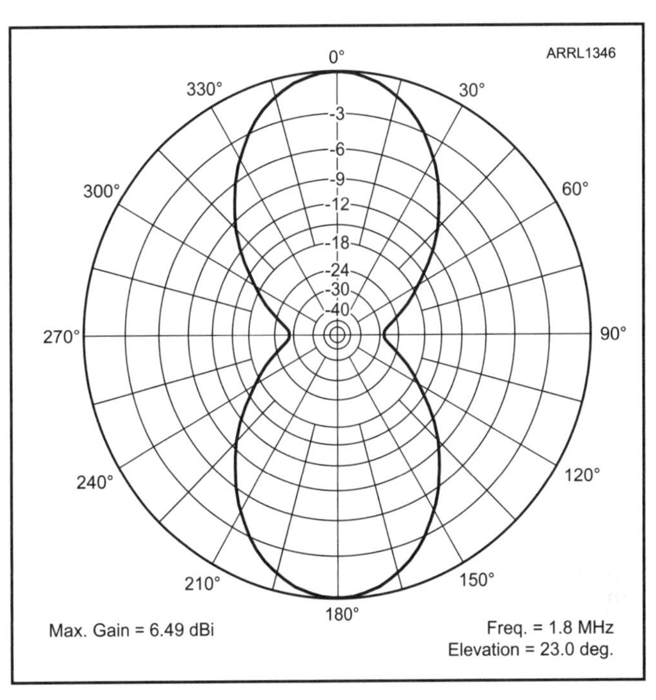


**Figure 7.21 — Azimuth gain plot of a full-size 160 meter Bobtail curtain showing the bidirectional pattern.**

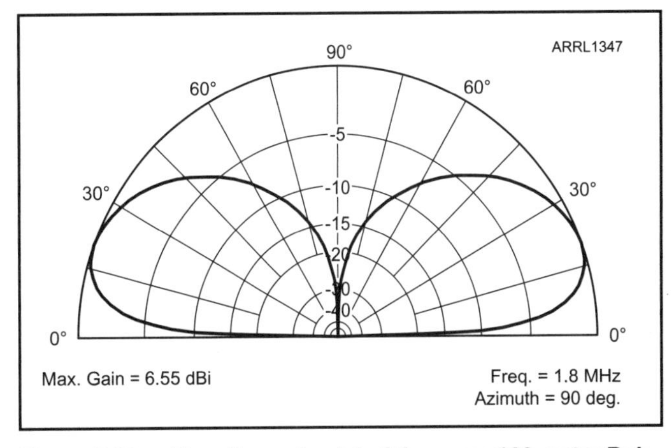


**Figure 7.22 — Elevation gain plot of the same 160 meter Bobtail curtain (Figure 7.21) broadside to the array (maximum gain).**

## Single-Element Vertical (Monopole Antenna)

A single ground-mounted vertical element has become the "standard" vertical antenna for a very wide range of applications. The ½ λ dipole is the only antenna that is more common. In this text, the first mention of these simplest of antenna configurations was in Chapter 1. The monopole's main advantages are its mechanical simplicity and its omnidirectional azimuth pattern (**Figures 7.23** and **7.24**). **Figure 7.25** shows a ⅝ λ vertical for comparison.

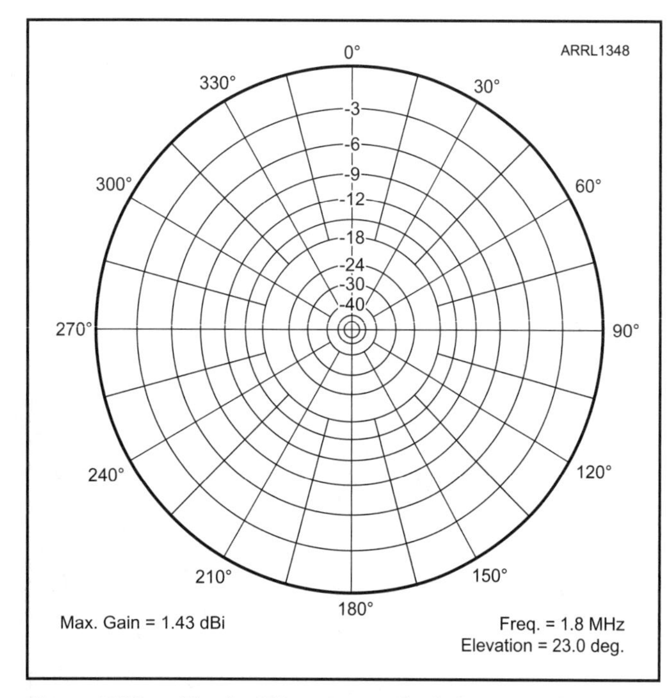


**Figure 7.23 — Single 160 meter vertical element azimuth plot at 23 degree elevation angle over average ground.**

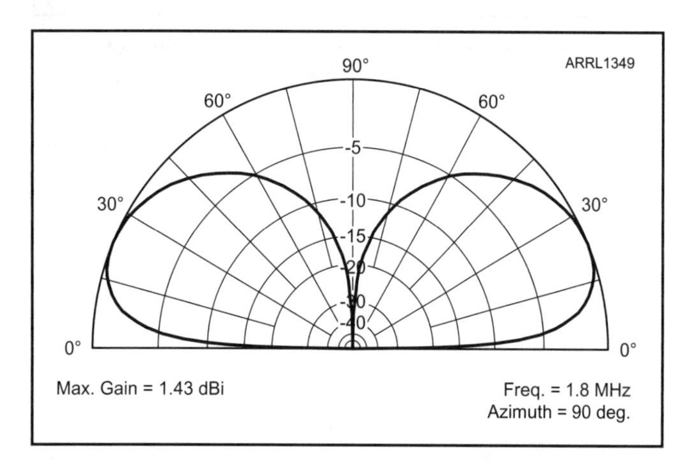


**Figure 7.24 — The same single vertical element, showing the elevation plot.**

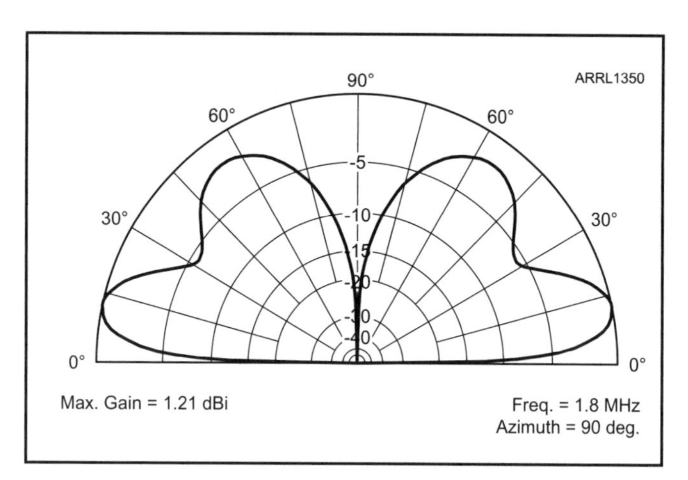


**Figure 7.25 — ⅝ λ vertical over average ground shows relatively poor gain. However, the gain maximum is at a very low 12 degrees. Slightly lower heights increase gain slightly and significantly reduce the very high angle parts of the pattern.**

## Multielement Vertical Arrays

As we discussed in Chapter 5, the addition of elements allows configuration of directional arrays. In AM broadcasting, nulls are created in antenna patterns to "protect" other AM station coverage areas, particularly at night. For amateur applications, the usual goal is to maximize gain in a particular direction, hopefully with many directions easily selectable.

In this section we will explore three multi-element configurations: 2, 4, and 8 element configurations. The plots shown in **Figures 7.26** through **7.31** are taken at 1.8 MHz, but the gain can be corrected for other frequencies with a single coefficient for a good approximation. For 75/80 meter applications, subtract 0.9 dB from these numbers. For 40 meters, subtract 1.5 dB (based on *EZNEC* modeling results).

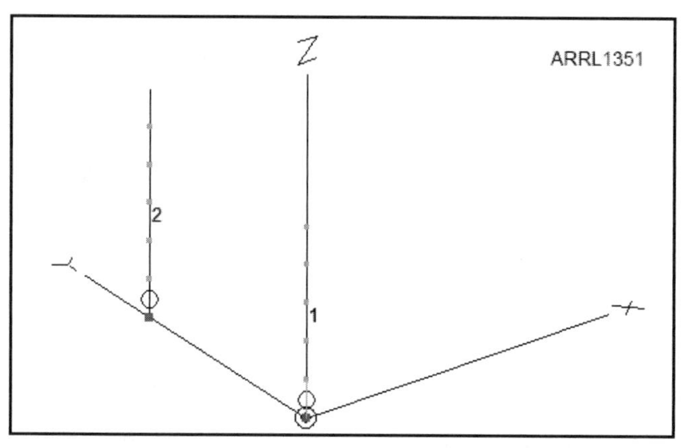


**Figure 7.26 — This diagram shows a 2 element cardioid antenna. The antenna height and spacing is ¼ λ, and the feed current values are equal. If element 2 is fed with the current lagging element 1 in phase by 90 degrees (–90 degrees), a mono-directional "cardioid" pattern results.**

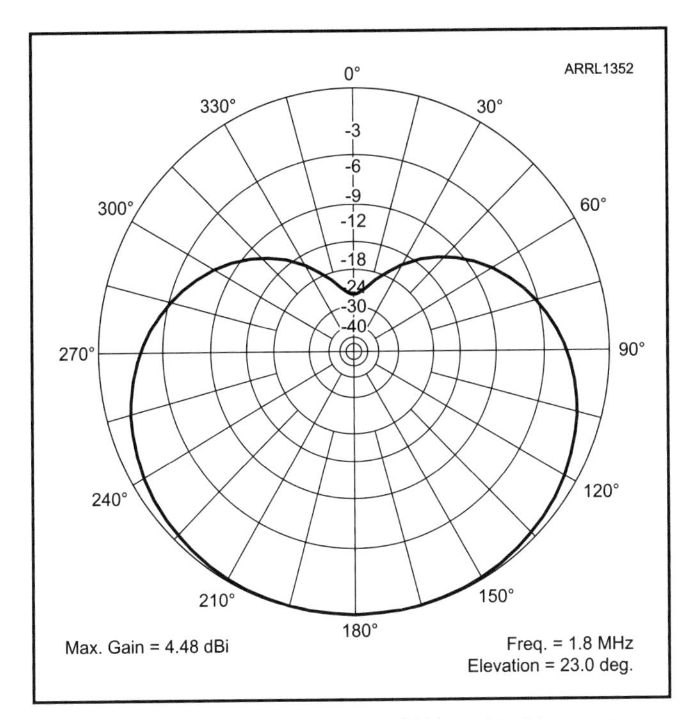


**Figure 7.27 — Two elements spaced ¼ λ and fed in quadrature, thus creating the heart-shaped "cardioid" azimuth pattern.**

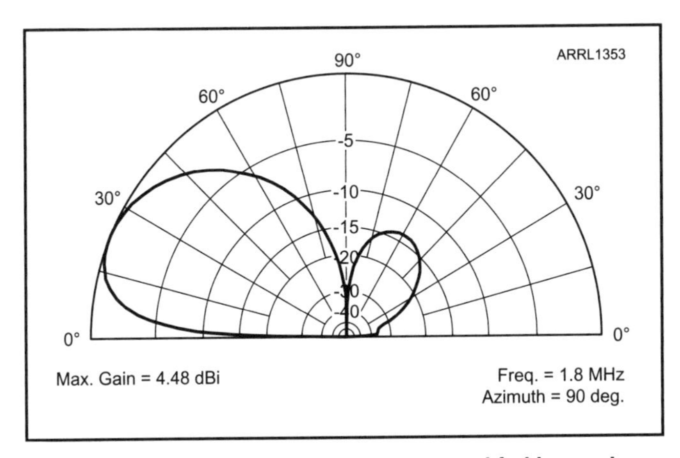


**Figure 7.28 — Two elements spaced ¼ λ and fed in quadrature showing the maximum gain azimuth angle (180 degrees).**

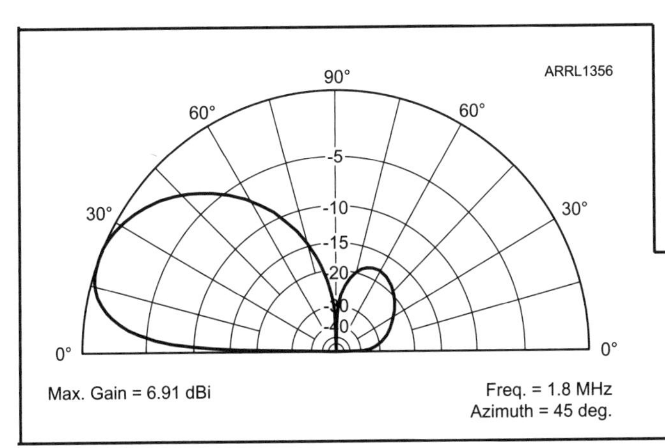


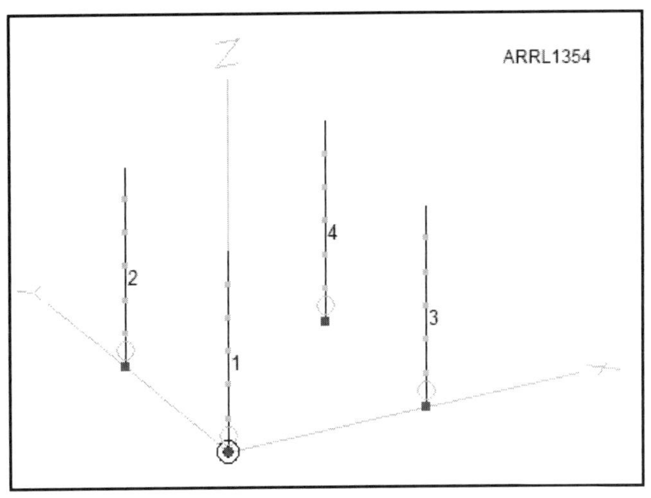


**Figure 7.29 — Four-square array. The elements are ¼ λ high, the element spacing is also ¼ λ. The elements are fed with equal current but with end-fire phasing. For example, to create a pattern that bisects the X-Y coordinate line direction "into the page," element 4 is fed at 0 degree angle, elements 2 and 3 are fed 90 degrees lagging element 4 (–90 degrees) and element 1 is fed 180 degrees out-of-phase with element 4.**

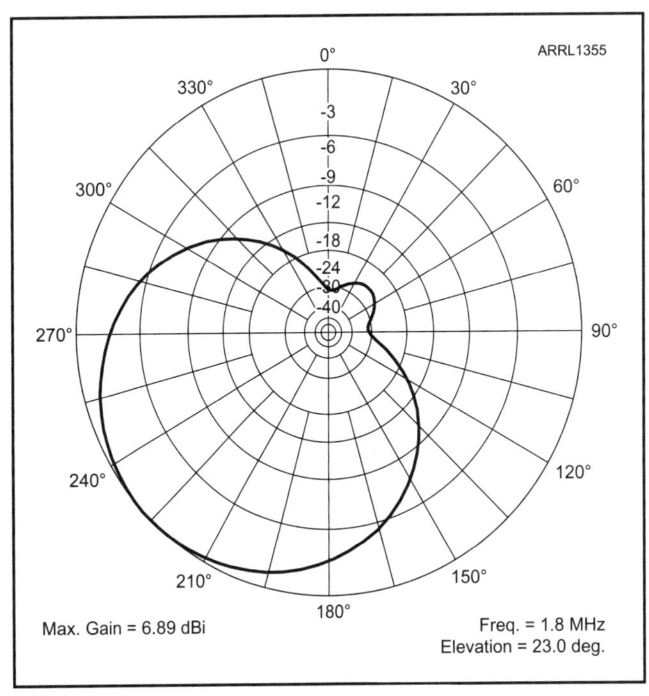


**Figure 7.30 —Azimuth coverage of the 4-square array at 23 degree elevation angle.**

**Figure 7.31 — Elevation pattern of the 4-square array at the maximum gain azimuth.**

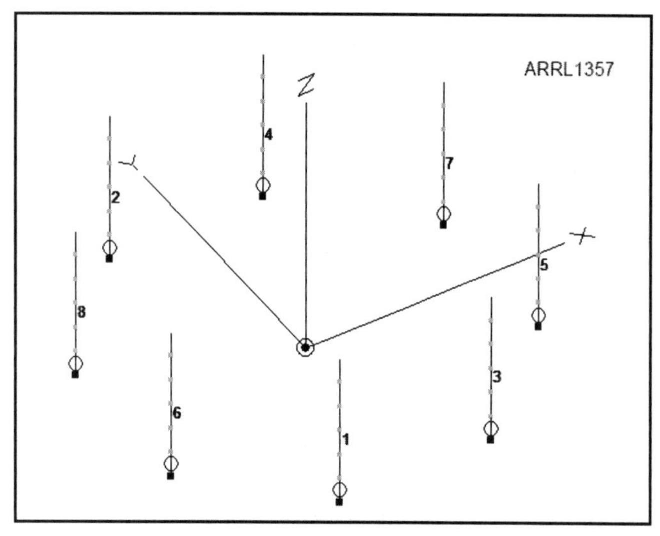


**Figure 7.32 — In this 8-circle array, only four of the eight elements are used for any of the eight directions. These four are at opposite sides of the "circle." Thus the 8-circle array is really several switchable 2-element cardioid arrays arranged in a broadside configuration. Four of the eight elements are fed for a chosen direction combining two 2-element end fire arrays into a single broadside array. For example, first we chose elements 2 and 4 for one cardioid array (as in Figure 7.26) and elements 1 and 3 for the other. Then we feed the two pair in-phase creating the broadside characteristics.**

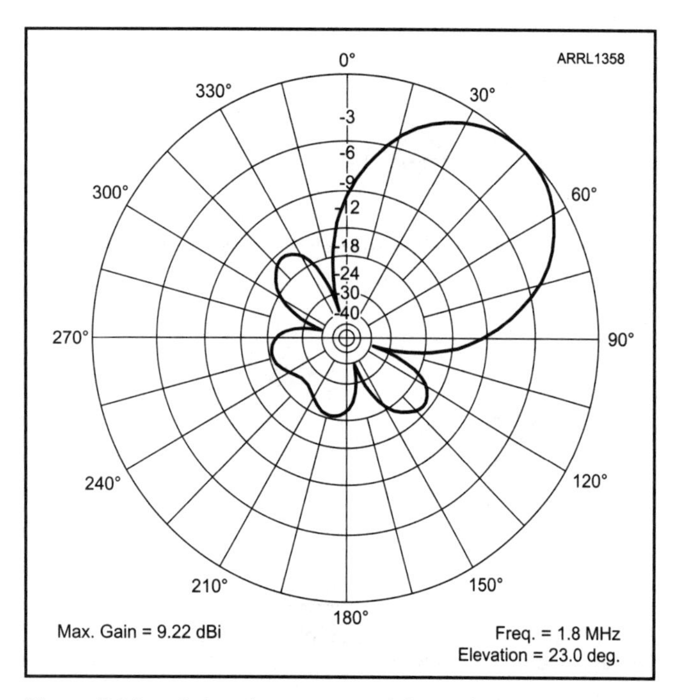


**Figure 7.33 — Azimuth coverage of the 8-circle array at 23 degree elevation angle.**

In 1994 while I was working up the first known 8-circle design, I noticed that with ¼ λ spacing of the elements around the circle, the resulting spacing of the two cardioid pairs would be very close to ⅝ λ, which is optimum for a broadside array! This results in a very effective configuration by virtue of the array's geometry. See **Figures 7.32** to **7.34**.

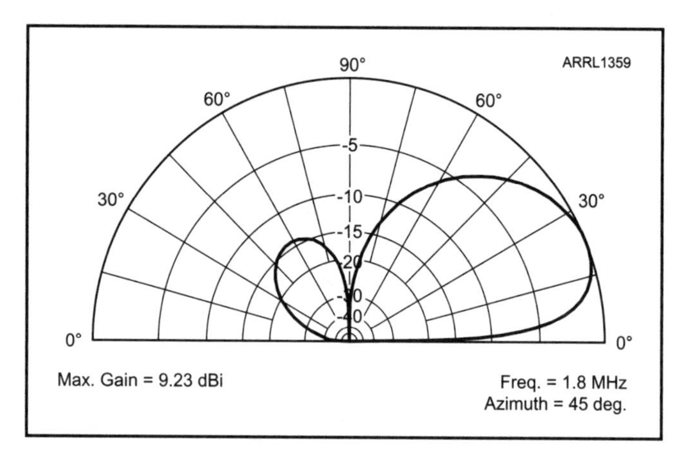


**Figure 7.34 — Elevation pattern of the 8-circle array at the maximum gain azimuth.**

## Gain Comparison of Vertical Arrays

We are now prepared to summarize the relative advantages and disadvantages of the various configuration presented in the following tables. The arrays detailed in **Table 7.2** are arranged by increasing gain.

When placed over "real ground," the frequency of operation *does* have an effect upon the absolute gain numbers. As the operating frequency rises above "average ground," the gain for the equivalent array decreases. The gain values given in **Table 7.3** are for 1.8 MHz. For all these antenna configurations, simply *subtract 1.5 dB for 7 MHz*, The actual range of deviation is about 1.4 to 1.6 dB. *For 3.6 MHz subtract 0.9 dB*. Also the elevation angle for maximum gain rises with frequency, but not more than 2 or 3 degrees. Since the ¼ λ ground plane in *free space* has no ground influences, its gain remains constant over all frequencies.

These *EZNEC* models show some interesting differences between the perfect and real ground systems. Although the ¼ λ vertical has a small gain advantage over the infinitesimal vertical in the idealized perfect ground circumstance, it actually shows *less* gain over real ground. The ¼ λ vertical begins to be affected by ground reflection cancellations, although it remains minimal. Unlike the ¼ λ vertical, the infinitesimal vertical far field ground effects are limited to only tangential dielectric absorption since there can be no ray cancelations. However, from a practical viewpoint, the $R_r$ of the infinitesimal (or very short) vertical will be very low and thus the efficiency will also likely be very low. Also, from a practical sense, losses in the matching network will also add to inefficiencies. Thus the slight advantage of a very short vertical over a ¼ λ vertical will in most circumstances will be more than offset by efficiency concerns. Appendix E shows plots for various vertical arrays, providing a graphic comparison of expected gain responses.

The top-loaded (capacitive hat) ¼ λ vertical shows a very slight compromise in gain due to the raising of the current maximum up the vertical height. However, the $R_r$ has increased significantly thus increasing efficiency significantly

**Table 7.2**

**Comparison of Vertical Antenna Configurations**

This table compares vertical polarized configurations over a perfect infinite conducting plane. There is no frequency dependence due to the perfect ground. Working against a perfect infinite ground the maximum gain is at the 0 degree elevation (on the horizon).

| *Antenna* | *Gain (dBi)* | *Elevation Angle (°)* | *Extensive Radials* | *Feed Location* |
|---|---|---|---|---|
| ¼ λ ground plane in *free space* | +1.25 | 0 | no | base |
| Near-ground inverted-V | +3.54 | 23* | no | apex |
| Infinitesimal (short) vertical | +4.77 | 0 | no | ground |
| Single ¼ λ vertical | +5.15 | 0 | no | ground |
| Single ¼ λ vertical top-loaded | +5.46 | 0 | no | ground |
| Phased near-ground inverted-Vs | +5.98 | 23* | no | apexes |
| Single ½ λ vertical | +6.96 | 0 | no | ground |
| Cardioid 2 element vertical | +8.20 | 0 | no | ground |
| ⅝ λ vertical | +8.14 | 0 | no | ground |
| 4-square | +9.55 | 0 | no | ground |
| Bobtail curtain (full-size) | +10.52 | 0 | no | ground |
| 8-circle | +11.88 | 0 | no | ground |

*The inverted-V arrays show a maximum gain elevation angle of 90 degrees (straight up). The elevation angle of 23 degrees is provided for easy comparison to other configurations at this low angle.

**Table 7.3**

**Maximum Gain for Various Vertical Configurations Over Average Ground**

The radial recommendations reflect a goal of about 90% efficiency and are used as a qualitative expression of the necessity for an extensive system.

| *1.8 MHz Antennas* | *Gain (dBi)* | *Elevation Angle (°)* | *Extensive Radials* | *Feed Location* |
|---|---|---|---|---|
| ⅝ λ vertical | +1.21 | 13 | medium | ground |
| ¼ λ ground plane in *free space* | +1.25 | 0 | 4 radials | base |
| Single ¼ λ vertical top-loaded | +1.32 | 21 | medium | ground |
| Singe ½ λ vertical | +1.32 | 16 | minimal | ground |
| Single ¼ λ vertical | +1.44 | 22 | yes | ground |
| Infinitesimal (short) vertical | +1.78 | 29 | critical | ground |
| Near-ground inverted-V | +2.38 | 23* | no | apex |
| Cardioid 2 element vertical | +4.48 | 23 | yes | ground |
| Phased near-ground inverted-Vs | +5.04 | 23* | no | apexes |
| Bobtail curtain (full-size) | +6.49 | 23 | minimal | ground |
| 4-square | +6.93 | 25 | yes | ground |
| 8-circle | +9.23 | 22 | yes | ground |

*The inverted-V arrays show a maximum gain elevation angle of 90 degrees (straight up). The elevation angle of 23 degrees is provided for easy comparison to other configurations at this low angle.

— especially over a compromised ground radial system.

With the ⅝ λ vertical, the effects from reflection nulling become dramatic. The current maximum on the ⅝ λ vertical is ⅜ λ above ground, thus providing enough height to generate multiple nulling reflections in the far field. The inverted-V configurations suffer comparatively little over real ground because the current maximum is at the antenna's apex, and the vertical polarized source is distributed over *both* the X- and Z-axis (sloping elements), thus "fuzzing" the reflection nulling.

The ⅝ λ gain maximum is at a very low 12 degree elevation angle. Shorter verticals will have significantly less gain at these angles. Therefore, the take-off angle desired for a given frequency and target area should be considered when comparing antennas. Usually for long-range F2 paths, higher frequencies typically require lower take-off angles. Thus 13 degrees may be *too low* for many 160 meter paths, but may be excellent at 40 meters. Again, from a practical sense, the slight advantages of the ⅝ λ ground-mounted vertical are usually offset by the fact it is more than twice as high as the ¼ λ option.

Also notice that the 2, 4 and 8 element phased arrays have nearly identical maximum gain elevation angles, near 21 degrees for 160 meters. This is also nearly identical to the monopole. Thus for a given vertical height, operating frequency, and ground conditions, there is a conservation of maximum elevation angle gain irrespective of the number of elements. This also holds true for the higher vertical configuration. Thus, for example, an array of ⅝ λ verticals would exhibit similar azimuth patterns to the ¼ λ arrays, but the maximum elevation gain angles would be about half, or about 11 degrees.

These real ground models in Table 7.2 and Table 7.3 assume a 100% efficiency ground radial system. Therefore this comparison must be tempered with compromised ground systems. For example, it is very unlikely that a practical comparison between the infinitesimal and ¼ λ vertical will show an advantage to the infinitesimal antenna, especially over compromised ground systems. Also a ½ λ or ⅝ λ vertical will have higher $R_r$ than the ¼ λ vertical and higher feed point impedance values, thus increasing the overall efficiency in compromised radial systems.

The 2, 4, and 8 element directive arrays provide a coinciding possible number of switchable azimuth directions. For example, the 2 element array inherently provides two switchable direction. The 4 element array provides 4 directions, and the 8 element provides 8 directions. It is also fortunate that in all cases these directional settings provide sufficient over-lap of azimuth patterns that result in only about 2 dB nulls for full azimuth coverage.

## Conclusion

When weighing the relative advantages and disadvantages of vertical arrays, Table 7.3 can represent a good "starting point" in that it compares:

1) Physical dimensions and necessary area for the antenna.

2) Gain that can be expected using an "adequate" radial system under the antenna when needed.

3) Comparative optimum take-off angles for maximum gain.

4) Relative radial requirements.

5) Feed point location.

In the case of a compromised radial system (when needed) the quoted gain will diminish as Equation 7.15. Since this equation is not closed form, for practical information the volumes of empirical data from amateur experiments can provide good approximations for real implications of Equation 7.15. However, Equation 7.15 is an attempt to actually quantify all the important terms into one unifying equation.

## EIRP Calculation for Ground-Mounted Low Frequency Vertical Antennas

Vertical antenna efficiency problems become particularly acute at very long wavelengths. Recent amateur bands below the AM broadcast band (the 630 meter band, 472 – 479 kHz) present new challenges for system design. Although the same issues apply for these frequencies as outlined above, added attention is required. EIRP is a power output specification where:

$$\text{EIRP} = P_i\,(G_{dbi}) \qquad \text{(Equation 7.18)}$$

where $P_i$ is the power input to the antenna.

For example, if the input to an antenna is 5 W and the antenna has a gain of 1.78 dBi, the EIRP in the direction of the 4.78 dBi gain is 15 W. A short vertical antenna (<⅛ wavelength) has a maximum gain of 4.78 dBi (linear gain of 3) with a perfect ground system and a perfectly conducting infinite ground. This gain figure is independent of the vertical's height for heights below about ⅛ wavelength. Unless ground losses in the far field are taken into account for EIRP calculations, this will be the coefficient in calculating EIRP values. For example if we are limited to 5 W EIRP and use a vertical shorter than ⅛ wavelength, our maximum power input to the antenna will be 5/3 or 1.6666 W.

At these frequencies, the typical amateur transmitting antenna will likely be very inefficient. A full size ¼ wave vertical is about 500 feet high! The efficiency will also be complicated by an almost certain compromised ground system. AM broadcast stations use 120 ground radials each ¼ wavelength long. The equivalent system at 475 kHz would require a circle of radial wires about 1000 feet in diameter (almost 20 acres).

The three primary efficiency concerns are as following for very short ground-mounted verticals:

1) Ground losses.

2) Vertical element losses.

3) Matching losses.

## Ground Losses

The short height and compromised radial system results in a very inefficient transmit antenna system. However, the FCC will be limiting the maximum EIRP to very low values, ≤5 W. Assuming 1.6666 W antenna input, this implies that an antenna only 1.6% efficient would require an antenna input power of 100 W. Wasting 98.3333 of 100 W is much more cost effective than the above extravagant solution. This is a huge advantage of regulating power in terms of *radiated power* rather than *transmitter output power.*

If we want to calculate the required antenna power input $P_i$, we use:

$$P_i = \text{EIRP}\frac{R_r + R_l}{R_r} \qquad \text{(Equation 7.19)}$$

where EIRP is the desired effective radiated power.

Now we need to find the proper values of $R_r$ and $R_l$.

$$R_r = 395\left(\frac{h_p}{\lambda}\right)^2 \qquad \text{(Equation 7.20)}$$

where $h_p$ is the physical height of the vertical (without the image) and λ is the wavelength, in this case about 2052 feet, or 632 meters. For the special case of verticals less than about ⅛ λ high, $h_e \cong h_p$ making the calculation easier. For example, a 50 foot high vertical ground-mounted antenna for 632 meters will have an antenna $R_r$ of 0.23 Ω. There is little advantage in attempting top loading since the loading structure would likely be very large, especially for verticals under about 100 feet.

The calculation of $R_l$ requires a direct measurement of the real portion of the feed point impedance.

$$R_l = R_{feedpoint} - R_r \qquad \text{(Equation 7.21)}$$

where again, $R_r$ is calculated by Equation 7.20. Now we have all the terms to calculate the needed antenna input power for a given EIRP (Equation 7.19). It is very likely that the measured $R_{feedpoint}$ will be much higher than the calculated $R_r$ indicating the losses will be very high and thus efficiency very low.

## Vertical Element Losses

For such very low values of $R_r$ it is also very important to lower the resistive value of the vertical itself in addition to ground losses. The losses along the vertical length will *not* be a linear function of $R_r$ and $R_{lv}$. *EZNEC* can estimate the losses for you, by applying the correct element diameter(s) and material(s). Using *EZNEC*, you can simply perform two models: one using "zero" for "wire loss" and the second using the appropriate metal and element diameter. The maximum gain difference will yield the dB loss due to the vertical element loss.

Some may think that vertical element loss will also affect $R_r$ which, in general, is an incorrect conclusion. The effective height will be reduced because there will be a reduction in average current along the vertical height. But also there will be a reduction in gain, and thus also aperture. These terms usually balance, and thus the calculated value of $R_r$ remains constant. The exception is when the loss resistance is "lumped" and of a high value. In effect this shortens the physical height of the vertical and simultaneously distorts the current distribution from its assumed linear function. However, when typical values of loss resistance are distributed over the length (especially the lower half of the vertical) and are less than about 1 Ω, the above assumption can be made. *EZNEC* makes short work of this calculation by the above technique.

The current distribution along the vertical element will be linear along its length with the maximum at the base. Therefore, the greatest effort should be to reduce resistance at the base of the vertical. For obvious reasons this is good news — heavier conductors are best placed closer to the ground.

## Matching Losses

Thus far we have limited the discussion to the antenna proper. However, in this case, short lossy verticals will exhibit relatively low real values at the feed point, perhaps tens of ohms. However, the imaginary part of the feed point impedance will be a high negative value (high capacitive reactance). This calls for the use of high Q matching, particularly the inductor. Typical inductive values for the tuner will be about 1 *millihenry.* An air core inductor with ¼ inch copper tubing would be about 6 feet long, 18 inches in diameter and 100 turns. It may be possible to build a comparable inductor out of ferrite cores, but this very high inductor value must also handle about 100 W.

## Summing the Losses and Determining Required Transmitter Power Output

Now we have all the terms we need to determine what input power is necessary for a given EIRP for a given system. Stated in dB losses and dBW power levels it is

$$P_t(dBW) = EIRP(dBW) + 10log_{10}\left(\frac{R_r + R_l}{R_r}\right) + Element\ dB\ loss + Matching\ dB\ loss \qquad \text{(Equation 7.22)}$$

where 0 dBW= 1 W and $P_i$ *(dBW)* is the power output from the transmitter. A detailed description of actual power levels using logarithmic notation is presented in Chapter 10.

This represents the basic equation for determining the proper transmit output power for a desired EIRP. The system design will be challenging if the FCC limits transmitter output power. For example, if the FCC restricts transmitter output power to 100 W and the maximum EIRP permitted is 5 W, implying an antenna input of 1.66666 W, a *total* of only 17.8 dB loss will be required to achieve maximum EIRP output.

Many of these complications will hopefully be settled with clear power rules. Aside from the EIRP calculation, will the calculation for $R_r$ require computer modeling? Will transmitter output power consider matching losses? Answers to these and other questions may change the coefficient values in Equation 7.22, or even change the equation itself.

# 8 Yagi-Uda and Cubical Quad Antennas

No antenna text can be considered complete without a discussion of parasitic antennas generally and Yagi-Uda antennas in particular. During the 1930s, Shintaro Uda (**Figure 8.1**) was a student of Professor Hidetsugu Yagi (**Figure 8.2**). Mr. Yagi acknowledged the Mr. Uda was indeed the inventor of the Yagi-Uda array. It continues to carry both names.

The Yagi-Uda has been the subject of intense investigation for many decades, resulting in highly optimized designs. There are many fine texts that detail Yagi-Uda and quad type arrays. This chapter will only review the basic computational methods used for optimized designs and link previous discussions in this book to those methods. Therefore, as one of the final chapters, the parasitic antenna can be considered as a specific culmination of the previous chapters.

**Figure 8.1 — Professor Shintaro Uda about 1960. [Courtesy of Tohoku University]**

**Figure 8.2 — Professor Hidetsugu Yagi about 1924. [Courtesy of Tohoku University]**

## Gain and Aperture of a Yagi-Uda Antenna

Most technical publications about the Yagi-Uda antenna stress F/B (front-to-back ratios) and forward gain as functions of boom length, thickness of the elements, element spacings, and lengths of the parasitic elements. I will not repeat this readily available and extensive information. Rather, we will explore the relationship of the Yagi-Uda to its effective aperture and continue the general discussion of parasitic arrays from Chapter 5.

The Yagi-Uda is a derivative of the simple dipole. There are two ways to increase the aperture (gain) of a ½ λ dipole antenna: increase the length, and/or create a multielement array (either by driving multiple elements, or by adding parasitic elements, or a combination of all these).

First we will explore the increased length dipole. We have already derived the broadside power gain of a ½ λ dipole as 1.64 (2.15 dBi). We have also defined broadside gain as a function of length (Figure 3.17 in Chapter 3).

Again, the aperture of a ½ λ dipole is

$$A_e\left(dipole\right) = \frac{1.62\lambda^2}{4\pi} \qquad \text{(Equation 8.1)}$$

If we double the aperture, the gain will also double and we achieve about 3 dBd gain. A 2-element collinear antenna is twice the length of a ½ λ dipole, but only achieves about 2 dBd gain. If we extend the length to 10/8 λ, we create an extended double Zepp. This is the length of a center-fed linear element that achieves the maximum broadside gain (about 3 dBd).

The boundaries shown in **Figure 8.3** are approximate. The actual boundary lines are a bit "fuzzy" but this figure serves as a good starting point to conceptualize what the aperture areas actually look like around linear antennas.

The second method of increasing the gain of a single dipole is to add elements. In Chapter 7 we showed many

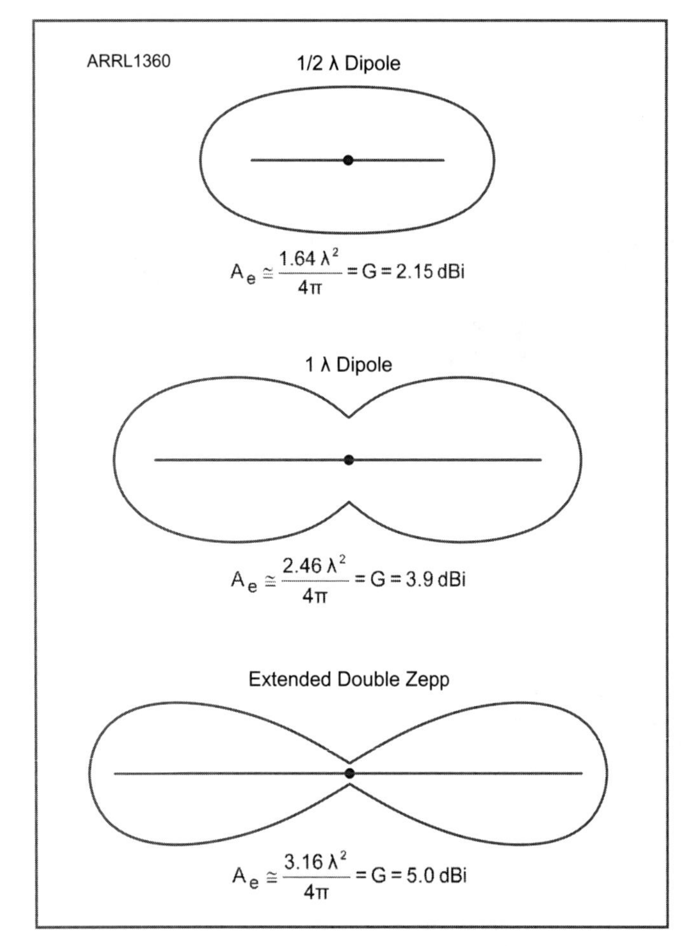


**Figure 8.3 — This figure shows the approximate boundaries of aperture for three linear single-element antennas. The 1 λ antenna does not provide 2× aperture over a single dipole because the spacing of the two dipoles is too close. Therefore part of the potential aperture has overlapping coverage from the two elements. The extended double Zepp, on the other hand, provides for greater separation of the current maximums on the two elements, thus minimizing "double coverage" and achieving near-theoretical broadside doubling of aperture from non-coupled arrays (about 3 dBd).**

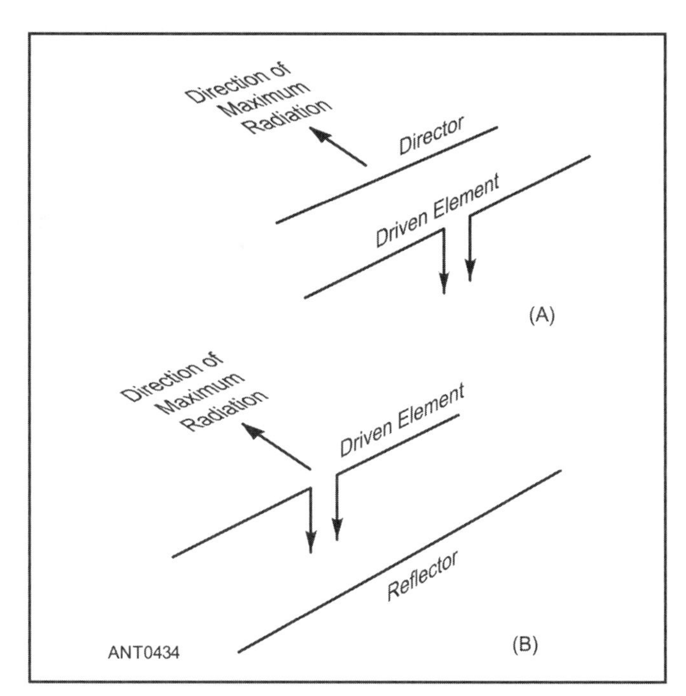


**Figure 8.4 — Two-element Yagi systems using a single parasitic element. At A, the parasitic element acts as a director, and at B as a reflector. The arrows show the direction in which maximum radiation takes place.**

multielement vertical driven arrays and their relative gains and patterns. The same basic techniques can be used for free space dipoles with similar results in gain. Thus free space driven arrays will not be covered here in detail to avoid repetition.

We begin the discussion with a simple two-element Yagi-Uda parasitic array (**Figure 8.4**). By making the parasitic element slightly longer than a resonant dipole and placed in back of it, the backward pattern can be nearly completely eliminated and aperture and gain increases in the opposite (forward) direction. By making the element slightly shorter than resonant, a different phase response is achieved, in that additional gain is realized in the forward direction when the *director* is placed in front of the driven element.

The very high scattering aperture (discussed in Chapter 2) of a shorted dipole makes possible the high gain characteristics of the Yagi-Uda antenna. For example, a typical 2-element Yagi has about 5 dBd of gain (5 dB higher than a simple dipole). Notice in Figure 2.21 in Chapter 2 that for a shorted ½ λ dipole the scattering aperture is 4× the maximum possible effective aperture. On the other hand, a 2-element quadrature driven array has a maximum gain of about 3 dBd, which is what would be expected by a simple doubling of a single dipole's aperture. See **Figure 8.5**.

By adding directors in the desired maximum gain direction, multielement Yagi-Uda arrays can be built. Antennas with 20 or 30 elements are practical for some applications. In the majority of configurations, there is one driven element and one reflector. Additional elements take the form of directors. This is the optimized design configuration because the reflector is more effective than directors at maximizing the front-to-back ratio. The addition of more and more directors is about proportional to the log of the gain. Therefore, adding more and more directors provides less and less gain addition per element. This is because, again, the additional directors begin to duplicate the available aperture around the antenna. On the other hand, the inherent advantages (electrical and mechanical) often outweigh this diminishing return.

Now let us assume a 3-element Yagi-Uda array. The addition of a single director is taken at a typical value of 2 dB more than a 2-element array. Therefore, we assume the Yagi-Uda has a gain of 7 dBd, or about 9 dBi, or a numerical power gain of 7.94 (in free space). Therefore the aperture of our 3-element Yagi-Uda is

$$A_e = \frac{7.94\lambda^2}{4\pi} = 0.632\lambda^2 \qquad \text{(Equation 8.2)}$$

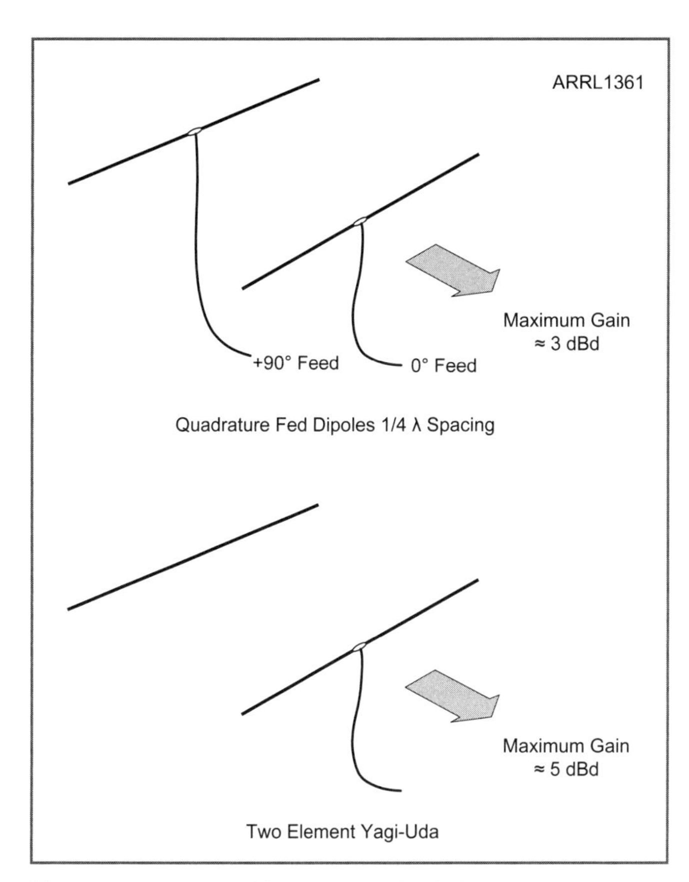


**Figure 8.5 — A parasitic 2-element Yagi-Uda antenna can provide about 2 dB more gain than a 2-element quadrature driven array. Furthermore, the Yagi-Uda requires only one feed.**

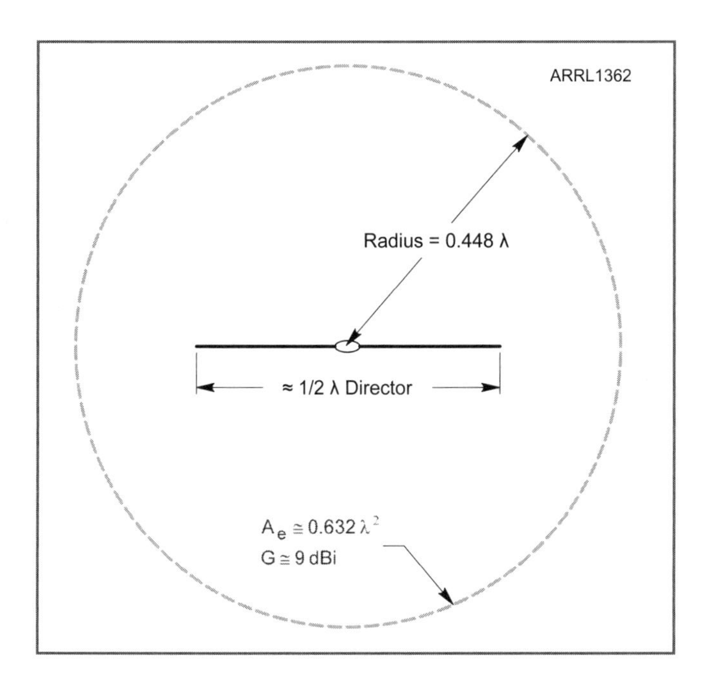


**Figure 8.6 — Circular aperture of a typical 3-element Yagi-Uda. The view is from the front of the Yagi-Uda looking "down" the boom with the director element shown. The circular shape is a good assumption, but not perfectly accurate. As in Figure 8.3 the actual boundary of the aperture is not distinct, but rather "fuzzy."**

In Figure 8.3 the shape of the apertures is elliptical. This is due to the horizontal current distribution along the wires. As the gain increases with a Yagi-Uda, the array proper begins to resemble a point source, and thus the shape of the aperture becomes progressively more "circular" as shown in **Figure 8.6**. In turn we can define the radius of this aperture circle by

$$radius = \sqrt{\frac{A_e}{\pi}} = 0.448\lambda \qquad \text{(Equation 8.3)}$$

## Stacking Yagi-Uda Antennas

We may increase the gain of a single Yagi-Uda by simply adding a second identical antenna. In principle this should double the aperture and thus double the gain (+3 dB). If yet more gain is needed, doubling the aperture again will yield

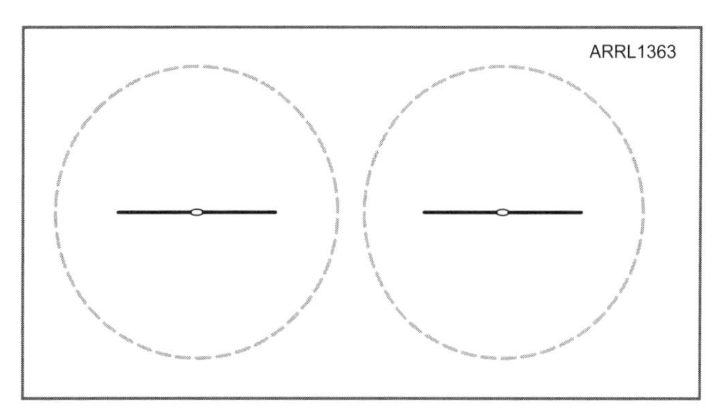


**Figure 8.7 — Proper side-by-side stacking of two Yagi-Uda antennas. The distance between the antennas assures no overlap of the antennas' apertures.**

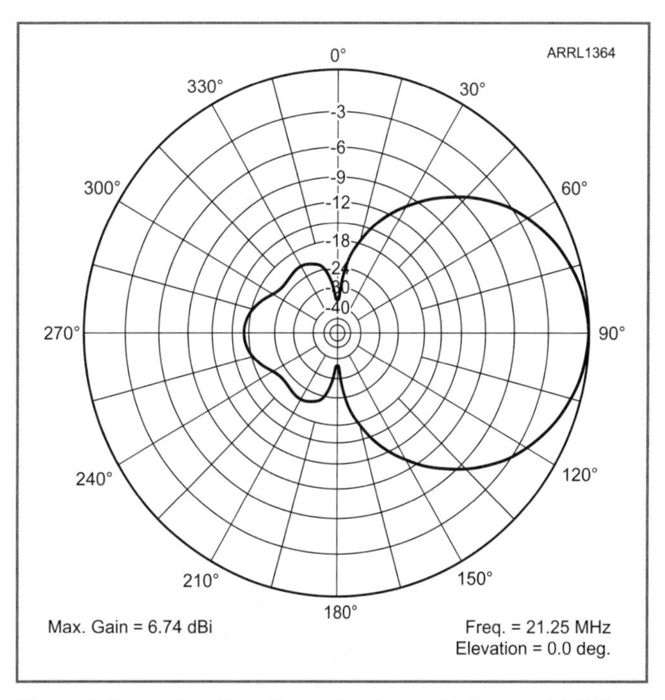


**Figure 8.8 — Azimuth pattern of a simple 2-element 21 MHz Yagi pattern in free space. The 3 dB beamwidth is 68 degrees.**

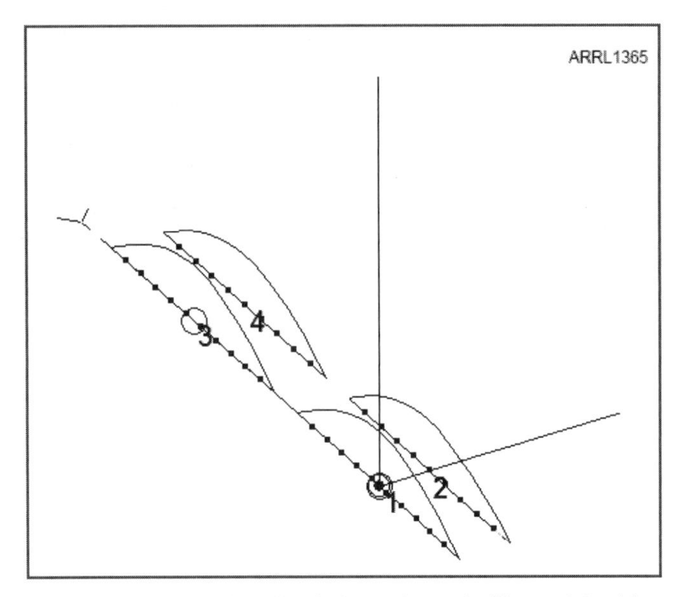


Figure 8.9 — Two of the Yagi-Udas shown in Figure 8.8 with close horizontal stacking.

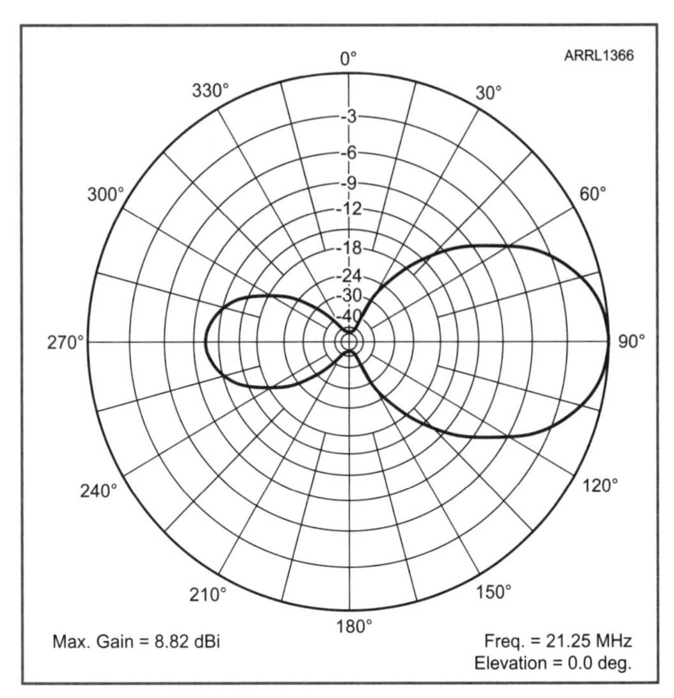


Figure 8.10 — Free space pattern of the closely spaced two Yagi-Uda array of Figure 8.9. Notice that the gain increase is about 2 dB, similar to the closely spaced dipoles. More significant is the beamwidth, which is now about 43 degrees.

(+3 dB) more (using four antennas). Another 3 dB can be realized by doubling the aperture again (now eight antennas), and so forth. In Figure 8.3 we showed that if two dipoles are spaced very closely (end-to-end), some of the potential aperture is covered by both antennas. This is analogous to our discussion of solar cells in Chapter 2. If one solar cell partially covers another (shading), the amount of aperture (gain) is reduced. The correct spacing among "stacked" Yagi-Uda antennas can be determined with fair accuracy by calculating the equivalent aperture (as in Figure 8.6) and then make sure the antennas are sufficiently separated to provide minimum "shading" of the other as shown in **Figure 8.7** for antennas stacked side-by-side.

For maximum gain there is no reason to increase the spacing further. Therefore, as we increase the number of elements of the Yagi-Udas, we increase the gain, increase the aperture, and thus the spacing among the elements must increase to avoid wasting the gain by duplicating the aperture coverage. Furthermore, if the spacing is made progressively larger, sidelobes begin to appear.

**Figure 8.8** shows the azimuth pattern of a simple 2-element 21 MHz Yagi in free space. We can stack two of these antennas horizontally as shown in **Figure 8.9**, and **Figure 8.10** is the resulting

Figure 8.11 — This array of eight long Yagi-Uda antennas is intended for 432 MHz "moonbounce" or EME (Earth-Moon-Earth) amateur communications. Each antenna has 27 elements. If each Yagi-Uda has a 20 dBi gain, then the *theoretical possible gain* from eight antennas is 29 dBi. [Noriyuki Yaguchi, JHØWJF, photo]

azimuth pattern for the stack. Figure 8.10 shows that the aperture of the array of Figure 8.9 is "distributed" more along the horizontal axis than the vertical axis. This results in a narrower azimuth response than the single antenna, but has little effect upon the vertical beamwidth.

We can also "stack" the two antennas vertically. The difference is how this will affect the directivity of the array. Vertical stacking of Yagi-Udas over real ground provides diversity of the maximum elevation angle of radiation. Low horizontal antennas will radiate at higher elevation angles than higher horizontal antennas (assuming the maximum gain).

**Figure 8.11** shows an array of eight Yagi-Uda antennas for 432 MHz using both horizontal and vertical stacking.

## Design and Optimization of a Yagi-Uda Array

The fundamental theory of parasitic arrays was presented in Chapter 5. We discussed the fundamental importance of the mutual impedances of the array. However, we only dealt with a 2-element array. Now we will present numerical methods for multielement arrays. In Chapter 7 we presented the concept of using the matrix as a method of linking multiple interrelationships of multiple functions. The modeling of a multielement Yagi-Uda antenna is most easily performed using such a matrix. Unlike the Chapter 7 discussion where we resorted to differential equations for the terms, with a Yagi-Uda we are only dealing with (in principle) relatively simple algebraic functions, in particular solving for Ohm's Law with complex impedances.

Consequently, matrix algebra is used to calculate the relative amplitudes and phases on each element that, in turn, determine the final directivity of the antenna. We begin by expanding the 2-element discussion in Chapter 5 to a 3-element Yagi-Uda. The first step is to compute the voltages that appear on each element as a result of all the self and mutual impedances:

$$\begin{aligned} I_1Z_{11} + \quad I_2Z_{12} + \quad I_3Z_{13} &= V_1 \\ I_1Z_{21} + \quad I_2Z_{22} + \quad I_3Z_{23} &= V_2 \\ I_1Z_{31} + \quad I_2Z_{32} + \quad I_3Z_{33} &= V_3 \end{aligned} \qquad \text{(Equation 8.4)}$$

The resulting voltages on each element are the sum of their corresponding currents and the *complex* impedances determined by the self and mutual complex impedances. It is these complex impedances that alter the relative phases of the currents on each element. Once again the phases, amplitudes, directions, and locations (PADL) of these currents, determine the final directivity of the antenna.

The matrix in Equation 8.4 actually represents three linear equations, with the currents being unknown. The mutual impedances can be computed analytically by knowing the spacings and electrical lengths of the elements. Thus if we apply a voltage to the driven element (element 2), we can derive the resulting voltages by this set of linear equations contained within the matrix.

The final step is to derive the currents by *inverting* the complex impedance matrix to yield

$$I = Z^{-1}V = \frac{V}{Z} \qquad \text{(Equation 8.5)}$$

which, of course is simply a special form of Ohm's Law

$$I = \frac{V}{Z} \qquad \text{(Equation 8.6)}$$

Once we know the complex currents in each element (amplitude and phase) and the spacing between the various elements, we can directly calculate the far-field of the array, again by the PADL method. This is the basic technique used by antenna modeling software such as *EZNEC*.

As with most antenna configurations, today computers are used to solve for these matrices and then calculate the resulting pattern. Furthermore, *optimization* for forward gain, front-to-back ratios, and other parameters involve hundreds if not thousands of calculations through trial-and-error algorithms to derive an optimized design. Again, today these computations require only seconds on a laptop computer.

## The W8JK Array

Although not a parasitic array, the W8JK array deserves review and may be compared to the performance of a Yagi-Uda. In 1937, John Kraus, W8JK, built a closely spaced driven array that resembled a Yagi-Uda. The array consisted of two ½ λ dipoles fed 180 degrees out-of-phase at 14 MHz. The result is a maximum bidirectional gain of 5.88 dBi, or about 3.74 dBd. This is about 0.74 dB higher than a mono-directional pair of dipoles, but about 1 dB lower than the mono-directional 2-element Yagi-Uda. The W8JK uses the 3 dB advantages of a doubling of aperture, but also both driven elements provide additional gain by virtue of their scattering apertures. This results in a slightly higher gain than a simple aperture doubling.

The W8JK can be thought of as yet another derived design from Chapter 5. See **Figure 8.12**. Rather than two

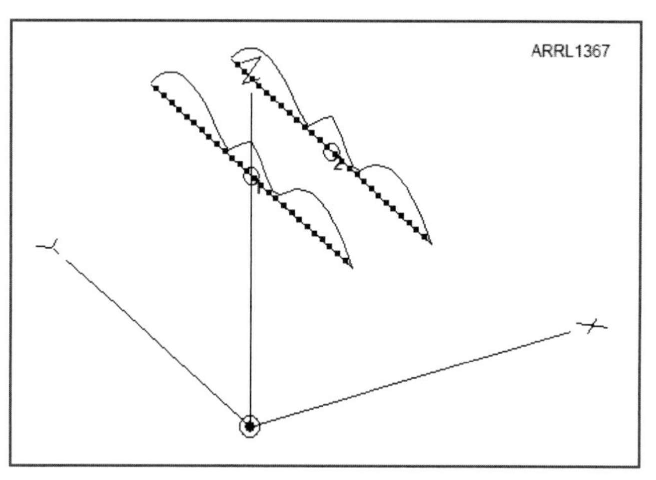


**Figure 8.12 — Two extended double Zepps (at 28 MHz) center fed out-of-phase. Each element is about 44 feet end-to-end. The array provides excellent gain from all frequencies down to about 10 MHz, where the array's $R_r$ drops to about 5 Ω, making matching more problematic.**

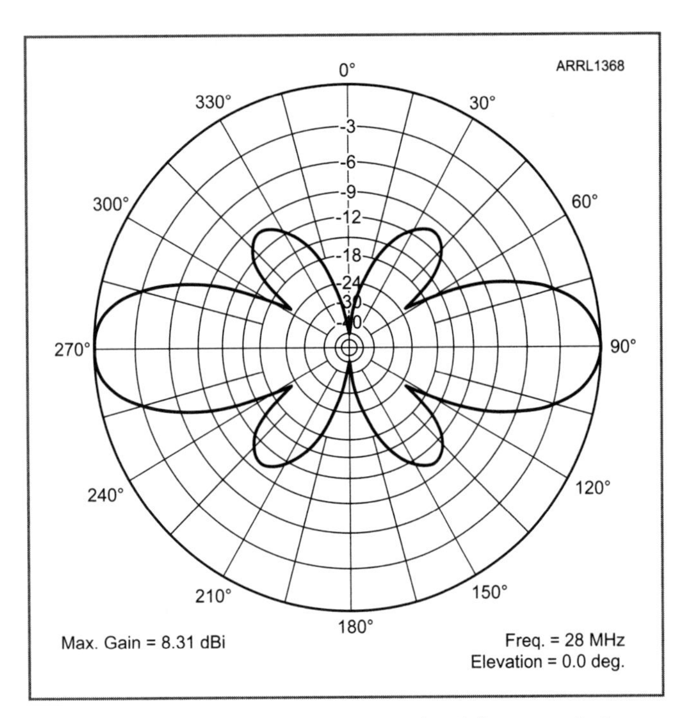


**Figure 8.13 — Free-space azimuth gain of the extended double Zepp/W8JK array. The modeled gain is 8.31 dBi, or 6.17 dBd, comparable to a 3-element Yagi-Uda in both directions.**

dipoles spaced ½ λ and fed 180 degrees out of phase, we now space them about ⅛ to ¼ λ. The flexibility of spacing is added to the flexibility of element length (both as a function of wavelength). Thus, the W8JK has the inherent advantage of providing broadside gain over one octave or more of usable spectrum (not mentioning the required matching for different frequencies!). If the element lengths are made ⅝ λ long, and use ½ λ spacing (let's say for 28 MHz), then appreciable gain and tolerable feed point impedances result as low as about 10 MHz (about 3⁄32 λ spacing). At the highest frequency, the two dipoles are extended double Zepps, and the *bidirectional* gain is comparable to a 3-element Yagi-Uda (in both directions simultaneously). The pattern is shown in **Figure 8.13**.

For all its simplicity, as the operating frequency changes, the feed point impedance moves "all over the Smith chart." Thus a practical broad-band W8JK should use open wire or ladder line to a tuner and balun for matching.

## Multiband Yagi-Uda Arrays

While the W8JK offers a very simple configuration for multiband operation, the Yagi-Uda does not. Yagi-Uda antennas are very effective when optimized for one narrow band of operation. There are several popular methods for designing a multiband Yagi-Uda.

1) "Traps" may be installed along the elements to electrically set the equivalent physical length of the elements. The traps are actually parallel resonant tank circuits placed in line with the element. At the resonant frequency, the trap effectively switches off the element length beyond its position. At lower frequencies the trap appears as a series inductor, thus making the element electrically shorter at lower frequencies. Good trap design can result in low $R_l$ loss, but there is always some loss. The shorter element lengths also reduce gain and decrease the VSWR bandwidth.

2) "Interlacing" elements simply places elements for different frequencies along the same boom. The main problem is inter-band interaction through mutual impedances of the various elements. Again, very careful design is needed to minimize these effects while maintaining acceptable performance. An interlaced Yagi-Uda array may have separate driven elements for each band, each with its own feed line, or it may use a broadband or multiband driven element with a single feed line. These may be a small log periodic cell or a set of closely-spaced elements that exhibits resonance at multiple frequencies (typically referred to as "coupled-resonator" or "open-sleeve").

3) Remote control of element length. This is an excellent technique used to change the length of the elements. Designs have included using air pressure to set the lengths against two stops for a dual-band configuration. Currently the most popular method utilizes stepper motors that can set the element lengths with high precision. The one disadvantage is shared with trap antennas: the element separation is fixed. However the design is inherently broad-banded in that it can be absolutely optimized for any frequency given the spacing constraint. Unlike the W8JK matching problem, the length of the driven element can be adjusted to resonance and a simple broadband balun-transformer can easily match to the popular 50 Ω coaxial cable feed line at any of the frequencies covered by the antenna.

4) Linear loading can involve many different configurations of multiple conductors running parallel to a "main" conductor along portion(s) of the element's length. Parallel conductors can provide parallel resonant dipoles or sections of open or shorted transmission lines that in effect, create multiple resonances at desired frequencies without the need for LC traps. This technique eliminates trap losses, but complicates the mechanical design.

5) In the January/February 2017 issue of *QEX*, I introduced a new technique for configuring Yagi-Uda parasitic configurations. Using traditional trap methods limits to available aperture of the parasitic element to that of a 1/2 λ or less. For example, a typical tri-band Yagi-Uda reflector only uses 1/2 of the available aperture on 10 meters. This configuration uses the entire length of the element on *all* bands. An actual antenna design is shown here to help explain this new concept in Yagi-Uda design.

An optimized 2-element Yagi-Uda antenna will yield about 5 dBd — 5 dB more than a half wave dipole — or about 7.15 dBi. Therefore, the goal will be to achieve 5 dBd over the five amateur bands (14 – 30 MHz) and also provides a bi-directional pattern on 10.1 MHz. A 7 MHz dipole is also possible, but the driven element length is marginal for this frequency. This will result in a six-band rotatable array, and possibly seven bands. Also, we want the boom length and number of elements to be minimum.

As shown in **Figure 8.14**, the design uses a 7-foot boom (about 2.5 meters) with 3 elements. Maximum gain from a center-fed linear antenna is achieved with a total length of

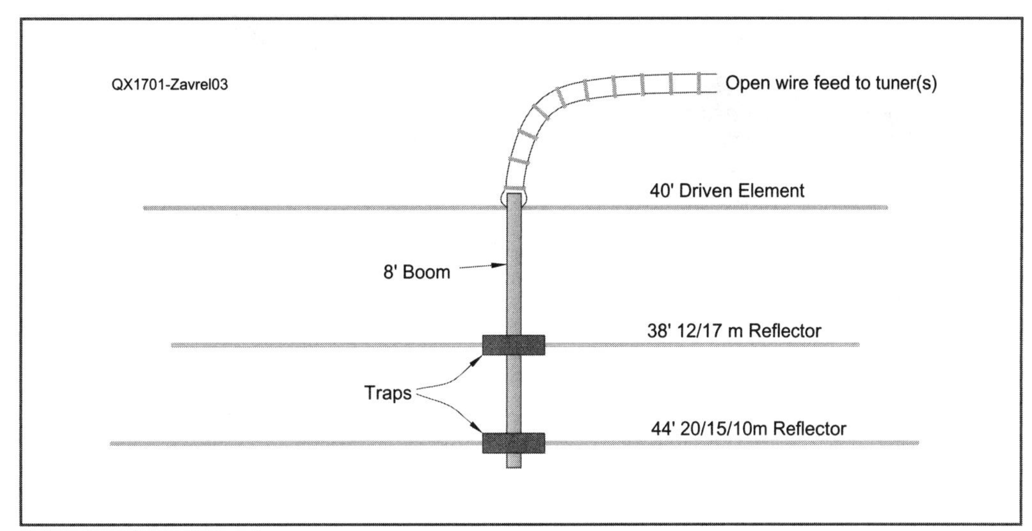


**Figure 8.14 — A new technique for configuring Yagi-Uda parasitic configurations can be understood through this design example. See text.**

1.25 wavelengths, the length of an extended double Zepp. Therefore, we begin with approximately this length for the highest operating frequency (the 28 MHz band). The three elements consist of the driven element, a reflector for 20, 15, and 10 meters, and a reflector for 12 and 17 meters. Only two traps are used in the array, one each at the centers of the parasitic elements; the driven element is a simple 40-foot center-fed dipole with open wire line. Calling these "traps" is a bit misleading. On one band each is indeed a trap, but on other bands they provide critical reactance values to affect optimized parasitic responses.

The advantages of this design are: (1) There are no traps or loading along any antenna element, (2) only two traps needed, one each at the reflector element centers, (3) the boom is short, (4) it offers a simple, single feed point, and (5) all bands take full advantage of the complete physical length of the elements (more gain, especially at the higher frequencies).

This configuration provides for reflector elements on 5 bands (10 through 20 meters) using two parasitic elements. It will also provide for a bi-directional pattern on 30 meters. The driven element is shown being driven by an open wire line that would connect to a tuner. The tuner could be "automatic" perhaps mounted on the antenna mast and then connect to a standard 50 Ω coaxial line. This technique can also include director elements with adjusted L and C values.

First, I will explain the simpler two-band trap for 12/17 meters shown in **Figure 8.15**. We can optimize a reflector by actually configuring two end-fire reflectors on the same element. The optimum reflector element length for the 12 m band is about 19 feet. So, we simply build an element twice this length or 38 feet. However, it is necessary to break this element in two, which would be simple if we only needed a reflector at 12 meters. Instead we place a resonant parallel LC circuit at the center and affect the break. We now have two collinear reflectors on 12 meters. On 17 meters, this element now looks too long for a reflector, so we simply add two capacitors on either side of the 12-meter trap. This "tunes" the longer element for optimum length at 18.1 MHz, and we now have a very effective 12/17-meter reflector element. Figure 8.15 shows the component values. However, the current remains distributed over a longer-than required physical length, thus increasing the antenna gain.

We can extend the idea of the dual-band trap into a tri-band trap, as in **Figure 8.16**. In this case we start with the two-element collinear at 15 meters, and as above, breaking the 44-foot length in two at 15 meters using another parallel LC tank, thus providing two collinear reflectors. At 14 MHz, this creates a center-loaded 44-foot reflector, too long for a traditional 20 meters parasitic array. At 10 meters, the element is more than one wavelength. We can take full advantage of this extra length at both bands by using a proper series LC circuit on both sides of the 15-meter trap, which tunes the element simultaneously for both 10 and 20 meters.

I won't duplicate the hand calculations I performed, but computer modeling tools such as *MATLAB* could be used to

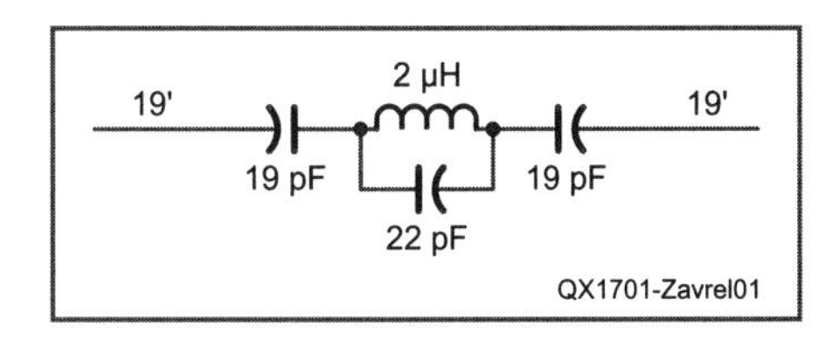


**Figure 8.15 — Parasitic reflector for the 12- and 17-meter bands.**

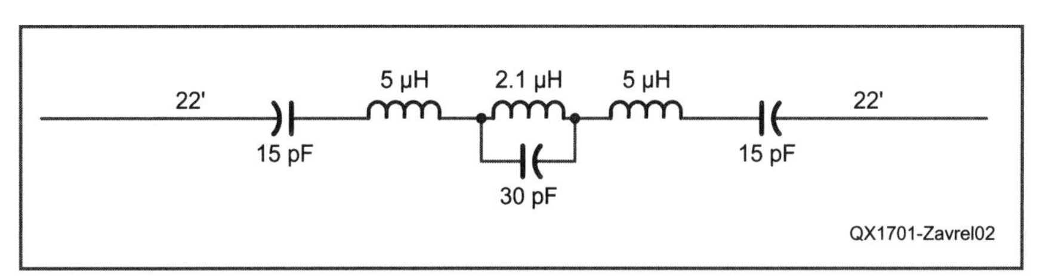


**Figure 8.16 — Parasitic reflector for 20, 15, and 10 meters.**

**Table 8.1**
**Overall Performance Summary**

| *Frequency (MHz)* | *Free space Gain (dBd)* | *Equivalent Yagi Elements* | *Feed point Impedance* | *VSWR on 450 Ω Line* |
|---|---|---|---|---|
| 7.1 | –0.18 | 1 | 17 –*j*472 | 54 |
| 10.1 | +0.26 | 1 | 46 –*j*153 | 10.7 |
| 14.1 | +4.4 | 2 | 46 +*j*239 | 12.7 |
| 18.1 | +4.88 | 2 | 77 +*j*700 | 20.1 |
| 21.1 | +4.66 | 2 | 1978 +*j*1525 | 7.0 |
| 24.9 | +6.1 | 3 | 786 –*j*2343 | 17.8 |
| 28.3 | +6.2 | 3 | 174 –*j*704 | 9.1 |

set up the necessary simultaneous equations and/ or a matrix to optimize the L and C values. I did not optimize, but I suspect these values are close enough. The front-to-back performance, in particular, may benefit more by such elaborate optimization efforts.

The first step in setting up the necessary equations is to determine the required reactance value for the given band and the given length of the element. This reactive value is most easily found through emperical modeling. I used *EZNEC* for the tool. The reactive value should be placed at the center of the element. *EZNEC* can then be used to maximize either the forward gain or F/B ratios (usually a compromise between the two). This is not necessary at 12 and 15 meters, since at these frequencies a simple two-element co-linear reflector is the goal. Then the values of the trap and additional reactive values must be determined.

The actual physical lengths of the reactive components must also be considered as the element lengths and reactive values shown here assume zero length for the reactive components.

**Table 8.1** shows the dBd (gain over a 1/2 λ dipole) in free space. The gain is then compared to an equivalent number of elements using a monoband Yagi-Uda. Feed point impedance and the resulting VSWR is also shown for first approximations for the required tuner.

The main disadvantage to this type of array is impedance matching for several bands. However, automatic antenna tuners using on-board computers are becoming common. If the driven element of such an array is driven with open wire line, an auto-tuner could be located at the tower base and provide a perfect 50 ohm match on all operating frequencies. Since the feed impedances are known, specific tuners could be built to get the match "in the ballpark", then an auto-tuner could be used to complete the process to exact matches.

## The Cubical Quad

The Yagi-Uda is a derivative of the linear antenna form: the dipole. Large loops (≅ 1 λ circumference) can be configured in a very similar manner to the Yagi-Uda by feeding only one loop (the driven element) and using parasitic elements slightly longer (reflector) and slightly shorter (director) to create more gain and/or front-to-back ratios. This is the basis of the cubical quad antenna such as the one shown in **Figure 8.17**. The calculations are identical to the Yagi-Uda.

**Figure 8.17 — This 2-element "quad" antenna for 50 MHz uses "cross" shaped "spreaders" to provide mounting for wire loops mounted in a diamond configuration. Quad antennas may also use "X" shaped spreaders with the loops mounted in a square shape. [Pete Rimmel, N8PR, photo]**

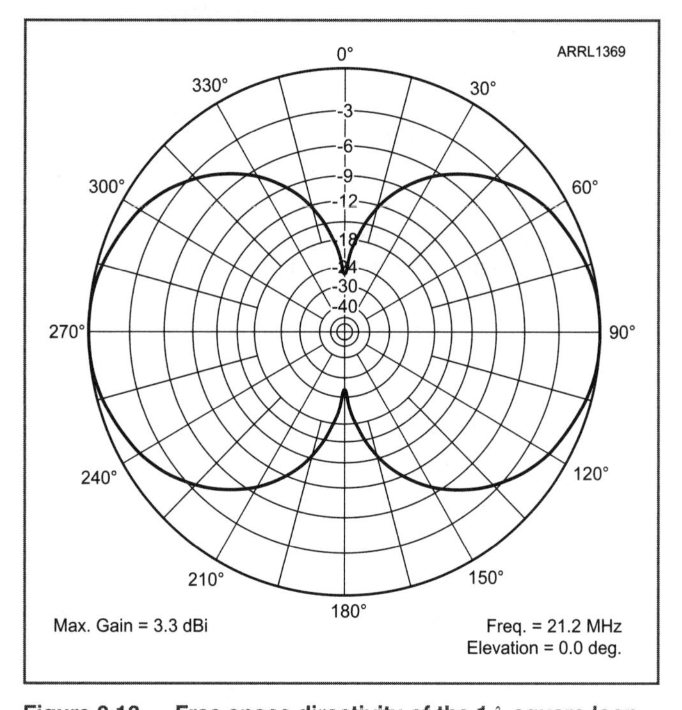


**Figure 8.18 — Free space directivity of the 1 λ square loop "1-element quad" with the plane of the loop represented by the Z-axis "in and out of the page." The gain is about 1.15 dB higher than a ½ λ dipole, or an aperture about 1.3x the dipole.**

**Figure 8.18** shows the directivity of a single 1 λ square loop, which has gain about 1.15 dB higher than a ½ λ dipole. The initial gain advantage of the quad over a dipole begins to diminish as more elements are added. Most models and empirical studies show a 2-element quad outperforms a 2-element Yagi-Uda (for forward gain) by about 1.8 dB, but as the boom length is extended for more elements, the Yagi-Uda shows at least a parity of gain with the quad. This advantage, coupled with the simpler mechanical structure of the Yagi-Uda, has made the Yagi-Uda far more popular. However, the quad has three big advantages in addition to the slight gain advantage of 2-element performance. First, it is much easier to build multifrequency quads. Smaller loops for higher frequencies are simply added to the same spreaders. Inter-band mutual coupling is also less with a quad, and there is no need for traps or very careful design of "interlacing" elements. Second, the quad loops inherently are lower in Q, thus providing a good match over a greater bandwidth than a dipole/Yagi Uda design. Third, a quad "element" may be used as a driven element on a band(s) and a parasitic element on other band(s). This is nearly impossible using a Yagi-Uda.

One final note: Good quality traps are expensive to buy or even build. High quality high voltage "doorknob" capacitors are usually required. Also, quality aluminum tubing is expensive. After the support, rotator system and boom, the quad requires only wire and bamboo poles, versus telescoping aluminum, expensive traps, and very low error tolerance in cutting and setting the aluminum. I built my first quad when I was 16 years old, I could not beat the performance for the required investment. Today I use Yagi-Udas, but would still consider a quad if my requirements changed.

## Quad Parasitic Elements in Vertical Arrays

We can also apply the advantages of scattering apertures to vertical polarization. Using ground mounted vertical elements as parasitic arrays is possible in principle. However, such parasitic arrays are also prone to the same $R_l$ losses as the driven element. Also, the fine tuning required to establish a parasitic element's parameters is difficult to maintain with changes in soil moisture and temperature particularly. This is, again, particularly true over compromised ground systems. Recalling our discussion from Chapter 5 on parasitic arrays, a parasitic element relies completely upon its mutual impedance with the driven element(s) to establish the necessary phase relationship for optimum gain in the desired azimuth direction(s). In contrast, a driven array's element phase relationships are primarily determined by the *driven* phase relationships, which are much easier to stabilize in an unstable environment (such as a typical ground-mounted vertical array).

An alternative is to use a quad loop parasitic element (**Figure 8.19**). In principle, a ½ λ parasitic element could be employed, but the element would be considerably higher. We can also bend the parasitic element, perhaps at the top and the bottom, to maintain the high current portion of the element in a vertical orientation. In effect, this forms half of the full quad loop! By completing the loop we add little to the required support, add additional gain, and provide for greater electrical stability (lower Q). **Figure 8.20** shows the gain pattern of a ¼ λ monopole for 7 MHz with a cubical quad loop director.

There are several important notes to this type of configuration.

1) The actual shape of the loop is not critical. We want a response to vertical polarization, so the vertical wires should

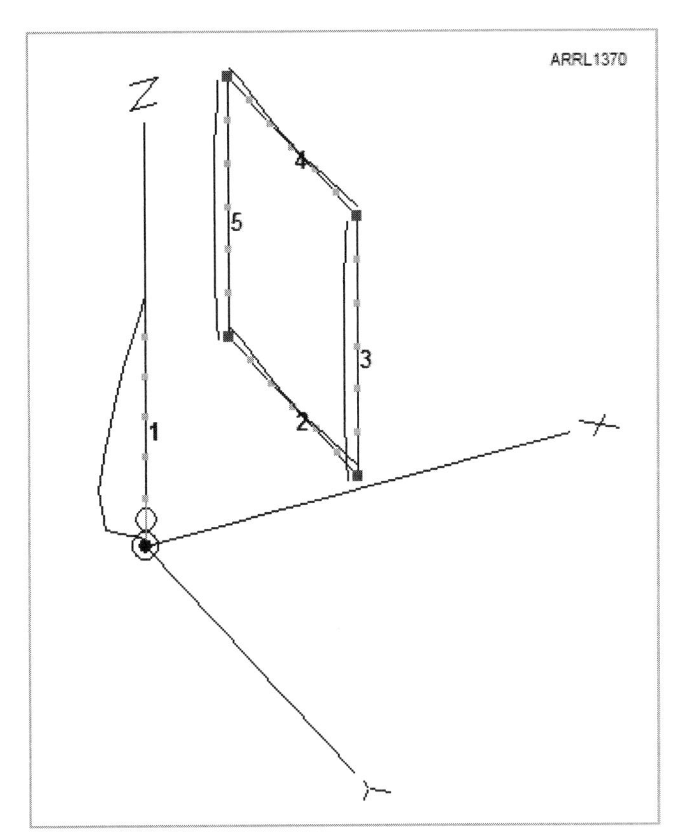


**Figure 8.19 — The configuration of a vertical monopole-quad array. The bottom wire of the quad is about 3 meters high.**

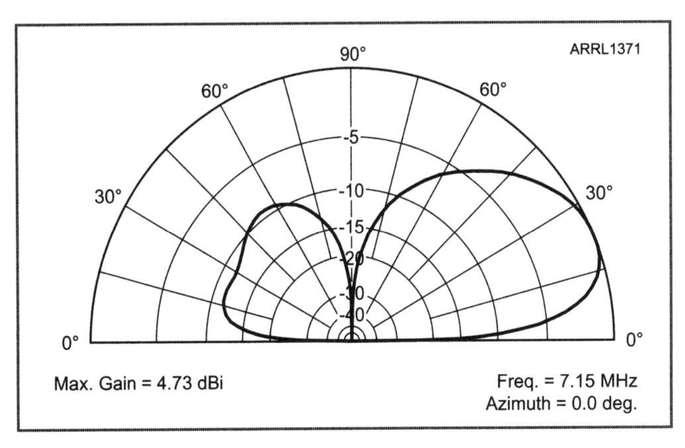


**Figure 8.20 — Gain pattern of a ¼ λ monopole-cubical quad loop parasitic element for 7 MHz. In this case the parasitic loop is a reflector. Notice the gain is approaching 5 dB over the monopole shown in Table 7.3 in Chapter 7. Here again, the parasitic element's scattering theoretical aperture is 4x that of an effective aperture, therefore we realize about 2 dB (single element) more than a 2-element cardioid array.**

be kept as long and "vertical" as possible.

2) We can change the loop from director to reflector by simply adding inductor(s) for the reflector. However, the inductor cannot be placed at the high voltage point (centers of the horizontal wires). They will have *no* effect. Rather, for a square loop, two inductors should be used, one each at the two bottom corners.

3) The lowest wire (wire 2 in Figure 8.19) will exhibit very high voltages and therefore should be kept well out of reach of people and animals *(as should all antennas used for transmission).*

4) The optimum length (circumference) for the parasitic element will be greatly affected by the overall height and the shape of the loop. Modeling will be necessary, but fine tuning can be accomplished by breaking the loop at a corner and resonating to the appropriate frequency above (for a director) or below (for a reflector). However, despite this sensitivity it is more inherently stable and efficient than a ground mounted vertical used as a parasitic element.

5) An inherent problem with any parasitic element is that it remains a parasitic element even when its associated antenna is not being used. For example, the presence of a loop reflector will also dramatically affect a receive antenna located nearby. This undesirable affect can be eliminated by making the loop's total wire length shorter than a full size reflector, about 0.9 λ or less (**Figure 8.21**). The loop is activated by switching in an inductance and closing the loop. The switch that breaks the loop is located adjacent to an inductor, so undesirable loading is avoided. **Figure 8.22** shows one method of creating a variable inductor that can be used to tune the loop.

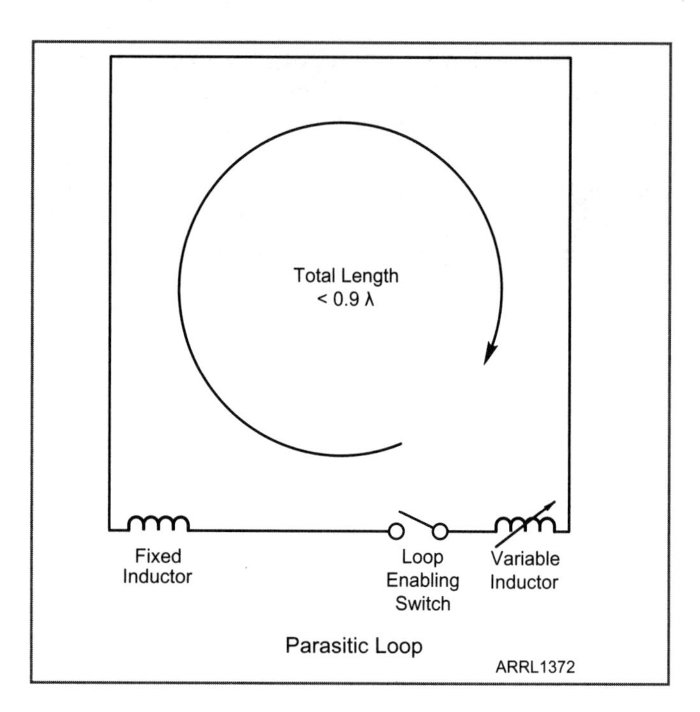


**Figure 8.21 — The loop can be effectively rendered a non-parasitic element by setting the loop length less than about 0.9 λ. Using two near-equal inductors at the bottom corners optimizes the loop for vertical polarization. The switch is placed adjacent to the variable inductor used for fine tuning. An alternative is to use a full-size loop, but two switches are required with careful placement.**

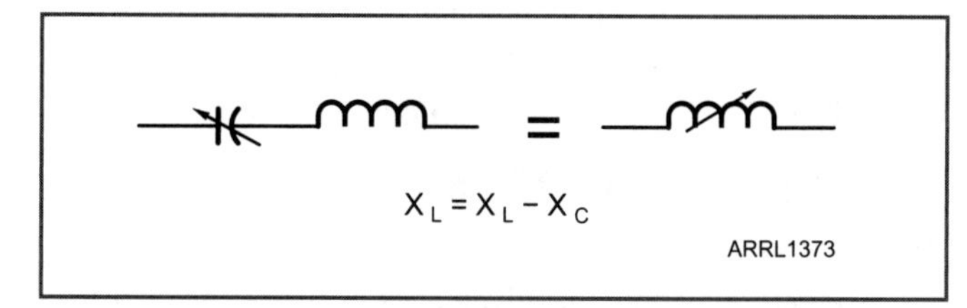


**Figure 8.22 — This circuit shows a convenient method of creating a variable inductor. Variable inductors are expensive, and remote control of such inductors using motors is very expensive. An alternative is to place a suitable variable capacitor in series with a fixed inductor. The variable capacitor is less expensive and far easier for remote motor control, especially since it does not require automatic motor "stops." A variable inductor can rotate indefinitely while the VSWR is observed at the control location.**

# 9 Specialized Antenna Configurations

Antenna design has evolved into many different configurations since the inception of radio technology. Thus far we have considered center-fed linear elements and derivatives, of which the ½ λ dipole is the simplest form. In this chapter we will investigate some of the other important configurations for professional and amateur applications alike.

## Fundamental Small Loop Antenna Characteristics

In Chapter 8 we described the cubical quad antenna. The quad is based upon elements about 1 λ circumference in size, making the quad a "large loop" antenna. Here we will describe the characteristics of "small loop" antennas. The shape of the small loop may be circular, square, triangular or any other shape as shown in **Figure 9.1**. It may also consist of one turn or more, and can range in size (circumference) from about ½ λ to a small fraction of a wavelength.

A loop's characteristics will change significantly as the diameter is reduced from a large loop to a small loop. Here we will treat the small loop in some detail, but will compare it to the large loop to gain insight into how the two configurations create their respective patterns.

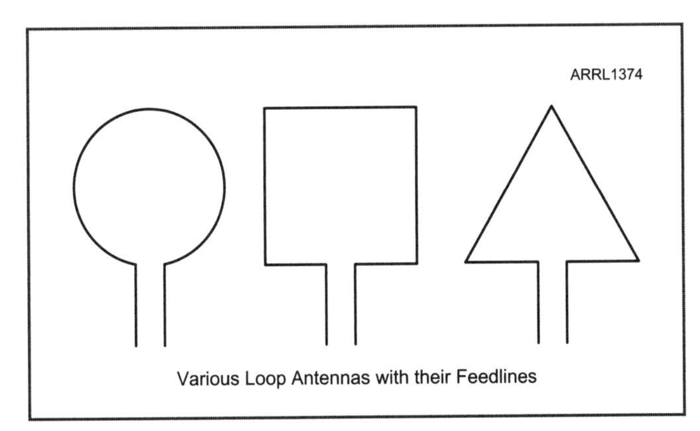


**Figure 9.1 — Various loop antennas with their feed lines.**

Let us begin with a very small loop. If we reduce the loop size to a very small size compared to the wavelength, we create an antenna that is "almost" a shorted transmission line. As expected, this shorting section will exhibit very high current and radiate very little power (very low $R_r$). As we make the loop larger, but still very small compared to the wavelength, the current begins to fall, and more power is radiated. However, since the circumference of the loop remains much shorter than the wavelength, the current value along the loop's conductor will remain essentially uniform as will the phase of the current.

We can once again employ the PADL principle introduced in Chapter 3 to gain an intuitive understanding of how a small loop's pattern is formed. In **Figure 9.2**, since the circumference of the loop is much smaller than the wavelength, both the phase and the amplitude can be assumed to

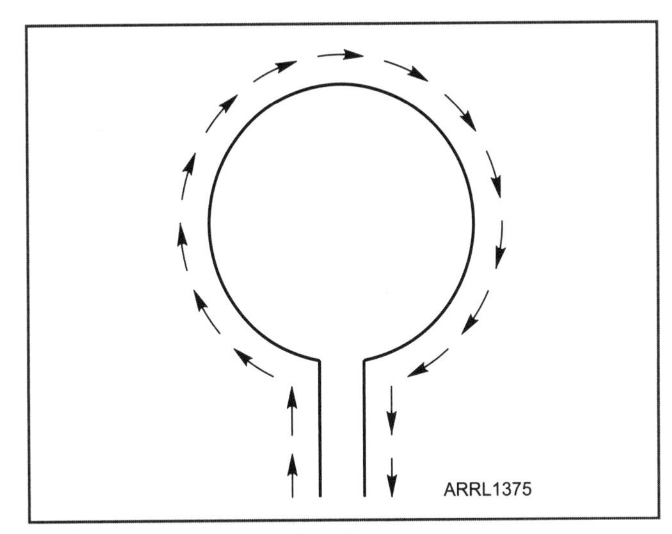


**Figure 9.2 — The RF current flowing around a small loop is effectively equal in amplitude and phase at all points around the loop, thus canceling the field at right angles ("broadside") to the loop.**

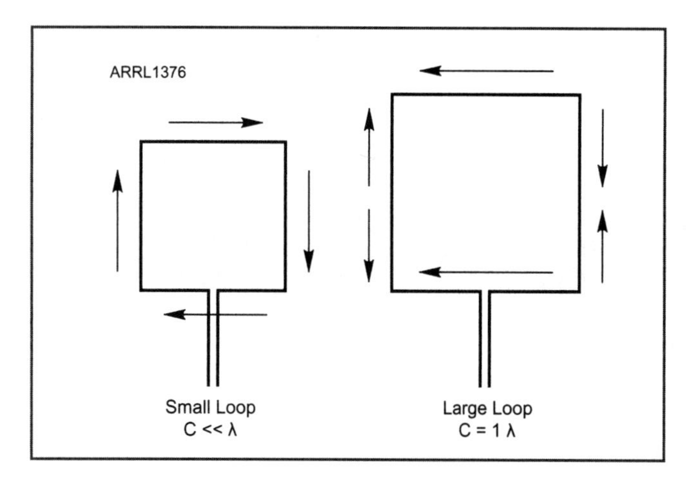


**Figure 9.3 — This figure illustrates the differences between current direction on the small and large loop (quad) antennas. As the loop is made larger, the current is no longer of near-identical phase and the distributed amplitude of the current becomes sinusoidal rather than constant.**

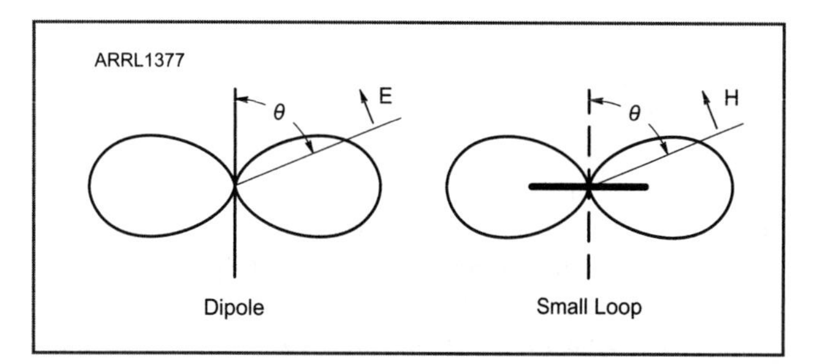


**Figure 9.4 — The patterns from an infinitesimal dipole are identical to that of a very small loop. However, the patterns are 90 degrees different relative to the broadside direction. Also, the *H* and E fields are "swapped." Here a "small" dipole is one less than about 1⁄10 λ in length, while a small loop is about 1⁄10 λ in diameter. The directivity of both small antennas is 1.5, or 1.76 dBi**

be constant around the loop. However, the direction of the current (accelerating charges) is equal *in all directions.* If we view the loop broadside to the plane of the loop as in Figure 9.2, the distance to all sections of the loop is equal from any point in line with the loop's axis. Therefore, there is perfect cancellation in the broadside direction.

In contrast, if we view the loop in Figure 9.2 from its side (from the right, left, up or down from the loop), one critical PADL parameter has changed. Again if we look at Figure 9.2, but this time from a distant point to the "left" or "right" of the loop, there is a slight difference from the distant point source to the two sides of the loop (the distance of the loop diameter). This difference is enough to provide a slight difference in the phase, and thus provides a slight non-cancellation of the fields, and thus some radiation maximum around the radius of the loop — but again, complete cancellation broadside to the loop. The very slight phase response normal to the loop's axis results in small loop efficiencies that are *very* low. **Figure 9.3** shows the current direction on small loop antennas versus large loop antennas.

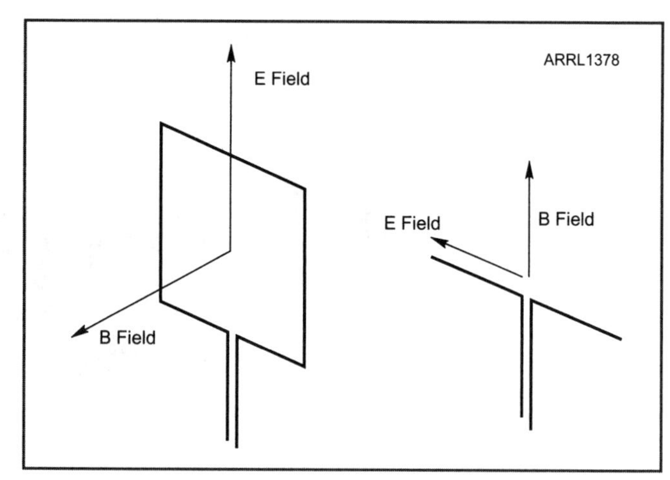


**Figure 9.5 — Comparison of the *E* and B fields in the far field of a small loop and a small dipole. Notice that a horizontal loop is vertically polarized and a horizontal dipole is horizontally polarized. Since AM broadcast ground wave signals are always vertically polarized, loop antennas in AM receivers are always mounted horizontally. A horizontally polarized small loop requires the axis of the loop to be mounted vertically to the ground, such as a ferrite core rod mounted vertically rather than horizontally.**

**Figure 9.4** shows the resulting figure 8 pattern with the null broadside to the loop. This is completely opposite the pattern of a large loop 1 λ in circumference, where the nulls are in the plane of the loop and maximum directivity is broadside to the loop. **Figure 9.5** compares the *E* and *B* fields of a small loop and a small dipole.

As we might expect from any antenna that is small relative to the operating wavelength, small loops tend to be rather inefficient. Indeed, the common expression for antenna efficiency also applies to loop antennas where

$$efficiency = \frac{R_r}{R_r + R_l} \qquad \text{(Equation 9.1)}$$

Loop antenna $R_r$ is calculated from the general form, Equation 3.37 in Chapter 3:

$$R_r = \frac{S(\theta,\phi)_{max} r^2 \mathbf{\Omega}_A}{I^2} \qquad \text{(Equation 9.2)}$$

Again where $S(\theta,\phi)_{max} r^2 \Omega_A$ represents power concentrated in the beamwidth defined as $\Omega_A$, and $I^2$ is the current. So, again, this is only a special form of the power law linking resistance to power and current. By integrating the power over the pattern (identical to the small dipole) over the sphere surrounding the loop, we find

$$R_r = 31,200\left(\frac{A}{\lambda^2}\right)^2 \qquad \text{(Equation 9.3)}$$

where *A* is the cross-sectional area of the single-turn loop.

Often the circumference of the loop is a more convenient measurement. If we use the circumference rather than the circular area of the loop, then the term *C/*λ becomes a coefficient that is a simple ratio applicable at any frequency, where $C$ is the circumference. Thus

$$R_r = 197\left(\frac{C}{\lambda}\right)^4 \quad \text{(Equation 9.4)}$$

It is strikingly apparent that as the circumference of the loop is made smaller, $R_r$ drops at the fourth power of the ratio! For example, a loop with a circumference of 0.1 λ will have an $R_r$ of about 20 *milliohms*. $C_\lambda{}^4$ is often used rather than $C/\lambda$ to make clear that this is now a coefficient. By increasing the circumference only to 0.15 λ, $R_r$ jumps to about 100 mΩ, a five-fold increase.

Furthermore, the dc resistance of the loop conductor (as in any application) is not adequate for computing $R_l$ in the efficiency equation. *Skin effect* increases the effective RF resistance of the conductor and must be taken into account when calculating loop efficiency, even with 20 milliohms of loop $R_r$.

Aside from using larger diameter wire, there are two techniques for raising the $R_r$ of a loop antenna: use more turns in the loop (as in **Figure 9.6**) and/or wind the loop on a material with higher permeability. These are precisely the same techniques used to increase the inductance of a coil. $R_r$ will increase as the square of the number of turns, so we simply add that coefficient to Equation 9.4.

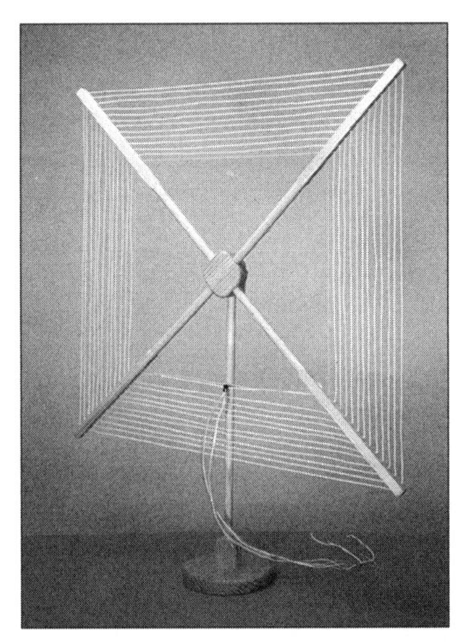

**Figure 9.6 — A small multiturn loop receiving antenna. [Courtesy Peter Jennings, AB6WM]**

**Figure 9.7 — Typical ferrite core antenna for the AM broadcast band as found in an AM/FM portable radio.**

$$R_r = 197n^2\left(\frac{C}{\lambda}\right)^4 \quad \text{(Equation 9.5)}$$

The other technique is to use a high permeability material, usually ferrite or iron core rods on which the coil is wound. Thus we add one last term to the equation

$$R_r = 197n^2\mu_{er}{}^2\left(\frac{C}{\lambda}\right)^4 \quad \text{(Equation 9.6)}$$

Consequently, for small loop antennas, we can increase $R_r$ and thus efficiency by winding many turns on a ferrite rod core, as shown in **Figure 9.7**.

Adding a ferrite core also increases the inductance, which must be tuned out for an effective impedance match. Therefore there is a limit on how big an inductor can be used in a real design. This becomes a practical limit for the number of turns (and efficiency) of small loops. Equation 9.6 makes clear the primary importance of the size of the loop relative to the wavelength.

There are also other issues with using a core material, such as core loss. Notice that we use $\mu_{er}$ rather than the usual specification of $\mu_r$ for ferrite rods. In practice $\mu_{er}$ will be somewhat smaller than $\mu_r$ due to demagnetization effects.

In recent years small loops have begun to be used as transmit antennas as well as receive antennas. In such cases extreme care must be used in design and construction to maximize efficiency to the extent possible given the severe constraints on size. The usual design technique is to use a resonating capacitor in series with the loop inductance to form a resonant circuit. This circuit would form a "short circuit" if not for the $R_L$ and $R_r$ of the loop. $R_r$ can be raised by increasing the size of the loop, and/or adding turns. This has a limitation if the simple LC design is used in that making the loop larger and/or adding turns increases the inductance of the loop. The inductance cannot be made so large as to preclude the use of "reasonable" values lumped capacitors. If "too large" a loop is desired or required, then other feed techniques must be employed.

The inductive near field of a small loop can become very high, thus extreme care must be used when using small loops as transmit antennas. Small transmit loops should never be used close to humans, pets, etc. at any power levels more than QRP.

In contrast, for receive applications at lower frequencies (i.e. $< 10$ MHz), the issues of efficiency can often become a secondary issue compared to directivity. This trade-off will be covered in detail in the next chapter. Once again, the simple equation (Equation 9.7) defines the terms for antenna efficiency.

Small loop antennas require careful design for transmitting due to their inherent low efficiency. However, under many circumstances they are very effective receive antennas. Small loops exhibit very deep nulls, making them useful for direction finding by establishing an azimuth bearing from the antenna. They also provide excellent nulling of undesirable signals and/or interfering noise.

$$efficiency = \frac{R_r}{R_r + R_l}$$ (Equation 9.7)

We are now prepared to calculate the efficiency and thus the gain and effective aperture of a small loop antenna. $R_r$ is a relatively straightforward calculation using Equation 9.6. On the other hand, $R_l$ is a function of several variables, all of which are difficult to predict. However, $R_r + R_l$ is a relatively simple measurement. If we tune the loop to series resonance with a capacitor we will measure a nonreactive total impedance equal to $R_r + R_l$. Then

$$R_l = R_{total} - R_r$$ (Equation 9.8)

The directivity of a small loop is 1.5 or 1.76 dBi. If the loop had no losses, its aperture would be

$$\frac{1.5\lambda^2}{4\pi}$$

Therefore we can calculate the efficiency and thus also gain and aperture by using Equations 9.6 and 9.7.

Finally, if we want to know how efficiency will change with more or fewer turns, we simply use Equation 9.9. We begin with the calculated and measured values of $R_r$ and $R_l$. These nominal values will be designated as $R_{rn}$ and $R_{ln}$. We know that the value of $R_r$ will change at a rate of $n^2$, where $n$ = number of turns, and $R_l$ will increase linearly by $n$. Thus

$$efficiency = \frac{R_{rn}\Delta n^2}{R_{rn}\Delta n^2 + R_{ln}\Delta n}$$ (Equation 9.9)

For example, we have a loop with 10 turns and $R_{rn}$ = 0.0001 Ω and $R_{ln}$ = 1 Ω. The efficiency of this antenna will be about 0.0001%, or 40 dB below 100% efficient (–38.2 dBi). Now we want to know the efficiency if we replace the 10 turns with 100 turns, thus $\Delta n$ = 10. The result is $R_r$ is now 0.01 Ω and the new $R_l$ is 10 Ω. Thus the new efficiency is about 0.001%, or 30 dB below 100% efficient (–28.2 dBi). Therefore, in the case of very low efficiencies where $R_l >> R_r$, efficiency increases nearly linearly with the number of windings.

### Small Loop Antenna Bandwidth

As with most small antennas which are small relative to the wavelength, the feed point of a small loop typically consists of a very low resistance and very high reactance. With a small loop the reactance is inductive therefore to resonate the loop, a capacitor is added thus creating a resonant series LC circuit. As with any LC circuit, the resulting circuit has a Q, where

$$Q = \frac{X}{R_r + R_l}$$

The inductor is usually the limiting factor for Q. Once Q is determined, the bandwidth can be derived through standard circuit theory equations. Clearly, the lower the loss resistance, the higher the Q, and the higher the radiation resistance lower the Q. These simply relationships allow derivation of antenna efficiency by calculating $R_r$ from the above equations and measuring the bandwidth (i.e. with an antenna analyzer). These two terms can then be used to calculate $R_l$ and thus also antenna efficiency.

## Magnetic(?) Loop Antenna

In recent years small loop antennas have become very popular among amateurs. If designed and built properly the small loop can be an effective vertically polarized antenna that can be mounted close to the ground (compared to wavelength) and provide similar efficiency to the more typical ground mounted vertical.

Unfortunately, along with this popularity a nearly unprecedented amount of misleading and false information has been published regarding the loop antenna. This section will begin with a description of what a small loop antenna *is not*, and then proceed to quantify what it *is,* and then offer some basic practical information.

The term "magnetic loop antenna" rather than simply "loop antenna" has crept into the amateur vocabulary. This unfortunate term has led to much misunderstanding of the operation of loop antennas, particularly small (relative to wavelength) antennas. Many otherwise very competent RF engineers and amateurs now believe that a small loop "only responds to the magnetic field" of an electromagnetic radiated radio wave. In reality, loop antennas behave exactly the same as *any* other antenna type. No receive antenna can "strip" either the magnetic or electric field away from the wave. *All* antennas must conform to Maxwell's Equations and thus must behave in essentially the same manner.

In Chapter 3 we discussed the differences among various types of fields. Part of that discussion dealt with RF non-radiating fields. A small loop antenna is obviously an inductor and small inductors do indeed create strong *non-radiating* magnetic "induction" fields in and around them. Small loops therefore, are also very effective at detecting magnetic induction fields for probing non-radiating magnetic fields in and around circuits and wires by inductively coupling to the inductive source of the field. Tank circuits in power amplifier output circuits also create strong magnetic fields, but are poor radiators. The small radiation that does occur from such applications is very small since there is very little "space" for the comparatively long wavelength fields to detach themselves from their source.

Induction fields from very close "noise sources" can be shielded from loop antenna elements by using coaxial shielded conductors on loops. These shields do not block the *radiated* wave, which is the usual goal for a receiving antenna but do offer rejection of undesired induction fields from electric motors, lights and other electric devices.

### Common myths about loop antennas:

• *The magnetic loop antenna only responds to the magnetic field of the radio wave.* If this were true, what happens to the proposed unaffected electric field? An electromagnetic wave's energy is split evenly between the electric and magnetic fields. If this statement were true, then *all* correct aperture models of loop antennas (over the past century) would be wrong by a factor of 2 since 1/2 of the power in an electromagnetic wave is in the **E** and the other half in the **B** waves. Also, what happens to the electric field? It would produce, if possible, a complementary magnetic field as it somehow passed the antenna without being affected, again, containing 1/2 of the power that should have been accepted by the antenna. If a small loop only responds to the magnetic field, then through reciprocity it must also only radiate only a magnetic field. This, of course, is impossible. A magnetic field cannot, by itself, detach itself from its source without violating Maxwell's equations. It cannot "radiate" away from the loop by itself and then Maxwell's equations apply and a "normal" wave appears.

• *Since magnetic loop antennas can be modeled by using magnetic dipoles this must signify it is the magnetic dipole that is the mechanism for the performance of a loop antenna.* Magnetic dipoles do not, in fact, exist. They are used to model loop antennas only as a mathematical convenience. Dipole antennas use "electric dipole" models and a theoretical equivalent is the magnetic dipole. Electric dipoles do exist, magnetic dipoles do not exist. This statement follows the obvious fact that you can only have an electric current (movement of electric charges). There is no such equivalent to magnetic current (movement of magnets in/on conductors?). Both dipoles and loops behave exactly the same. Both radiate and receive both fields, Maxwell explicitly states (through his equations) you cannot have one field without the other in a radiating wave and/or in a radio wave interacting with a conductor (antenna).

• *The near field of a loop antenna is dominated by a magnetic field, therefore a loop antenna radiates a magnetic field.* Again, the magnetic induction field dominates the near field because the loop is an inductor, both in the receive and transmit modes. It also re-radiates a *radiated* field by virtue of its scattering aperture as discussed in Chapter 2. A similar situation exists on short dipoles and short vertical antennas. There is a very strong electric induction near field, but no one seems to say dipoles only respond to electric fields! For this point, we should remember that near fields are dominated by *reactive* fields and radiated fields by fields with *real* power (**W = E × H**).

*No* major textbook on antennas (Kraus, Balanis, LaPort, Schellkunoff/Friis, Terman, Stutzman/Thiele and every other credible source) ever states any of the above and *none* ever use the term "magnetic loop." The best explanation of this confusion comes from *Fundamentals of Electric Waves* by Hugh Hildreth Skilling, John Wiley & Sons, second edition, 1948, 9th printing, 1960.

I will directly quote Skilling's description from page 190:

"A question that very commonly arises in reference to receiving antennas is: Is the antenna voltage produced by the electric field of the passing wave, or the magnetic field, or both? This is a natural question, but the answer is clear when it is considered that *anywhere in space* the electric field is the result of a changing magnetic field. The electric field induced in an antenna is likewise the result of a changing magnetic field, and whether one wishes to consider the electromotive force as the integral of the electric field of the wave in free space (which it is) or as produced by the change of magnetic field (which it also is) is immaterial. The above question is analogous to asking whether a cork rising on the crest of a water wave is lifted by increasing pressure or by the higher water level: in wave motion there cannot be one without the other."

Through reciprocity, the transmit case is similar; the RF current flowing on a transmit antenna creates both an electric and magnetic field, using Skilling's bouncing cork as an analogy. It should also be pointed out that any antenna's radiation and/or reception of radio waves can be explained exactly by solving for the electric *or* magnetic wave

## Calculation of Efficiency of Small Loops Using *EZNEC*

The extremely low $R_r$ values of small loops require that the loss resistance be made as low as possible to achieve any worthwhile efficiency for transmit applications. Consequently, large conductors (usually copper pipe) are commonly used for the loop element. This is the same technique used for mobile whip antennas for low frequencies (i.e. 80 or 40 meters). Although the above equations provide an easy method to calculate $R_r$, these equations assume a circular loop. Small circular loops are difficult to model on *EZNEC* since each wire in the loop must have a minimum of one segment resulting in segments that are too small. Therefore, it is necessary to substitute the circular loop for a square loop using *EZNEC*. This can provide a reasonable estimate for $R_r$ and then also the average gain number. The use of a square loop (using *EZNEC*), with sides about 1.5 times the diameter of a circular loop, provide comparable $R_r$ values (calculated $R_r$ for the circular loop and measured values (*EZNEC*, Src Dat) for the square loop.

Using *EZNEC*, $R_r$ values for the square loop are taken with wire loss set to Zero. The resistive part of the feed point impedance (Src Dat) will give the $R_r$ of the square loop. Switching to copper for the wire loss, the resistive part of the feed point will increase, now $(R_r + R_l)$. You can now calculate the efficiency of the antenna using the efficiency equation or simply take the average gain number from *EZNEC*. The two values will be very close.

In Chapter 4 we introduced a simple technique for deriving the radiation leakage of an open wire transmission line. In this case, the process is much easier in that we can read the antenna efficiency *directly* using the "average gain" feature in *EZNEC*. To access this feature, you simply select "3D" for the Plot type, as described in Chapter 4. To eliminate the effects of ground, you use the "free space" option for "Ground Type"

The graph in **Figure 9.11** shows that the efficiency of a loop is quite satisfactory for 1/2 λ circumference using even smaller (i.e. #14 or 16 copper wire), and even higher efficiencies are implied for full-wave circumference quad loops. However, as the circumference decreases to less than about 0.2 λ, choice of conductor diameter becomes critical. A single-turn loop becomes almost useless at circumferences less than about 0.1 λ no matter what the size of the loop's conductor. For receive applications, however, –30 or even –40 dB losses are acceptable, especially at low band frequencies, especially when using high impedance amplifiers. Very small multiple winding "loop stick" ferrite core antennas with losses exceeding –30 dB are adequate for AM broadcast receivers where relatively low receiver sensitivity is adequate for even small signal applications. Note that if you extend the graph at $C/\lambda = 1$, you have a full-wave loop, or a "quad" element.

## Larger Small Loops

The coefficient

$$\left(\frac{C}{\lambda}\right)^4$$

from Equation 9.6 is a great incentive to build larger loops for greater efficiency. As we will see, efficiency for weak signal reception is critical but other factors come into play. As operating frequencies decrease, the usual trend is for both natural and man-made noise to increase. On the other hand, the desired signal strength is often considerably higher than even the high noise levels at MF frequencies. Because of the high noise levels at these frequencies, the most important parameter becomes the received signal-to-noise ratio (SNR).

Many types of receiving loops have evolved over the years. The K9AY loop shown in **Figure 9.8** provides excellent over-all performance for its physical size.

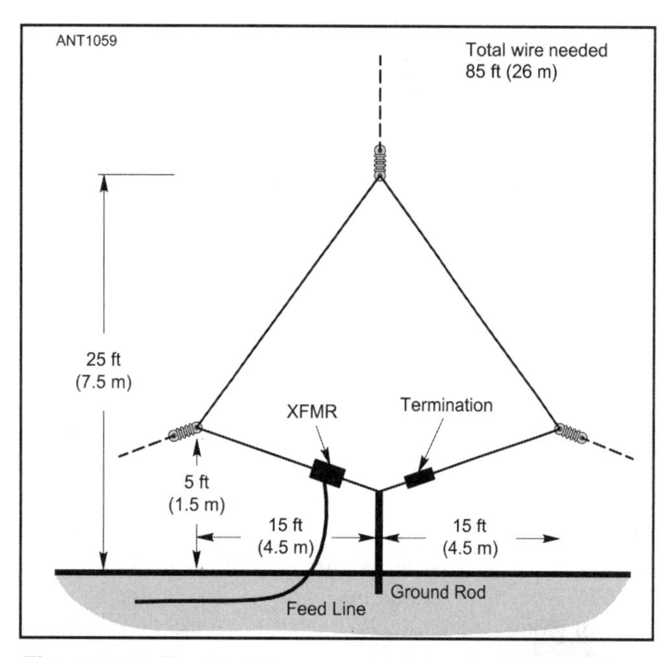


**Figure 9.8 — The K9AY loop provides a mono-directional pattern (in this case to the left). Owing to its modest size and support system, multiple loops can be installed and each can be switched to "front and back" directions for full-azimuth coverage.**

The directivities of the Beverage as well as a short monopole are shown here for easy comparison to the K9AY loop (**Figures 9.9** to **9.11**). Together with the plots in Chapter 7, a comprehensive set of patterns for MF and low HF are presented. The interpretation of these plots requires some knowledge of noise in systems, and we will explore noise in Chapter 10. In addition, we will present a discussion on alternative methods of defining antenna parameters at a system level. These techniques will be shown to be critical for understanding the applicability of small loop antennas to frequencies about ≤ 5 MHz.

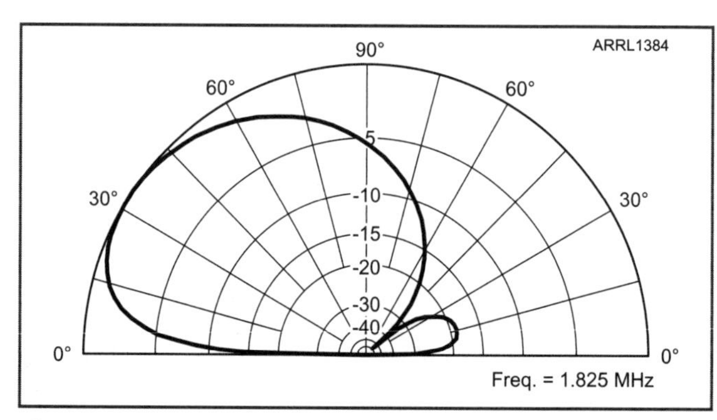


**Figure 9.9 — Directivity of the K9AY loop (elevation angle).**

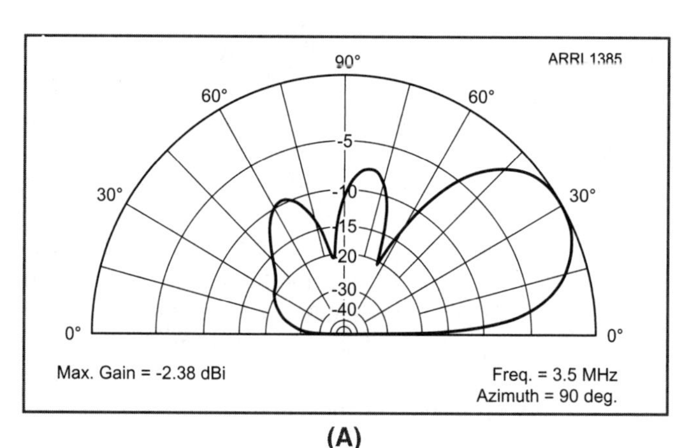


**(A)**

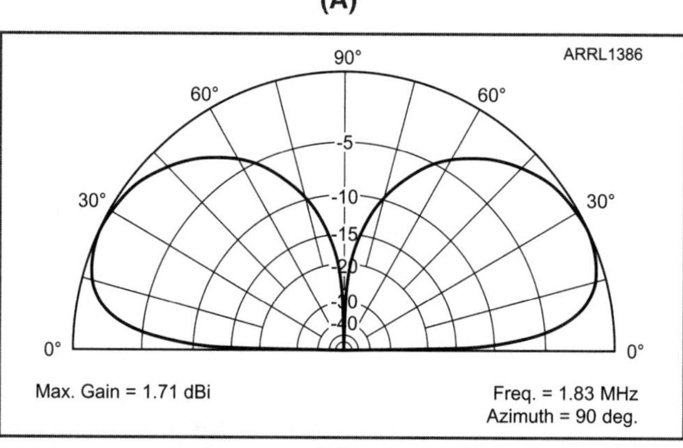


**(B)**

**Figure 9.10 — At A, directivity of a 2 λ Beverage (elevation angle). At B, directivity of a short ground-mounted vertical (elevation angle).**

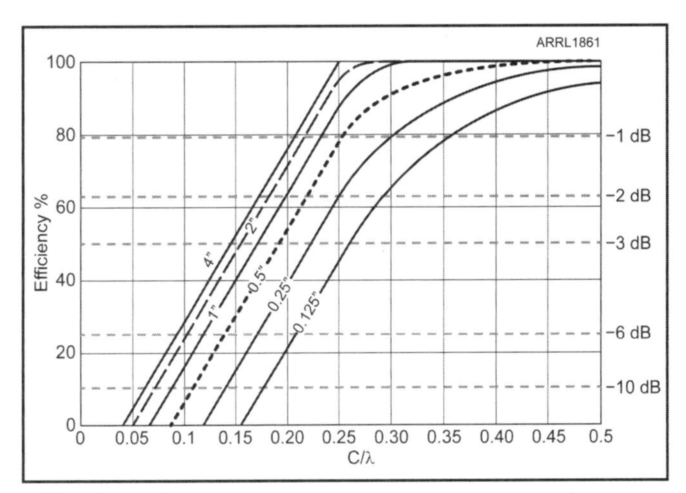


**Figure 9.11 — This graph plots efficiency of square loop antennas (in numerical % and –dB compared to a lossless loop) vs. the circumference of the loop (4 × side length)/wavelength, (C/λ) for six different diameters of copper: (4, 2, 1, 0.5, 0.25, 0.125 inches). These numbers are a good approximation for loops in the HF and low VHF bands.**

## Helix Antennas

As a small multiturn loop antenna's size is made larger (relative to the wavelength), other directivity characteristics emerge, creating a *helix* antenna. In 1946, John Kraus was the first to discover the remarkable characteristics of the helix (**Figure 9.12**). The gain of the helix increased as the length (number of turns) increased. It is inherently broadbanded and is relatively easy to match to common 50 or 75 Ω feed lines. It is also circularly polarized, making it preferable for a wide range of application. It's particularly useful for space applications, where "horizontal" and "vertical" polarizations become meaningless.

**Figure 9.12 — John Kraus, W8JK, was the inventor of the W8JK antenna, the helix antenna, and the corner reflector. Dr Kraus made substantial contributions to antenna design and radio astronomy, and he was professor emeritus at Ohio State University. Here he is shown with a massive array of helix antennas used for early radio astronomy research. The photo dates from 1956. [Courtesy of Ohio State University]**

The helix is an excellent antenna with which to visualize the subtleties of circular polarization. If we imagine the helix like threads of a screw, we find the "left hand" and "right hand" circular polarization depends on the direction of the "threading."

For example, when we look down on a standard thread screw, we rotate it clockwise to tighten and counterclockwise to loosen. If we have a threaded rod, we can turn the rod to the other end and the same directions apply to tightening and loosening. Helix antennas are similar in that no matter which direction you view them from, they are either "left" or "right" handed. A reverse-threaded screw will tighten when turned in a counterclockwise direction, and thus has reverse threading. It is a common mistake to assume a right-hand helix will become a left-hand helix if it is turned over. The "threading" does not change by flipping it over.

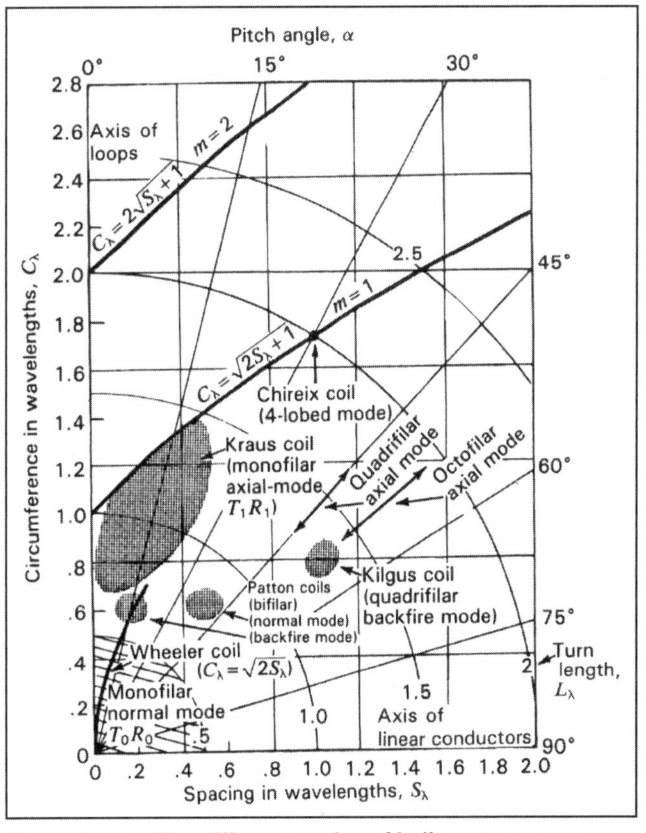


**Figure 9.13 — The different modes of helix antennas as a function of loop circumference and the spacing of the windings. The length of each turn and the pitch angle of each turn are also shown but are simply functions of the other two parameters. Notice that for small, single wire helixes, the helix is operating in the "normal mode." In effect it is a small loop antenna. As the helix is made larger (in particular a larger loop diameter) the currents flowing around a single turn are no longer in-phase, resulting in directivity off the ends of the helix — more like a quad antenna. However, in the helix the current forms a continuous sinusoidal distribution rather than the closed form of a quad. This results in circular polarization instead of the quad's linear polarization. [From John D. Kraus, Antennas, 2nd ed, 1988, reprinted with permission of McGraw-Hill Education]**

## Small Helix Antenna

A small helix antenna may be considered identical to a small loop antenna of multiple turns. When the loop circumference is less than about 0.5 λ, the resulting antenna responds as a small loop, the maximum directivity being normal to the axis of the loop. This is defined as a "normal mode" helix. As the loop is made larger, it begins to show a broadside pattern to the loop, as does the 1 λ circumference quad. However, the characteristics of the antenna become very interesting if we treat the one-element quad as the first winding of a helix, and begin to add more windings. This is the basis of the helix architecture. **Figure 9.13** shows the different modes of helix antennas.

As **Figure 9.14** shows, the gain increases and the beamwidth decreases with more helix turns. There is also a decrease in bandwidth as the array becomes longer.

When a circular polarized (CP) wave is reflected, its "hand" changes. For example, if you use a right-handed helix to drive a parabolic surface, the transmitted wave off the parabola will be left-handed. Similarly, a left hand CP wave reflected off the Moon's surface will return as a right hand polarized wave. Sometimes this is a desirable effect. For example, GNSS navigation signals use right hand polarization. A major problem with GNSS accuracy is ground reflections that distort the time of arrival to the receiver from the satellite. CP antenna inherently reject opposite polarized signals, so they will attenuate ground reflections greatly. However, for ultra-high performance GNSS, even the slightest ground reflected signal can cause location errors. No "real" antenna is perfect, and therefore a right-hand CP antenna will never have "infinite" rejection of a left-hand signal.

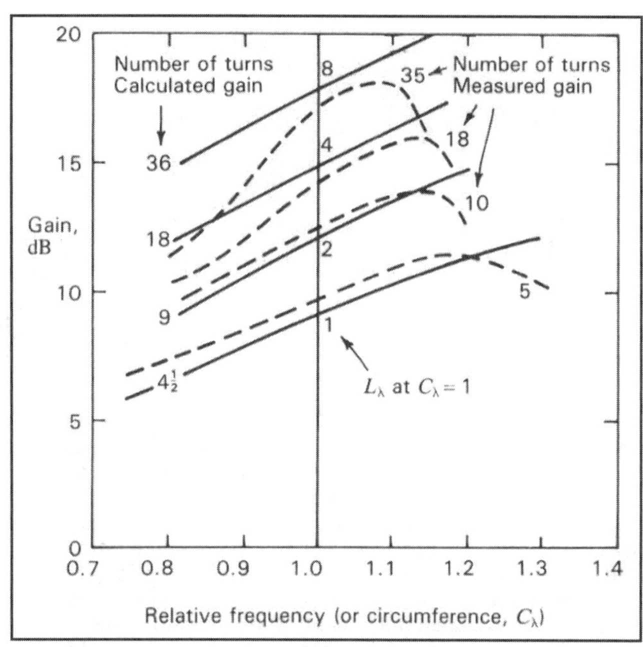


**Figure 9.14 — Theoretical gain (dBi, solid lines) and measured gain (dotted lines) of a helix antenna as a function of the circumference of the windings. For reference when $C_\lambda \approx 1$ the loop is about the same size as a quad element at the same frequency. [From John D. Kraus, Antennas, 2nd ed, 1988, reprinted with permission of McGraw-Hill Education]**

## Aperture Antennas

The parabolic antenna and corner reflector antennas are often termed "aperture antennas." The reflecting surfaces of such antennas are not strictly "part" of the antenna, but rather form passive reflection rather than taking advantage of mutual coupling as in a true parasitic element.

The parabolic antenna derives its name from the fact that the surface is shaped into a parabolic curve. Parabolic surfaces will focus a plane wave onto a "focal point" where a true "antenna" is mounted to receive or transmit power from or toward the parabolic surface (**Figure 9.15**).

The antenna mounted at the focal point ideally will have a directivity that results in a radiation pattern that uniformly coincides with the surface of the parabolic reflector. Too

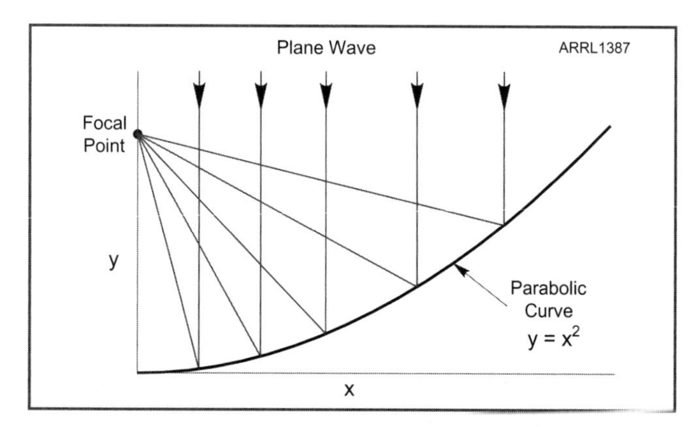


**Figure 9.15 — The principle of a parabolic reflector. A plane wave striking the surface parallel to the Y-axis (the boresight of the antenna) will be reflected to a "focal point."**

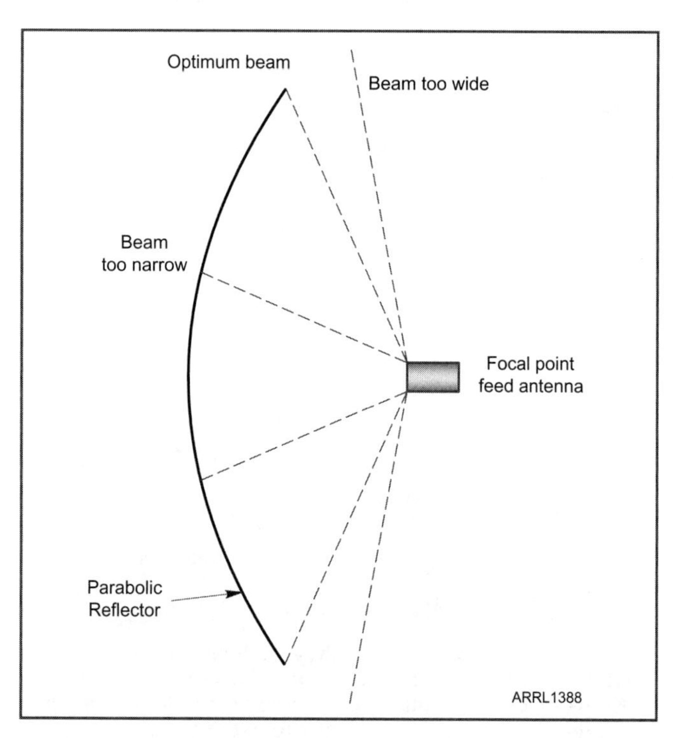


**Figure 9.16 — Feed point directivity of a parabolic antenna. It is critical for the focal point feed antenna to properly illuminate the surface of the parabolic reflector.**

much gain will "waste" some of the parabola's surface area (aperture), while too little gain will expose the antenna's response to directions outside the parabola's response. See **Figure 9.16**.

The aperture of a parabolic antenna is perhaps the simplest aperture calculation of all antenna forms. For a perfect parabolic antenna $A_e$ = cross-sectional area. For example, if the diameter of a circular "dish" antenna is 10 meters, then $A_e$ = 78.5 m$^2$. At 1 meter wavelength, such an antenna will provide a gain of 986.5 or about 30 dBi. A unique characteristic of the parabolic "dish" is that the aperture is *constant* for all wavelengths. However, since gain is a function of both aperture and wavelength, the gain (dBi) of a parabolic antenna changes as the square of the wavelength. **Figure 9.17** shows the Cassegrain antenna, a variation of the parabolic dish.

The corner reflector configuration (**Figure 9.18**) has found wide applications, particularly at VHF and UHF. Construction is easier than a parabolic structure in that it is easier to build a planar reflecting surface rather than one formed as a parabola. Its advantages are also significant for linear polarization as the reflector need not be "solid" but can be constructed of linear elements in the plane of desired polarization.

The feed antenna of a corner reflector is not as critical as the feed antenna for a parabolic antenna. The corner reflector's driven element is typically located "inside" the reflecting surface, thus a simple dipole can be used without the concern illustrated in Figure 9.16.

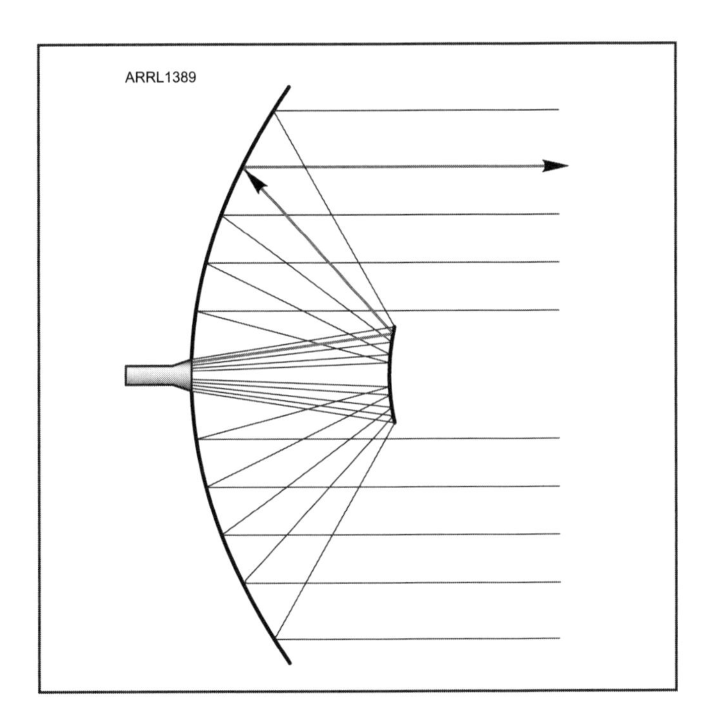


**Figure 9.17 — The Cassegrain antenna consists of two reflecting surfaces — the primary parabolic reflector and a hyperbolic-shaped second reflector. The Cassegrain has the advantage of placing the focal point antenna at or "below" the primary reflecting surface. Large feed antennas and other equipment are often large, heavy, and bulky — complicating the mechanical support of the feed and/or "shading" or scattering the desired incoming waves.**

On the other hand, the maximum gain available from a corner reflector is limited to about 13 dBd, as shown in **Figure 9.19**. This is due to the fact that there is no "focal point" in a corner reflector antenna, setting a limit on the potential aperture (**Figure 9.20**). Therefore if more gain is required, then a parabola antenna is necessary where the aperture, in principle, can be expanded indefinitely

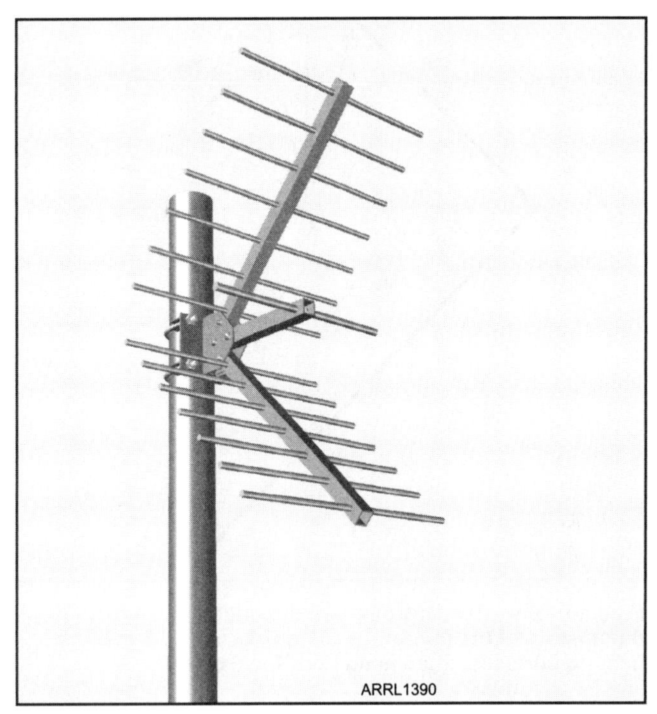


**Figure 9.18 — A horizontally polarized corner reflector.**

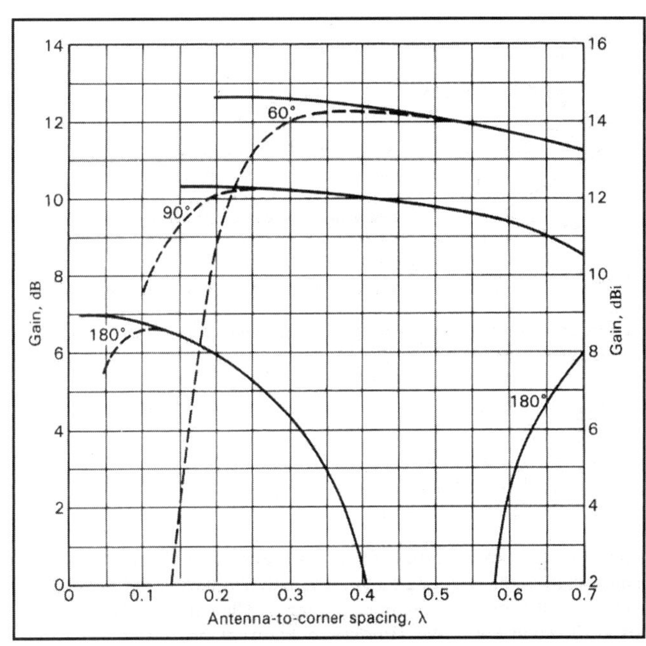


**Figure 9.19 — Possible gain from a corner reflector antenna as a function of the distance between the driven dipole and the corner of the reflector and the angle formed by the reflecting surface. Here 180 degrees is a "flat" reflecting surface. [From John D. Kraus, Antennas, 2nd ed, 1988, reprinted with permission of McGraw-Hill Education]**

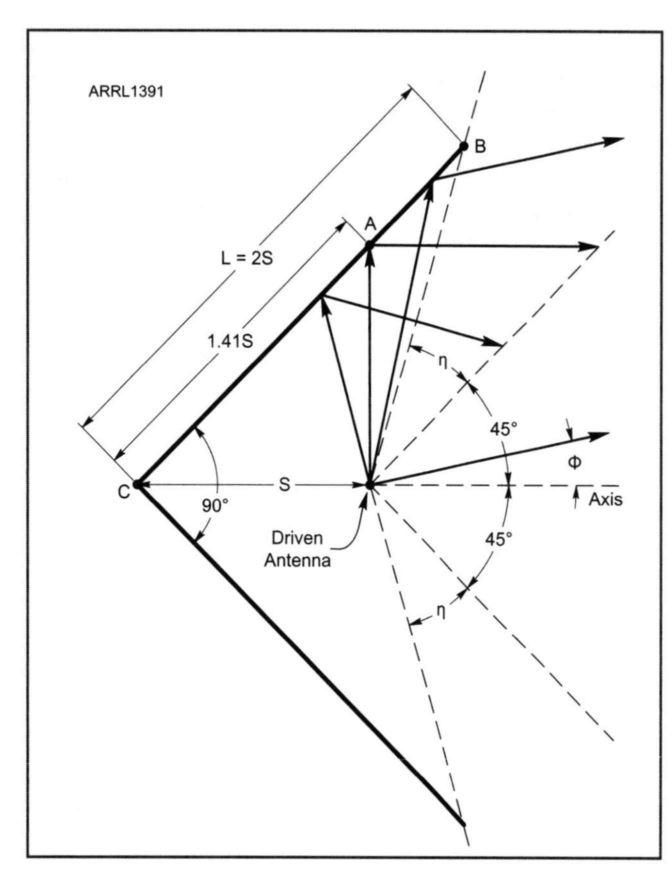


**Figure 9.20 — This figure shows the critical dimensional parameters of the corner reflector. Because of the flat reflecting surfaces, there is no true "focal point."**

The corner reflector can be modeled using the mutual impedance methodology introduced in Chapter 8.

## Small Antennas

In Chapter 1 we discussed the general trend in wireless communication toward higher frequencies and smaller form factors (size of device). Design problems arise when the physical dimensions of the form factor become small with respect to the wavelength of the wireless device(s). We also discussed how antenna efficiency relates to antenna gain and how both relate to the physical size of the antenna. Therefore, a fundamental question arises: How small can an antenna be made and still be effective?

In Chapters 2 and 3, some points we made hint at some physical limitations on antenna size. In particular, as an antenna size shrinks relative to the wavelength, $R_r$ also begins to drop. One may recall from ac circuit theory that efficient LC matching circuits require greater Q to match higher ratios of impedances. This also holds for matching from the typical 50 or 75 Ω transmission lines to the resistive portion of a small antenna's feed point impedance which is typically very low. A simplified equation for the necessary Q of an LC matching circuit is:

$$Q \approx \sqrt{\frac{R_{higher}}{R_{lower}}} \qquad \text{(Equation 9.10)}$$

where $R$ is the real component of the two impedances to be matched.

In this equation, either the source or load real value can be either the higher or lower value. If the load or source has a reactive value, the equation becomes more complicated. However, Equation 9.10 shows how the bandwidth of an antenna becomes narrower as its load resistive value becomes lower.

As the reactive-to-resistive portion of the antenna proper is made larger, the Q of the antenna also rises. This shrinks the bandwidth but also increases the antenna's *stored* power versus the radiated power. In Chapter 4 we described the basic modes of electromagnetic waves as they relate to transmission lines. These modes also begin to play a significant role in the behavior of small antennas, particularly in the near field.

Therefore, one effect of making an antenna small is that higher Qs become necessary and thus the bandwidth of the antenna is reduced. The second major problem is that very low values of $R_r$ tend to drop the antenna efficiency as in Equation 9.1. See **Figure 9.21**.

A "small" antenna can be defined as an antenna whose maximum dimension(s) are confined to a spherical volume which has a radius of

$$kr < 1 \qquad \text{(Equation 9.11)}$$

where

$$k - \frac{2\pi}{\lambda} \qquad \text{(Equation 9.12)}$$

and $r$ is the radius of the sphere, where

$$r < \frac{\lambda}{2\pi} \qquad \text{(Equation 9.13)}$$

McLean quantified the minimum *antenna Q* (assuming zero matching loss) for both linear and circular polarized antennas as:

$$Q_l = \frac{1}{k^3 r^3} + \frac{1}{kr} \qquad \text{(Equation 9.14)}$$

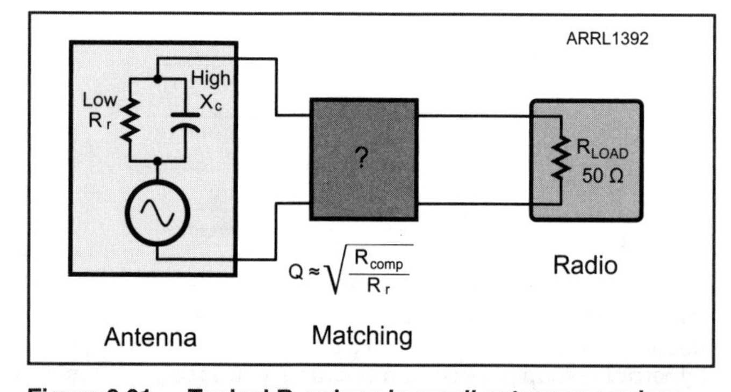


**Figure 9.21 — Typical $R_r$ values in small antennas are low, while the reactive value is typically a high value of capacitance. Complex impedance matching is required to "tune out" the reactive value in addition to matching the two real portions of the impedances.**

for a linear polarized antenna, and

$$Q_c = \frac{1}{2}\left(\frac{1}{k^3r^3} + \frac{2}{kr}\right) \quad \text{(Equation 9.15)}$$

for circular polarized antennas.

For example, when $\lambda = 2$ m, then $k = \pi$. If we set $kr = 1$ (maximum size of a "small" antenna) then $r = 1/\pi$. Therefore a *maximum size* small antenna for the 2 meter band must be confined within a sphere of radius $1/\pi$ meters or less. The maximum size dipole that could fit inside this sphere size would be $2/\pi$ meter (the diameter of the sphere), or 0.64 meter long, while a full-size ½ λ dipole would be 1 meter long. Since $kr = 1$, then by Equation 9.15 the minimum Q of a dipole of this length is 2.

These limitations are derived from the assumption that equal power is excited into the TM and TE modes (Chapter 4) in the immediate vicinity of the antenna. A detailed analysis of these "higher modes" requires an advanced mathematical treatment which can be found in some of the texts referenced in Appendix C.

As the size of the antenna decreases, maintaining efficiency and a reasonable bandwidth becomes increasingly more difficult. As a practical matter, impedance matching also becomes more problematic, which adds to the overall system efficiency. Figure 9.22 dramatizes the challenges facing antenna designers with ever-smaller devices. However, as the operating frequencies have increased (λ decreases) handheld devices can accommodate reasonably efficient antennas.

An intuitive understanding of the small antenna problem can be taken from our discussion on vertical antennas (Chapter 7). As the size of the antenna is reduced (compared to the wavelength), the effective height $h_e$ of the antenna decreases. Again we can see this effect in

$$R_r = \frac{h_e^2 Z_0}{4A_e} \quad \text{(Equation 9.16)}$$

which is the same as Equation 7.9.

In effect, a longer length (and therefore the size) of an antenna distributes the current over a larger distance and thus requires less current to create the required field. As $h_e$ shrinks, larger current values (lower $R_r$) are required to generate the same field strength. The same principle applies for a linear vertical, a planar antenna, and a 3D structure.

Finally the Q of the antenna is a function of its self-impedance. As antennas are made smaller, $R_r$ decreases and thus decreases the damping factor on the increasing capacitive reactance. See **Figure 9.22**. Similar to circuit theory this results in a higher antenna Q, including an increase in stored energy around the antenna.

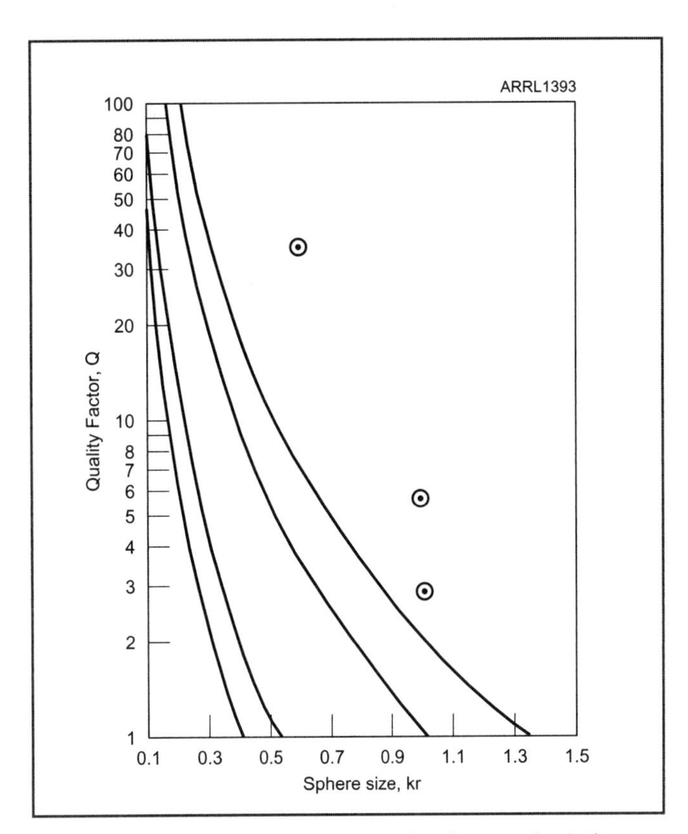


**Figure 9.22 —This figure shows the fundamental relationships among Q (bandwidth), volume of an antenna, and maximum efficiency (ignoring matching losses). [After Balanis; see the References in Appendix C]**

## Traveling Wave Antennas

As we discussed in Chapters 6 and 7, as a vertically polarized wave travels along the Earth's surface, the velocity of the wave nearer the ground is slower. The source of these waves can be a ground-mounted vertical antenna with its wave propagating over the ground (ground wave), or it can be a sky wave arriving from a distant transmitter. This effect results in a tilting of the wave "forward." This forward tilting effect can be used to advantage in creating a traveling wave antenna very close to the Earth. A wire (about 1 – 4 λ in length) placed near the ground surface and laid out in a straight line will receive waves from the direction of the wire. Such an antenna is called a Beverage, after its inventor Harold Beverage, who developed it in the 1920s. See **Figure 9.23**. There are cases where a combination of directivity, along with better efficiency is needed, especially for very weak signal detection in the MF and low shortwave bands. The Beverage will exhibit better efficiency than small loops, but at the expense of laying out a very long wire in the desired direction(s).

The traveling wave antenna differs from most other antenna types in that the waves travel in only one direction along the wire due to the presence of the terminating resistor. In effect, received signals from the unwanted direction are simply dissipated as heat in the terminating resistor. There is no reflection of the signal as at the element ends of dipoles and verticals. Therefore, the RMS voltage level will be constant at all points along the Beverage.

Another important type of traveling wave antenna is the rhombic. The rhombic is based upon directive patterns obtained from wires about 1 λ or longer. We have already described the patterns of shorter wires, including the extended double Zepp, where broadside gain is maximum. As the wire

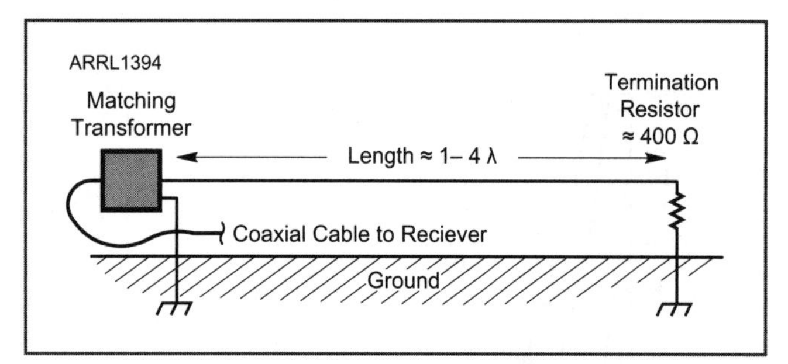


**Figure 9.23 — A typical Beverage receive antenna. Vertical polarized waves (with some small horizontal component) arriving from the right side induce a self-reinforcing voltage that "travels" along the wire from right to left. Finally it is matched into a lower impedance to be fed to the receiver. Waves arriving from the left side also form a reinforcing signal that travels to the right, where waves arriving from the left are dissipated as heat in the terminating resistor. The result is a mono-directional, vertically polarized antenna. Efficiency and directivity are increased by making the Beverage longer. Height of the wire can actually be zero (laid on the ground) to about 3 meters, although efficiency is also increased by elevating it above ground. The main advantage of the Beverage compared to a loop is its higher efficiency and thus gain at the cost of much larger size.**

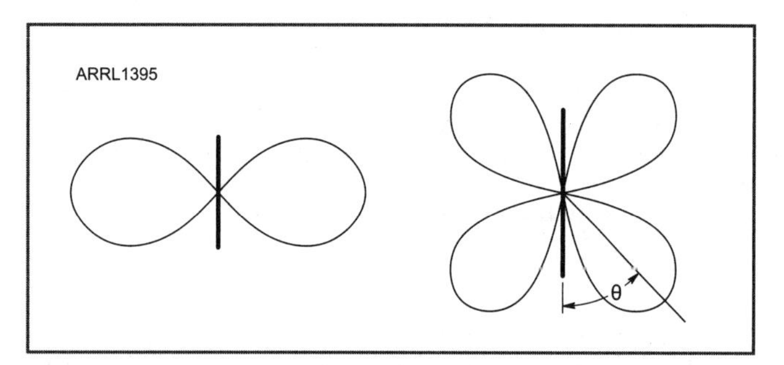


**Figure 9.24 — The familiar "figure 8" azimuth pattern of a ½ λ wire compared to a "4-leaf clover" pattern of wire greater than about 1.5 λ. As the wire is made longer, then angle θ decreases.**

is made longer, the azimuth pattern quickly changes from a "figure 8" to a "4-leaf clover" pattern (**Figure 9.24** and **Table 9.1**).

The rhombic is constructed from four wires, each longer than about 1 λ at the antenna's lowest operating frequency (**Figure 9.25**). Opposite the feed point is a terminating resistor that performs the identical function as the terminating resistor in the Beverage. Without this terminating resistor, the rhombic would be bidirectional. With the resistor, half the power (from the rear direction) is dissipated in the resistor. Unlike the Beverage, the rhombic efficiency is typically very high. Therefore, if used as a transmitting antenna, the resistor must be able to dissipate half the transmitter's power output. This is a major disadvantage of the transmitting rhombic.

**Table 9.1**
**Angle θ as a Function of the Length of an End-Fed Wire**

| Length of Wire (λ) | Angle θ (°) |
|---|---|
| ½ | 90 |
| 1 | 51 |
| 2 | 34 |
| 3 | 28 |
| 4 | 24 |
| 5 | 21 |
| 10 | 15 |

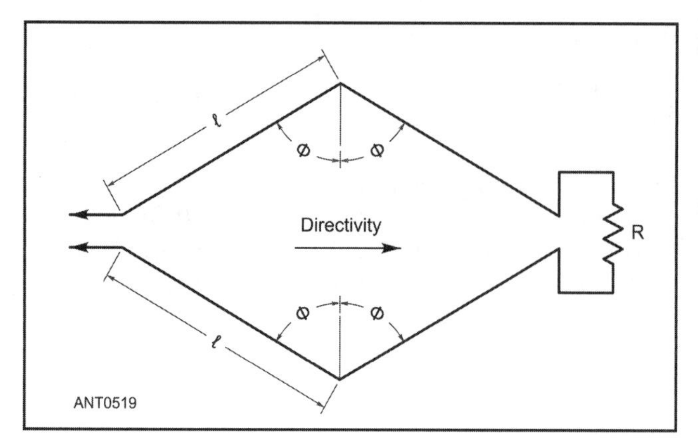


**Figure 9.25 — The layout for a terminated rhombic antenna.**

## Antennas for Navigation and Location Determination

The sharp nulling effects of a small loop can also be used as effective direction-finding and thus navigation aids. By nulling a radio "beacon" of known location, an azimuth line may be drawn through the point location of the beacon and the user's unknown location. With two beacons of sufficient directional difference, two azimuth lines are drawn. The intersection of the two lines yields the unknown user's location, as shown in **Figure 9.26**.

Prior to the advent of GNSS (global navigation satellite system), medium wave (1.6 – 2.0 MHz) and LF wavelengths were used. These wavelengths have the advantages of excellent ground-wave propagation and also small portable loop antennas can provide the sharp nulls necessary to provide a high precision azimuth.

More advanced navigation systems such as LORAN utilized multiple beacons with time (and thus phase) synchronization. These types of systems provided excellent 3D location for the users. These older MF wave systems have been largely replaced by GNSS (global navigation satellite system) which includes GPS (US), GLONASS (Russia), Galileo (European), and others that use microwave signals from satellites. The instantaneous satellite locations are known with high precision (despite their high velocity and altitude). Very precise location determination (within centimeters) is possible using sophisticated algorithms for error correction. However, even these very sophisticated satellite systems rely upon simple position and timing for these calculations.

## Patch Antennas

With the nearly universal use of PC boards in radio construction and the general trend toward higher operating frequencies, it was only inevitable that antennas would soon be fabricated onto the PC board. Also, low profile antennas suitable for working off a ground plane were obvious choices for

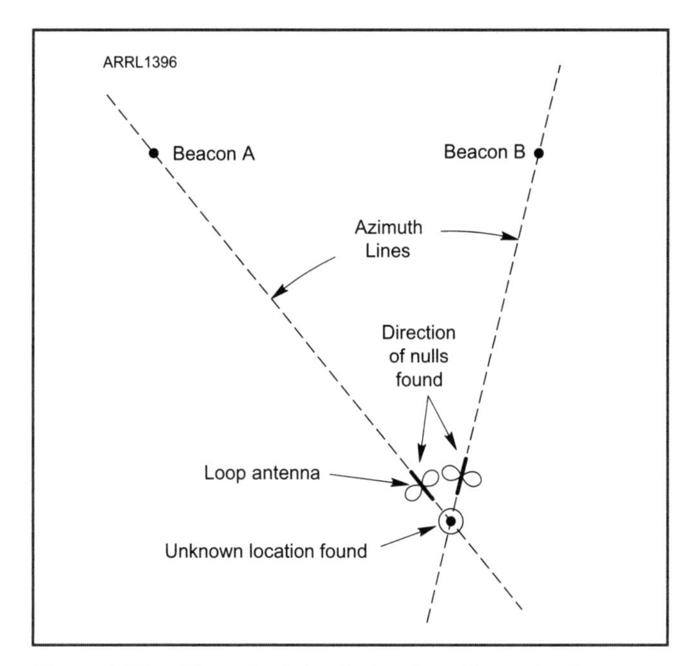


**Figure 9.26 — The principle of a basic radio navigation or location-finding system. If the locations of the beacons are known, two azimuth lines can be found through antenna nulling when using small loop antennas. The intersection of these lines yields the unknown user's location. On the other hand if a radio transmitter's location is unknown, two separate users' points finding two corresponding azimuths will yield the unknown transmitter's (beacon) location.**

aircraft and space applications. In response, a wide variety of antenna forms have evolved. The simplest PC board uses two sides of thin copper separated by a dielectric of fiberglass. Since PC boards use at least two metal layers, transmission lines, matching networks as well as the antenna proper can be combined into one planar structure.

At frequencies above about 1 GHz, fiberglass (fire-retardant FR4 material used in PC board construction) becomes a comparatively high loss material. Therefore, for higher performance RF circuits, other substrate materials are often used. However, some compromised performance is often accepted to keep manufacturing prices down.

At first, simple dipoles were etched onto the PC board, followed by more complex geometric forms including spirals, planar "patches," and even fractal shapes. The various form factors may either be placed above a ground plane or emulate "free space" by keeping other metal traces away from the antenna proper, or some combination of both.

An alternative approach has been developed that permits the use of fiberglass PC board by creating a separate antenna structure using much higher performance ceramic as the antenna's substrate (**Figure 9.27**). The ground plane is simply included on the PC board's top layer, thus making the shortcomings of fiberglass mute. A rectangular, circular or other shape metal plane with small spacing to the ground plane forms most of configurations in use. The above discussion on small antennas is directly applicable to patch antennas since they are often quite small relative to the wavelength in use.

**Figure 9.27 — An assortment of ceramic patch antennas now in common use above about 600 MHz. Only the large patch supported by my little finger has a ground plane.**

PADL parameters become very complex since the patch uses two-dimensional planes rather than linear wires or thin cylinders for elements. Both transverse electrical and magnetic fields appear between the patch and the ground planes as well as outside the dielectric.

The patch antenna has gained wide acceptance for a variety of applications from UHF well into the microwave spectrum. They are typically mounted on a conducting ground plane and thus form an approximation of a hemispherical pattern. Depending upon the feed point within the structure, either linear or circular polarization may be achieved. Because of their small size and relatively good efficiency, they find wide application in handheld equipment, particularly for GNSS and UHF RFID applications.

## Phased Array

In Chapter 5 we introduced the concept of a phased array with an example that expands the usefulness of a 4-square vertical receive array. Here we expand to show how a phased array can provide steering of both the azimuthal direction (as in the 4-square) as well as elevation angles to cover the hemispheric sky.

In its simplest form, a phased array consists of multiple antenna elements (sometime hundreds of individual antennas) distributed over a plane (**Figure 9.28**). The resulting array may be steered to any direction above the plane by changing the phase and amplitude relationships among the elements.

If all elements are fed in phase, the pattern is broadside to the array or "straight up." If we progressively delay the phase to elements on the left as in this figure, we can establish directions that are dependent upon these phase relationships.

Changing the phase relationships among all the elements to any desired phase delay/advance is difficult as the number of elements grows. With the introduction of digital

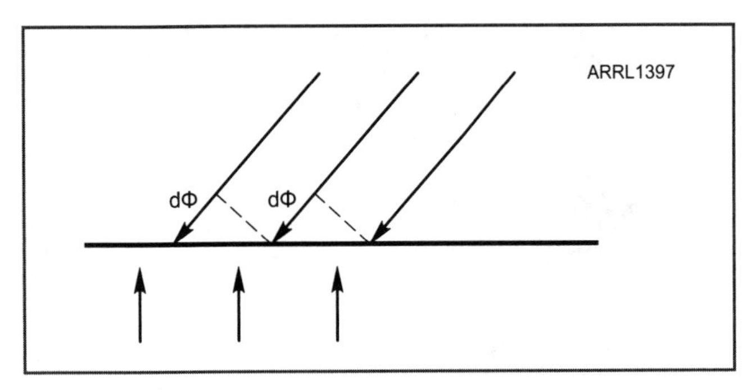


**Figure 9.28 — Three elements of a phased array can point to any angle above the mounting plane.**

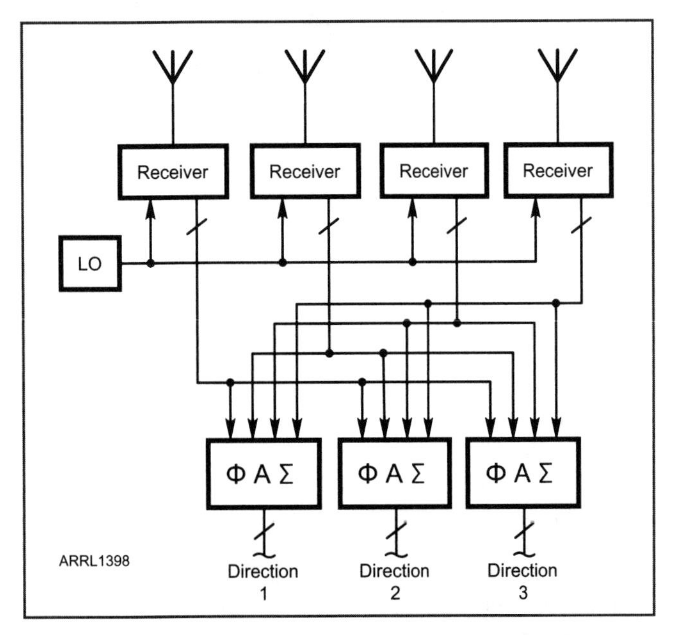


**Figure 9.29 — Multiple receivers/phase steering. In this figure, four antennas and receivers use the same local oscillator to phase lock their signals. The receiver outputs are digital representations of the IF. Three digital processors are used to adjust the phase and amplitude responses from the four receivers and then sum the four (φ, A, Σ). The outputs of these processors represent three simultaneous directive array outputs from the same antenna array. Of course such a system is scalable for more antennas and more simultaneous directions.**

signal processing, an alternate method of steering becomes possible. We use separate receivers for each of the elements, as shown in **Figure 9.29**. The receivers use the same local oscillator so that the signals from each receiver will be phase-locked (identical), but the location (as in PADL) is different. After analog-to-digital conversion we feed the digitized signals into digital phase shifters. Therefore the beam steering is accomplished at the digitized IF frequency. In addition, multiple directions may be received by simply increasing the number of digital phase shifters. In Figure 9.29 three simultaneous directions may be realized.

## Aperture Distribution

No matter what the antenna configuration, sidelobes begin to appear in the main pattern as the aperture becomes greater than about 1.5 λ. See **Figures 9.30** to **9.32** for the following discussion.

The simplest example is to plot the resulting pattern along one axis as a result of the aperture distribution along that axis. The aperture can consist of multiple individual antennas (as with a phased array) or a large single aperture (as with a parabolic antenna). Again, the calculation uses PADL for its solution, but with a rather sophisticated pair of integral functions.

The calculation of the sidelobe response from the aperture distribution utilizes a *Fourier transform*. The Fourier transform is a special form of integral that is extremely useful for a wide variety of engineering problems. In electronics it can provide the spectrum (frequency domain) of a signal captured in the time domain. Time domain displays are most often shown on oscilloscopes, while frequency domain displays are shown on spectrum analyzers. If you have time domain data (for example a voltage versus time) the spectrum of that signal (frequency versus voltage or power) can be calculated using the Fourier transform. On the other hand, if you have spectrum data you can use the inverse Fourier transform to derive the time domain signal. The simplest example is the Fourier transform of a pure sine wave to a single spectrum line, and vice-versa.

Readers who are familiar with transitions between time and frequency domains may recognize the similarity of the relationships shown in Figure 9.30. If the uniform aperture distribution shown in Figure 9.30A, were a "square wave" on an oscilloscope, we would expect to see *Fourier series* of the fundamental frequency and a series of odd-order harmonics (3rd, 5th, and so on) decreasing at a

$$\left(\frac{\sin X}{X}\right)^2$$

rate. This is exactly the decreasing rate in the sidelobe power shown in Figure 9.30A.

A second example is the Gaussian distribution, which is the unity function for the Fourier Transform. For example, if we have a time domain signal that traces a Gaussian curve on the oscilloscope, the frequency distribution of that signal will be the same Gaussian spectrum. Thus the Fourier and inverse Fourier transforms of a Gaussian function are the same Gaussian function relationship as shown in Figure 9.30E.

The trace in Figure 9.30F shows the sidelobe response of an inverse taper in the aperture. Such aperture distortion occurs when a large focal point structure is placed over a parabolic antenna. Thus careful design of the aperture distribution must be taken if sidelobe response is to be minimized.

## Polarization Diversity

In principle, an infinitely thin perfectly horizontal antenna will have zero response from an infinitely thin, perfectly vertical wire in free space. Since linear wire or tubular antenna elements have non-zero diameters, there is some response from one such antenna to the other.

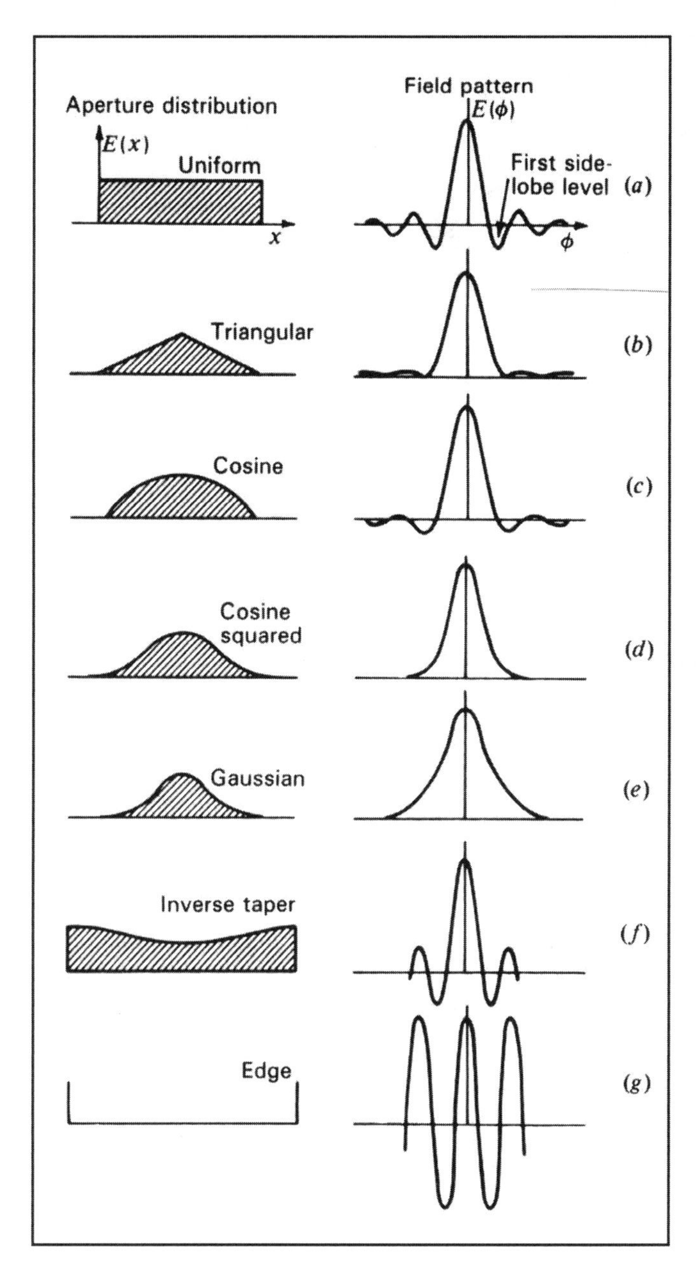


**Figure 9.30 — Different distributions of apertures along the X-axis (line distribution) form different sidelobes. Thus a phased array with a uniform aperture distribution will produce sidelobes in addition to the antenna's main lobe response. The sidelobes are shown distributed along the X-axis, but actually are relative to the azimuth direction ϕ. [From John D. Kraus, Antennas, 2nd ed, 1988, reprinted with permission of McGraw-Hill Education]**

Furthermore, reflections from the Earth and/or from the ionosphere often change the polarization characteristics of the radiated wave.

This difference in response between polarizations can be put to use in transmitting one channel of information using vertical polarization and a second channel using horizontal polarization but using the same frequency band simultaneously.

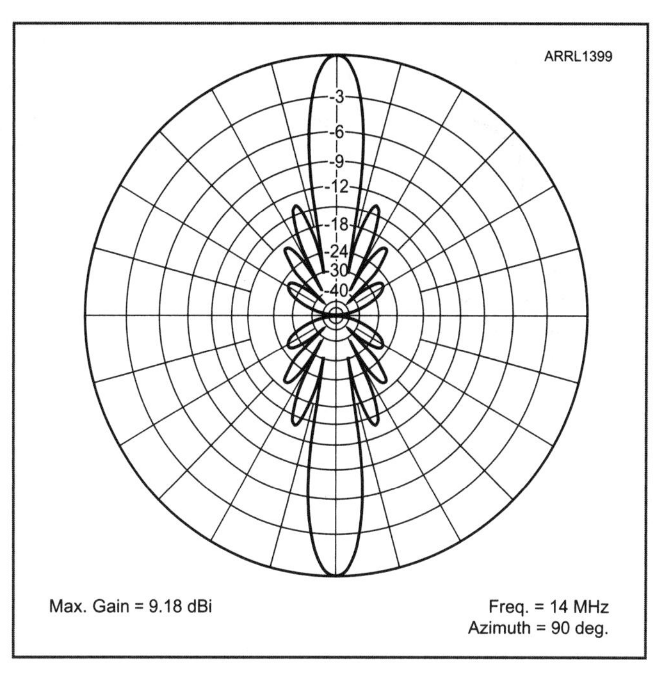


**Figure 9.31 — An array of nine co-linear dipoles along the X-axis in free space. This represents an approximation the aperture distribution shown in Figure 9.30A Near-uniform aperture is achieved by driving each of the identical dipoles with equal current. Notice the significant sidelobes in the pattern, close to the center of the plot.**

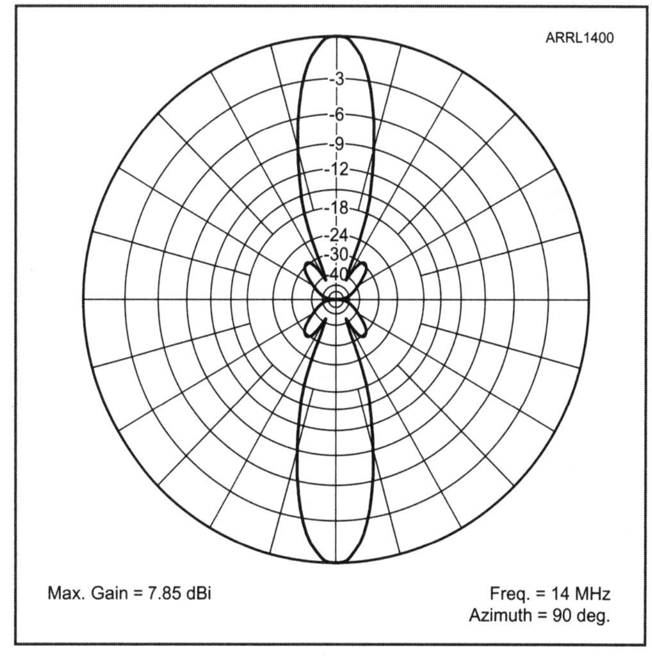


**Figure 9.32 — The same nine dipoles in Figure 9.31. Sidelobes can be greatly attenuated by proper adjustment to the aperture distribution. Here, for simplicity of modeling the aperture is made triangular by tapering the nine feed current values: 0.1, 0.3, 0.5, 0.7, 0.9, 0.7, 0.5, 0.3, 0.1 to form a triangular aperture distribution along the X-axis as in Figure 9.30B.**

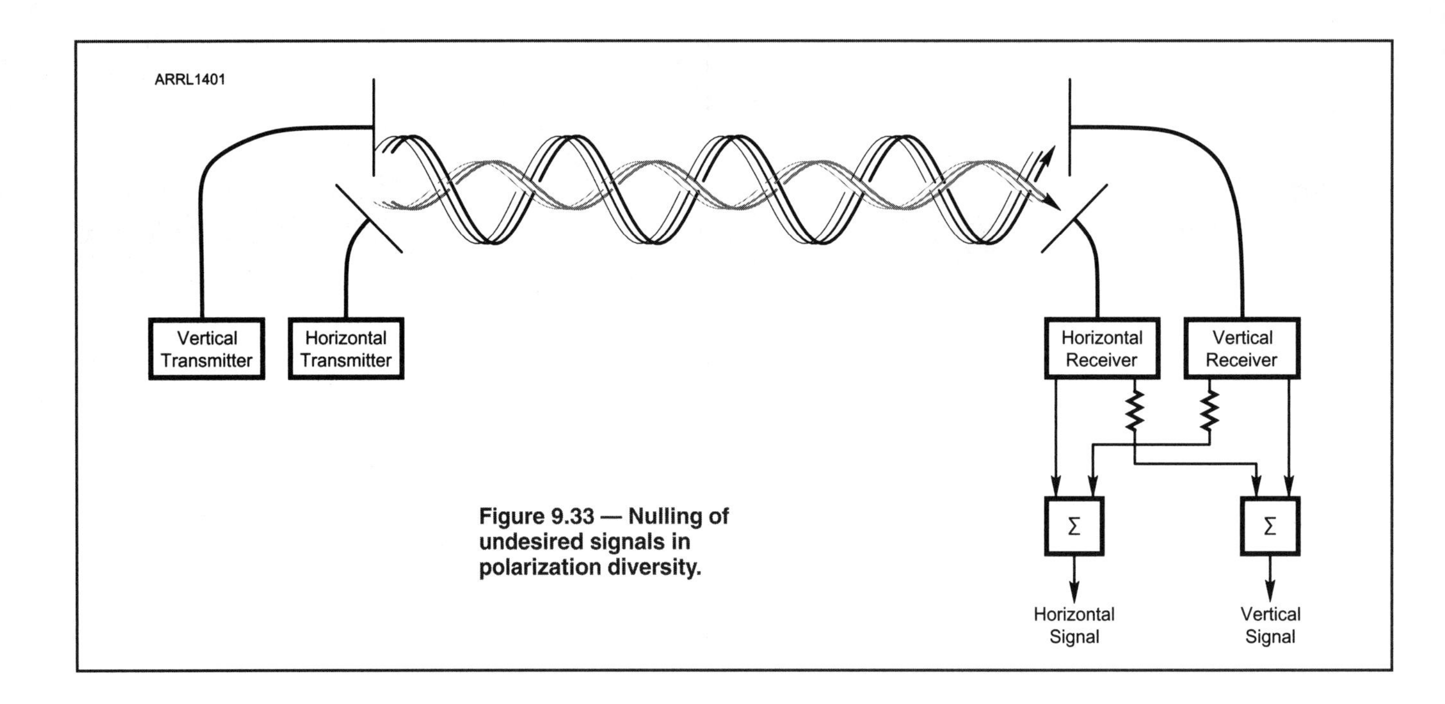


Figure 9.33 — Nulling of undesired signals in polarization diversity.

This technique has the potential of doubling the capacity of the frequency channel. Left hand circular combined with right hand circular can provide the same effect.

In such systems additional nulling can take place in the receiver to offer better polarization isolation since both polarization signals are received as shown in **Figure 9.33**. Since the error signals bleed-over from the other polarization, we can simply sample the undesired signal, correct for amplitude and phase and provide additional nulling.

## MIMO Antennas

"MIMO" is an abbreviation for "multiple input, multiple output." Consider the system is **Figure 9.34**. The transmit power is divided between two antennas of some distance apart. The receiver receives these two signals on two receive antennas also separated by some distance. Therefore there are now four possible paths for the signal to travel thus greatly increasing the probability that the system can avoid multipath problems. A second advantage is that since there are different time delays among the four separate paths, it is possible to transmit different information along the separate links, or a combination of the two.

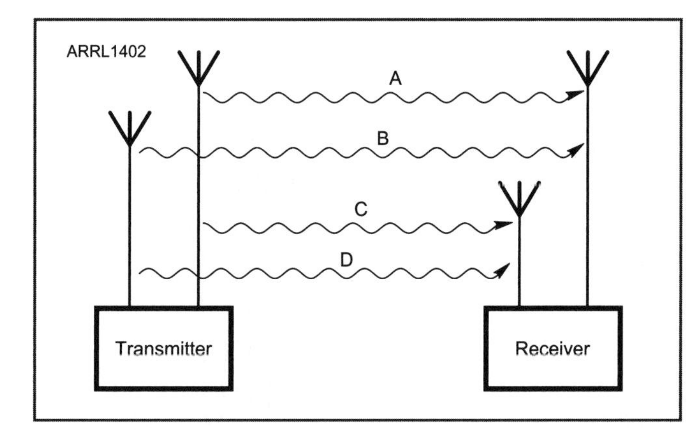


Figure 9.34 — This figure shows how four paths can be created with two antennas at each end of a link.

## Other Configurations

The above examples of specific antenna configurations are but a small sample of current development work in antennas. The last few examples point to a progressively closer relationship between the antenna and the receiver. The development of digital signal processing and software defined radios (SDR) has opened new frontiers for improving antenna performance. In particular the difficulty of dealing with expensive passive components for phase and amplitude adjustments can now be performed using inexpensive yet very powerful digital processors. In effect, critical adjustments to antenna parameters are now being performed in the digital domain. However, the physics of the antenna proper remains constant.

# 10 Noise, Temperature, and Signals

Thus far this text has presented some of the key concepts of antenna physics. The physics of antennas sets physical and theoretical limitations (and opportunities) for radio links. The purpose of this final chapter is to link antenna performance to the overall theoretical limitations of the radio link.

Of critical importance is the omnipresence of *noise*. The first part of this chapter deals with the origins of noise and how these mechanisms relate to the RF spectrum. Second, we discuss how noise affects the radio link. Third, we introduce a new parameter for antennas: *antenna temperature*. Aside from the electromagnetic noise limitations to a radio link, there is one other critical physical parameter that, together with the electromagnetics, completes the physical limitations of a radio communication system: information science. This will complete an introductory treatment of *all* the theoretical factors that place physical limits upon a radio communication link. A discussion of medium-frequency (MF) atmospheric noise serves as a practical application of many concepts treated earlier. As an aside, we will show how an antenna was the indispensable sensor for the first direct detection of the beginning of the universe, the Big Bang.

## Boltzmann's and Planck's Constants

Noise is a type of waveform present in all communication systems. It also sets fundamental limitations upon all communication links. Therefore an understanding of what causes noise and how it imposes limitations can set the stage for dealing with it most effectively.

For radio systems, noise has numerous sources, and each can become the limitation for a given application. We begin the discussion by looking at thermal or "Boltzmann" noise. We will see that thermal "radio noise" is closely related to temperature, as its name implies, but this relationship is often confusing, particularly when its effects upon communication systems are considered.

**Figure 10.1 — Max Planck.**

**Figure 10.2 — Ludwig Boltzmann.**

*Thermal noise* originates from the fact that any atom or molecule with a temperature above 0 Kelvins or 0 K (absolute zero) will vibrate. The classical definition of *absolute zero* (0 K) is the point where all atomic vibration stops. Some experiments as of this writing (2014) have achieved temperatures only a small fraction of a Kelvin higher than absolute zero. (Note that unlike the Fahrenheit and Celsius scales, *Kelvins* is the unit of measure for this scale — without "degrees." So while we would normally write 50 °F or 50 °C, when using the Kelvin scale, 50 K or 50 Kelvins is the correct notation.)

The absolute position of a vibrating atom is impossible to predict at any given instant. However, the distance from a "norm" position is a matter of probability. This physical vibration of atoms also has another effect: it produces a random electrical voltage. Since the voltage is random, we call it "noise." If a voltage is "random" it must therefore also be "alternating." Since this is a random alternating voltage, we can also deduce that it is *broadbanded* and (to the point of this text) extends into radio frequencies. Furthermore, we recall from Chapter 3 that *radiation* can only occur when an electrical charge is accelerated. Atoms contain electrons (charges) that therefore also vibrate, which in turn is a random acceleration of charges, which creates a random electromagnetic radiation. A random RF voltage is produced within a resistor at some temperature above absolute zero. This random voltage will be *conducted* and thus appear across the terminals of the resistor. The source of this RF voltage will also *radiate,* thus causing thermal radiated noise according to the dynamics of the antenna PADL (Phase, Amplitude, Direction, and Location — introduced in Chapter 3) of the resistor.

As we might suspect, the random movement of atoms will increase with increasing temperature. This is indeed the case and was first quantified by Max Planck (**Figure 10.1**). In 1900 Planck introduced the concept of *black body radiation.* This was follow-on work of Ludwig Boltzmann (**Figure 10.2**), who, 23 years earlier had linked entropy (from theory of gasses) and probability. The mathematics dealing with probability functions is of primary importance in the study of quantum mechanics. In turn, black body radiation derives its theoretical basis in quantum mechanics and is also described in terms of probability functions.

As temperatures increase, the *total* power emitted from a black body increases very quickly, proportional to $T^4$! The spectral distribution of this power is *not* linear across the electromagnetic spectrum, but limited to RF and microwave frequencies a linear relationship can be assumed. The peak power typically occurs well into infrared wavelengths and shorter. Even for very low temperatures near absolute zero, the noise temperatures at RF frequencies change linearly with temperature defined by Boltzmann's constant. For example, the peak spectral power of a very cold 3 Kelvin black body is centered at about 160.2 GHz. Although this can be considered a very short microwave wavelength, technically it is still in the RF spectrum. So for radio noise considerations, the Planck constant need only be considered at superlatively very high (RF) frequencies and measuring at very low temperatures. The general equation across the entire electromagnetic spectrum for black body radiation is:

$$\frac{E}{m^2} = \frac{8\pi hf^3}{c^2}\frac{1}{e^{\frac{hf}{kT}}-1} \quad \text{(Equation 10.1)}$$

where $E/m^2$ is the energy being radiated per $m^2$, $h$ Planck's constant, $6.626 \times 10^{-34}$ J-s (Joule-seconds) and $k$ is Boltzmann's constant (given in Equation 10.3). It was Planck who first calculated (within a few percent) the now-known value of Boltzmann's constant.

Fortunately for radio engineers, when $f$ is made comparatively low (RF frequencies), this equation can be replaced with a much simpler linear equation taken from classical physics. When an object is very hot, the peak frequency becomes very high. For example, the Sun's highest power

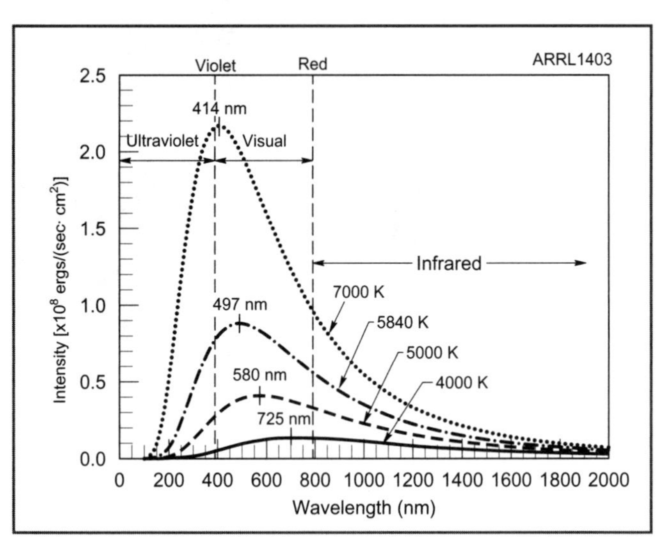


**Figure 10.3 — This is a plot of thermal noise power plotted at four different temperatures. Notice that as wavelengths become progressively longer (closer to RF frequencies) the spectral power becomes more constant (versus wavelength). Also, the noise at RF frequencies becomes a linear function of temperature.**

spectral output occurs in the visible light spectrum. The Sun is about 5000 K. If it were made hotter, the peak power would increase, and the frequency of maximum output would also increase.

**Figure 10.3** shows thermal noise power plotted at four different temperatures. Note that Figure 10.3 plots black body radiation. The Sun, for example emits higher power on the spectral lines of hydrogen and helium, which make up the bulk of the Sun's mass. Few "real" objects emit a "pure" black body spectrum. If we look closely at Figure 10.3, we see the noise power (Y-axis) converges into a mostly linear function at lower frequencies (RF). Therefore, for the majority of considerations concerning RF frequencies, Equation 10.1 converges to a much simpler linear equation:

$$P_{noise} = kTB \quad \text{(Equation 10.2)}$$

for RF frequencies, where $T$ is temperature (Kelvins), $B$ is bandwidth, and $k$ is Boltzmann's constant:

$$k = 1.38 \times 10^{-23}\ \text{J/K} \quad \text{(Equation 10.3)}$$

for 1 Hz of bandwidth at a rate of 1 second, where J is energy in Joules, and K is the Kelvin temperature. It is interesting to note that $k$ is an expression of *energy/temperature*. How do we derive a *power* term from $kTB$?

"Frequency" was defined by "cycles per second" (*cy/sec.*) until the ISU (International System of Units) redefined "frequency" as "Hertz" where today

$$1\ Hz = \frac{1}{sec} \quad \text{(Equation 10.4)}$$

The old "cycles" are now dimensionless and therefore

simply become the coefficient for Hertz — and thus in a noise calculation, also power. So if

$$k = 1.38 \times 10^{-23}\ \text{J/K} \quad \text{(Equation 10.5)}$$

and $J = W \times sec$ (energy = watts × seconds), $T$ = absolute temperature, and $B$ (bandwidth) = $X/sec$, therefore,

$$P = kTB = \frac{W \times sec}{T} T \frac{X}{sec} = X \text{ watts} \quad \text{(Equation 10.6)}$$

Thus $kTB$ is in watts, with temperature ($T$) and bandwidth ($X/sec$). Also, $W \times sec = energy$.

## Boltzmann Noise in Circuits

**Figure 10.4 — Harry Nyquist. [Courtesy of Alcatel-Lucent USA]**

During the 1920s, John Johnson at Bell Labs measured a "noise voltage" across a resistor that was not connected to any power source. Harry Nyquist (**Figure 10.4**), who was also working at Bell Labs, was able to explain the mechanism of the source of this thermal noise by linking it to Planck's constant. Suitably, "resistor noise" is today referred to as Johnson-Nyquist noise. Like black body radiation, noise voltage is directly related to the absolute temperature (Kelvin) of the resistor and the value of the resistor, where:

$$V_n = \sqrt{4kTR} \quad \text{(Equation 10.7)}$$

This is the noise *voltage* value in 1 Hz of bandwidth produced by a resistor ($R$) at an absolute temperature ($T$). To derive the noise *power* we can use the simple equation:

$$P_n = \frac{V_n^2}{R} \quad \text{(Equation 10.8)}$$

or

$$P_n = \frac{4kTR}{R} \quad \text{(Equation 10.9)}$$

or

$$P_n = 4kT \quad \text{(Equation 10.10)}$$

for power in 1 Hz of bandwidth.

Adding the bandwidth term, we get

$$P_n = 4kTB \quad \text{(Equation 10.11)}$$

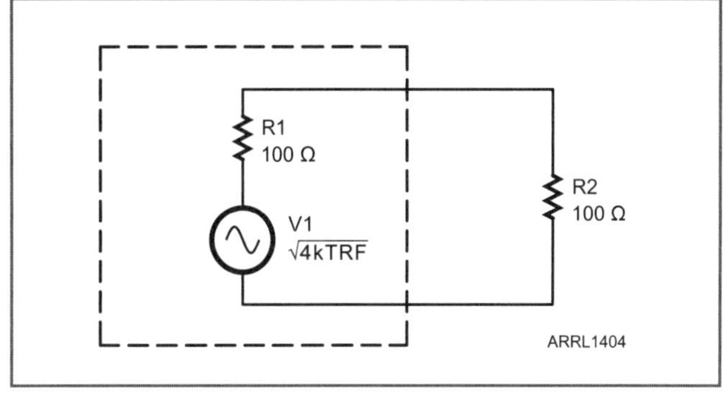


**Figure 10.5 — Transfer of power to an external matched load.**

There are two important points to be made regarding this equation for thermal noise:

1) In the noise power equation the $R$ term cancels, indicating that the *noise power* generated in a resistance is *independent* of the resistor value, whereas noise voltage and current values are dependent upon the actual $R$ value.

2) The coefficient 4 is a common source of confusion. This is the voltage seen across the thermal resistor source. However, if we want to transfer the power to an external matched load as in **Figure 10.5** the power becomes:

$$P = kTB \quad \text{(Equation 10.12)}$$

By the following derivation (from Equation 10.7):

$$V_n = \sqrt{4kTR}$$

Then the voltage found across R2 is

$$V_n(\text{R2}) = \frac{\sqrt{kTR}}{2} \quad \text{(Equation 10.13)}$$

or,

$$V_n(\text{R2}) = \sqrt{kTR} \quad \text{(Equation 10.14)}$$

And, simultaneously adding the bandwidth correction

$$P(\text{R2}) = kTB \quad \text{(Equation 10.15)}$$

This is the derivation of Boltzmann's noise.

A final point which is very important when considering noise in circuits and antennas is that although the power generated in a resistor is independent of the resistor value, the noise source will have an impedance equal to the value of the source resistance. Figure 10.5 shows a Thevenin equivalent circuit which is a good approximation for source impedances of finite value. However, in the case of near-zero or zero ohm source impedances in thermal noise modeling, the signal noise source is actually a short circuit. Thus the very low resistance noise source provides a "short" across the input impedance of the load, and no power is transferred. In effect, the random thermal noise power generated in a non-connected resistor simply re-dissipates itself as heat. Therefore we have a noise-voltage/current-noise cycle inside all resistors. So, a simple carbon resistor sitting on your bench contains a frenzy of electromagnetic activity! Extracting a portion of that noise power is another matter. As we will see, for antenna noise analysis, this source impedance is simply equated to $R_l$ for efficiency and noise calculations. For circuit noise calculations, including amplifiers, the mathematics becomes significantly more complicated.

## Random Processes

Equation 10.12 provides a simple definition of thermal noise, either radiated or conducted. At first glance we might expect to measure a *constant* voltage or power due to Johnson-Nyquist noise. This is not the case since thermal noise is created by a *random process* and thus the *instantaneous* power value must be governed by the laws of probability. Equation 10.12 defines a *mean* value. The instantaneous value of a black body radiator is defined by a *Gaussian distribution*.

Note that the *theoretical* probability curve *never* quite reaches zero even for an infinite power level! Using the Boltzmann constant we can compute an actual mean power level at any temperature. This is the reason that Boltzmann's constant is represented as an energy term. The energy term, by virtue of its time integration, also finds an average power over time. In this case it represents the average of the Gaussian function over time, or the peak of the Gaussian probability curve as in **Figure 10.6**.

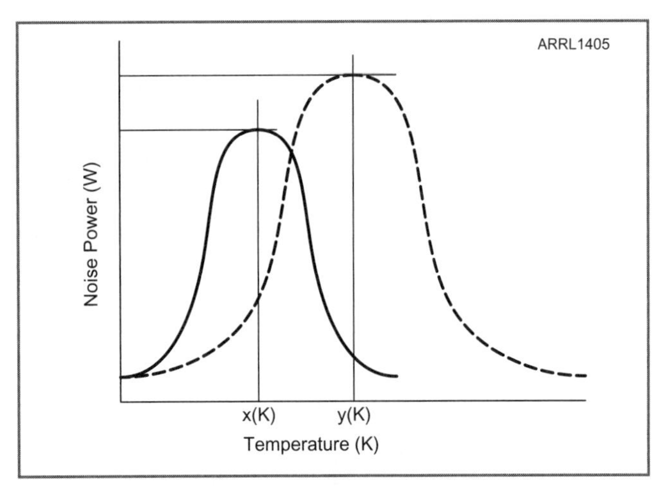


**Figure 10.6 — This is the shape of a Gaussian Distribution (sometimes also referred to as a "bell curve"). For our purposes, the X-axis represents the mean noise power value (a function of the temperature K), while the Y-axis represents the probability of an instantaneous power. The center represents the mean value determined by the temperature and Boltzmann's Constant. The probability is maximum at this power level, but can vary above and below this value with Gaussian Probability. As the temperature moves higher, the peak mean value increases by Equation 10.12, but the probability around the mean value remains governed by the Gaussian distribution.**

At room temperature (290 K), the power in 1 Hz bandwidth is

$$1.38\times10^{-23}\,\frac{\text{J}}{\text{K}}\,(290) = 4\times10^{-21}\,\text{W} \qquad \text{(Equation 10.16)}$$

This is about 204 dB below 1 W. We have found that antennas are usually specified in dBi or dBd. These are related to *power gain* and express ratios, not absolute power levels. When we get into radio systems engineering, we need to specify *real power levels*. In Chapter 7 we introduced such an absolute power term; now we will formalize it relative to RF system engineering. The usual convention in radio engineering is the use of dBm, which uses the reference:

$$0\ \text{dBm} = 1\ \text{milliwatt (mW)} \qquad \text{(Equation 10.17)}$$

Therefore, Boltzmann noise at 290 Kelvins, also +16.85 °C and +62.3 °F (or "room temperature") is –174 dBm/Hz. The noise power will be directly proportional to the bandwidth, therefore we can simply add the term

$$10 \log (B) \qquad \text{(Equation 10.18)}$$

to derive

$$P_{noise} = -174\ \text{dBm} + 10 \log (B) \qquad \text{(Equation 10.19)}$$

at 290 K, where $B$ is the bandwidth in Hertz.

## Radio Receiver Sensitivity

There are two more terms needed to derive the critical specification of *receiver sensitivity.* Any radio receiver will generate its own noise. The process of amplification, attenuation and mixing all contribute to the *receiver noise figure* usually expressed as $NF$ (dB). The noise figure (specified in dB) is the log of the *noise factor (F)*, which in turn is the ratio of the input to output *signal-to-noise ratios (SNR)*. Therefore the noise factor is a ratio of two ratios. The noise factor of a circuit (active or passive) is simply defined as the degradation of the signal to noise ratio when a signal passes through the device. Therefore,

$$F = \frac{SNR_{out}}{SNR_{in}} = \frac{S_{out}/N_{out}}{S_{in}/N_{in}} = \frac{S_{out}}{S_{in}}\frac{N_{in}}{N_{out}} \qquad \text{(Equation 10.20)}$$

for actual power.

Noise figure is defined in dB. Therefore, for the NF expressed as power,

$$NF\ \text{(dB)} = 10 \log (F) \qquad \text{(Equation 10.21)}$$

For example, a "perfect amplifier" will not degrade the SNR of the amplified signal, therefore, $NF = 0$ *while the* $F$ *is 1.* When we refer to a "device" we are referring to amplifiers, mixers, filters, and other components of a *radio*. Other sources also must be considered, including noise power that is received by the antenna. Consequently, we can divide "noise" into two basic sources: external to the receiver, and internal to the receiver. Both must be considered in the calculation of the quality of a communication link. However, the first step is defining the *receiver noise figure*.

Finally, the *minimum performance level* of a receiver requires some level of SNR for audio, video, or digital bit error ratio performance. In effect, the quality of a link is directly related to the ratio of signal power (or energy) to noise power (or energy). In analog systems, we usually refer to the SNR, which implies a *power* ratio. However, in digital systems it is more appropriate to specify an *energy* ratio since the *symbol rate* of a digital communication system implies that the symbol value (amplitude, frequency, and/or phase) stays constant for the *time* that symbol is transmitted.

$$T_{symbol} = \frac{second}{r_s} \quad \text{(Equation 10.27)}$$

$$r_s = \frac{r_{bit}}{bits\,/\,symbol} \quad \text{(Equation 10.28)}$$

where $r_s$ is the symbol rate and $r_{bit}$ is the bit rate. Thus, a receiver performs a time integration ($T_{symbol}$), which results in an energy value in contrast to an analog signal's power value SNR.

$$\frac{E_b}{N_o} \quad \text{(Equation 10.29)}$$

where $E_b$ is the "energy per bit" and $N_o$ is the power in a 1 Hz bandwidth.

Therefore, at room temperature receiver sensitivity is defined as

$$\text{Sensitivity} = -174\text{ dBm} + 10\log(B) + \text{NF} + \text{SNR}(\text{or }10\log\frac{E_b}{N_o}) \quad \text{(Equation 10.30)}$$

The conversion factor between SNR and $E_b/N_o$ is simply:

$$\text{SNR} = \frac{E_b}{N_o}\frac{R_b}{B} \quad \text{(Equation 10.31)}$$

where $R_b$ is the channel data rate (bits/second) and $B$ is the channel bandwidth.

To summarize, Equation 10.30 simply states that at 290 K, we first calculate thermal noise power per Hertz, then correct for the bandwidth of the channel, degradation from the receiver NF, and finally the required ratio of signal to noise (power or energy) to maintain a predefined quality of the link. This is the definition of receiver sensitivity.

For example, let us assume that we have a system that requires 1 MHz of bandwidth, the noise figure of the receiver is 5 dB, and the required SNR is 10 dB. What is the sensitivity of the receiver?

$$\text{Sensitivity} = -174\text{ dBm} + 60\text{ dB} + 5\text{ dB} + 10\text{ dB} = -99\text{ dBm} \quad \text{(Equation 10.32)}$$

This receiver requires a –99 dBm *minimum* of signal power assuming that only thermal noise is present at the receiver input.

## Calculating a Minimum Antenna Gain Specification

In Chapter 2 we described the physics of a radio link. Now we can use that physics coupled with a radio sensitivity specification to calculate how much antenna gain will be required to span a free-space distance using some restricted transmitter power. See **Figure 10.7** for the following example.

Let us assume that we need –99 dBm of power to provide the minimum signal level at the receiver input. Let us also assume we have a transmitter that provides 1 W output (+30 dBm). Through a simple calculation (–99 dB – 30 dBm = –129 dBm) we also know that we can lose 131 dB between

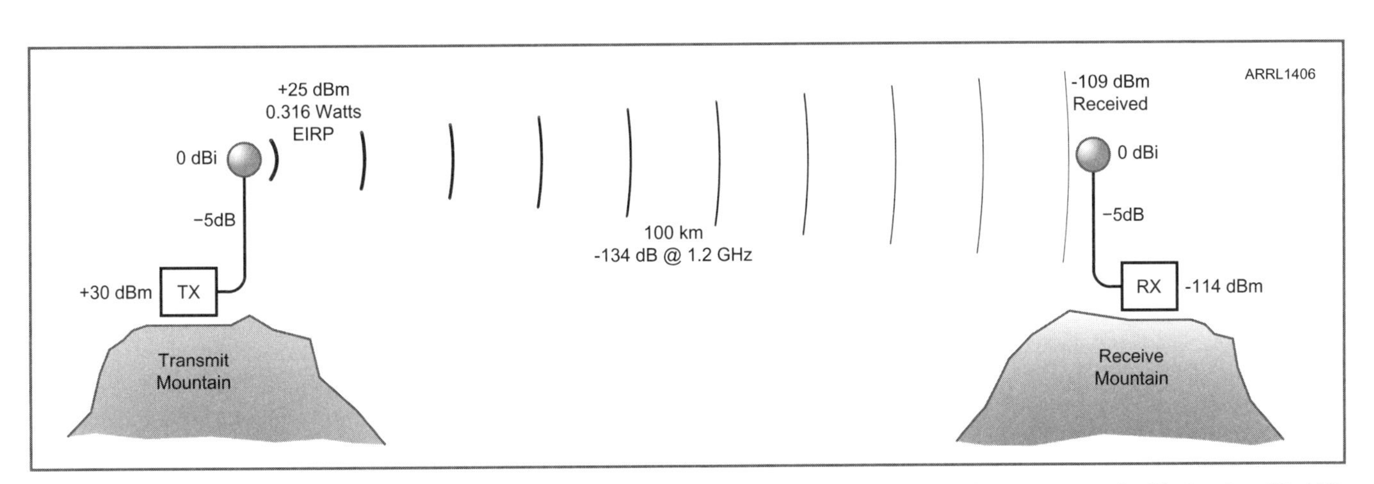


**Figure 10.7 — This figure shows a graphical representation of the radio link described in the text example. We begin with 1 W (+30 dBm) at the transmitter, lose 5 dB in the transmission line, radiate +25 dBm, lose 134 dB over the 100 km link, and another 5 dB in the receiver transmission line, thus presenting –114 dBm to the receiver. Notice the serial nature of the link analysis and the inherent possibility of increasing the power input to the receiver by decreasing loss (increasing gain) with all the parameters (including improving receiver sensitivity)!**

the transmitter output and the receiver input. This is termed the total link budget and like all budgets, we can scrimp and save some places and splurge in others. However, at the end of the day we have to stay within our budget.

Now we also know that we are using a frequency of 1200 MHz, or a wavelength of 0.25 meters. We need to link two mountaintop sites 100 km apart, but we have a free-space link. We also assume that we have 5 dB of transmission line loss at both sites, for a total of 10 dB loss. How much *total* antenna gain will we need to provide minimum quality to the link?

From Equation 2.6 in Chapter 2 we know that the antenna-to-antenna link budget (assuming isotropic antennas at both ends) to be:

$$P_r = P_t\left(\frac{\lambda}{4\pi r}\right)^2 \quad \text{(Equation 10.33)}$$

Now we can plug in the numbers and calculate the power received:

$$P_r = 1\ \text{W}\left(\frac{0.25}{4\pi(100{,}000)}\right)^2 = 39.6 \times 10^{-15}\ \text{W} = -104\ \text{dBm} \quad \text{(Equation 10.34)}$$

After we subtract 10 dB from the transmission line losses, we find –114 dBm at the receiver, or 15 dB lower than we require. We can *gain* this 15 dB by adding a 7.5 dBi gain antenna at each end; or a 10 dBi antenna at one end and a 5 dBi antenna at the other; or use lower loss transmission line with lower gain antennas. Any of these changes will result in an increase of 15 dB at the receiver. We could also increase the power of the transmitter, or use any combination of these approaches. This rather complex problem simply is a matter of adding and subtracting decibels to the *actual* power (dBm) levels of Boltzmann noise, transmit power, and necessary received power. All the other terms are simply dB loss or gain, but we begin with dBm (transmit power) and end with dBm (signal power) at the receiver input.

Now suppose we have *transceivers* at both ends of the line, and we assume both are identical to the transmitter and receiver specifications stated above. Since links are *reciprocal* we will find the power to the receivers at both ends to be identical.

## Noise Temperature

In the above example, we assumed that the noise at the input to the receiver was –174 dBm/Hz since the assumption was a system temperature of 290 K. Therefore NF is a temperature-dependent value related to the reference of 290 K. Suppose by Equations 10.20 and 10.21 we measure and calculate a NF of 3 dB, the F is 2 and let's define the gain as 10 dB. Let us drive the amplifier with a 1 W signal with an accompanying ¼ W noise. The output will contain 10 W of signal, but 5 W of noise. We can see the input SNR is 4/1 and the output SNR has been reduced to 2/1, therefore F = 2 and NF = 3 dB. We have assumed an input noise temperature of 290 K and the amplifier has contributed an equal amount of noise (before amplification). Therefore the noise temperature of the amplifier must also be 290 K.

For NF or F, we can calculate the equivalent noise temperature of any device by

$$T_e = T_0\,(F - 1) \quad \text{(Equation 10.35)}$$

where

$T_e$ is the equivalent noise temperature
$T_0$ is 290 K (since $F$ is a function of 290 K)
$F$ is the noise factor

Good quality low-noise amplifiers can exhibit noise figures approaching 1 dB, even operating at 290 K. This corresponds to a noise factor of 1.26 and by Equation 10.35 an equivalent noise temperature of about 75 K. Thus the noise temperature of an amplifier may be considerably below the ambient temperature at which it is operating. This very beneficial but somewhat paradoxical feature is explained by relationships in semiconductor physics.

Since NF is taken as a log function of F, we must use an inverse-log function to derive the NF.

$$F = 10^{NF/10} \quad \text{(Equation 10.36)}$$

We can also lower the $T_e$ and thus also the F of an amplifier by physically cooling it down. The relationship is linear where:

$$T_{e\,4.22} = T_{e\,290}\,\frac{4.22\ \text{K}}{290\ \text{K}} \quad \text{(Equation 10.37)}$$

Here we will begin with the same amplifier as above with a NF of 1 dB, or a $T_e$ of 75 K. If we cool the amplifier down to the liquid helium temperature of 4.22 K, we derive a new $T_e$ for the device of 1.09 K. In Equation 10.37, our $T_{e290}$ is 75 K. (The $T_e$ of the amplifier at 290 K.) Thus the $T_{e4.22}$ is a very low 1.09 K.

The effort of cooling an amplifier to such extreme low temperatures is worthwhile is some specific applications which we will soon address. With this background information we are ready to describe an antenna specification critical to many system-level design concerns.

## Antenna Temperature

A resistor can be used to measure temperature by simply measuring its noise voltage by Equation 10.7. However, a resistor cannot measure temperatures below its own physical temperature, while it is possible for an antenna to do so. In a "perfect" antenna the "resistance" is radiation resistance ($R_r$), which cannot be specified by an ambient temperature. The concept is quite simple. If we measure the power level (dBm) being received by an antenna, then we can calculate an *equivalent* temperature using Boltzmann's constant as the relating term. However, now the noise power source is the $R_r$ of the antenna instead of a "real" resistor generating Johnson-Nyquist noise. $R_r$ cannot have a physical temperature.

Therefore in a "perfect" antenna there can be no Johnson-Nyquist noise produced by the antenna itself as with a resistor. Antenna noise temperature is only a function of the RF noise power received by it. And the noise power is only a function of the equivalent temperature of the matter in the antenna's steradian gain response. This "temperature" is coupled to the antenna by the antenna's $R_r$. The "matter" might be the ground, the moon, or a distant galaxy. Here again, the *voltage* received in 1 Hz bandwidth will be:

$$V = \sqrt{kTR_r B} \quad \text{(Equation 10.38)}$$

However, in this case, $T$ is the equivalent temperature of the noise/signal being received by virtue of the antenna's $R_r$, *not* the physical temperature of $R_r$. Therefore,

$$V^2 = kTR_r \quad \text{(Equation 10.39)}$$

and since

$$V^2 = PR_r \quad \text{(Equation 10.40)}$$

then

$$P = \frac{kTR_r B}{R_r} \quad \text{(Equation 10.41)}$$

and

$$T = \frac{P}{kB} \quad \text{(Equation 10.42)}$$

Thus the noise power *(dBm)* received by an antenna is directly proportional to the *antenna temperature* in Kelvins. Note: this assumes that the antenna is 100% efficient (lossless). In the case where the antenna is not 100% efficient, we *must* consider the relationship between the radiation resistance and the loss resistance $R_l$ (actual temperature of the antenna's loss resistance). This is similar to the efficiency calculation of ground-mounted vertical antennas in Chapter 7.

See **Figure 10.8**. We have already explained the mechanism for a simple noise measurement of a resistor (Figure 10.8A). If we place an antenna whose radiation resistance is the same as the resistor in Figure 10.8A in an anechoic chamber at the same temperature, we will measure an *identical* voltage and thus power level (Figure 10.8B). Finally, in Figure 10.8C, we point the antenna at a part of the sky with an *equivalent noise temperature* and again, we read the same noise power.

In the third example above, we can see how an antenna can become a remote temperature sensor. This is somewhat of a peculiar circumstance. *The distance to the "sky" temperature is cancelled by the antenna aperture, thus neither affect the antenna temperature!* This is *provided* that the portion of the sky (measured in steradian solid angle) has uniform thermal temperature over the beamwidth of the antenna.

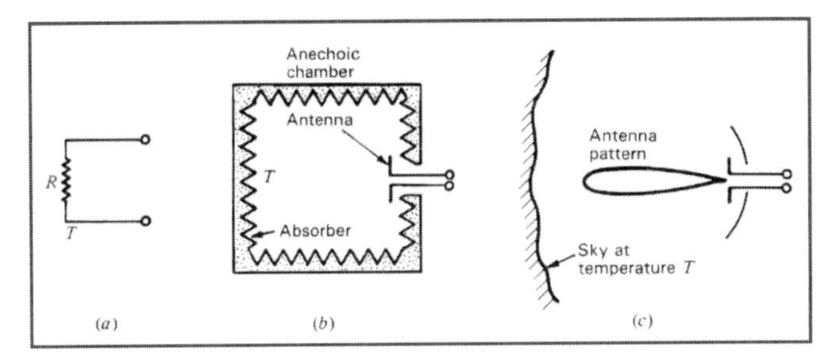


**Figure 10.8 — Three examples of noise measurement systems. [From John D. Kraus, *Antennas*, 2nd ed, 1988, reprinted with permission of McGraw-Hill Education]**

Until this point in this text we have assumed that a distant signal source is a point source. With noise as the source, we must now consider it as a source with some steradian "area" of the sphere.

For example, if we use a 20 dBi gain antenna, we are looking at 1/100 the area of the full spherical sky and measure X Kelvins. Now we replace the 20 dBi antenna with a 23 dBi antenna, so we are only viewing 1/200 of the spherical sky. Therefore we might conclude that we are only receiving ½ of the thermal power, therefore the temperature should be ½ X. However, we just increased the antenna gain by 3 dB so the smaller area's power is now increased by 2. This leaves us with a temperature again of X Kelvins (an identical noise power)! Also, the distance (even to the edge of the universe) does not matter by the same cancellation. Of course the sky temperature is *not* constant, so if within the beamwidth of our antenna there are different source temperatures, we need to adjust our measurement technique. However, in many directions, sky temperature *is* constant, even for lower-gain antennas.

Now let us assume that the antenna is not 100% efficient. If the antenna is receiving the equivalent of 50 K of thermal noise and the antenna is 99% efficient operating at 290 K, we can assume that this resistive loss (due to antenna inefficiency) will contribute thermal noise, and thus raise the antenna temperature. Thus in this example antenna noise would be:

$$T_{ant} = 50 + 290\left(\frac{1}{0.99} - 1\right) \quad \text{(Equation 10.43)}$$

or, about 53 K.

Therefore we suffer a 3 K rise in noise temperature due to the antenna's inefficiency. The equivalent would be to place a resistor with a value of ½ Ω at 290 K in series with a radiation resistance of 50 Ω at 50 K. Unfortunately, if we point the antenna toward the Earth, we will see an antenna temperature close to 290 K, the ambient temperature of the Earth. However, if we point the antenna at the sky, we must also minimize all thermal radiation from the Earth and keep our system noise temperature very low. Only then can our antenna become an extremely sensitive heat detector for man-made and natural objects. Even a small sidelobe response to the 290 K Earth will overwhelm an attempt to measure very weak extraterrestrial thermal noise sources.

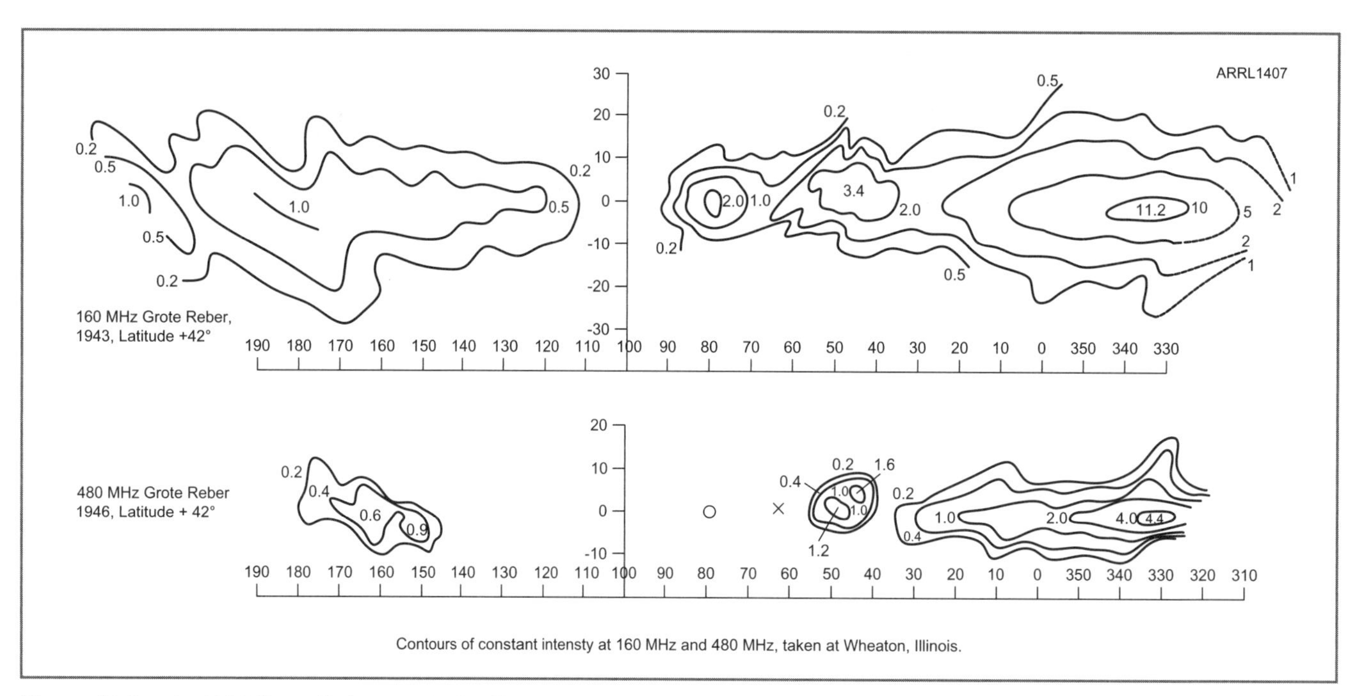


**Figure 10.9 — In 1944 Grote Reber made the first radio map (temperature of the sky). He found that the temperature was highest along the plane of the Milky Way and peaked toward the Milky Way's center. He also discovered the high radio emissions coming from Cygnus A and Cassiopeia A.**

**Figure 10.10 — Grote Reber is the acknowledged first radio cartographer of the universe. He extended the work of Karl Guthe Jansky who discovered radio waves emanating from the Milky Way. Reber held the amateur call sign W9GFZ.**

**Figure 10.11 — Reber's home built antenna in his Wheaton, Illinois, back yard. Reber used this antenna to make the first radio maps of the sky.**

However, if we *know* $R_r$ and $R_l$ of the antenna (usually an easy calculation), we can subtract the contribution from $R_l$ and determine the temperature of a remote object, or even space itself. But if we point our antenna to "empty" sky, it is reasonable to assume the antenna should record zero power since there is no matter emitting thermal radiation from "empty sky." This would be true in a "static" universe, but our universe not static, so what will empty sky provide?

**Figure 10.9** shows the first radio map detailing the temperature of the sky. This map was made by Grote Reber, W9GFZ (**Figure 10.10**), who used the home built antenna shown in **Figure 10.11** for his early radio astronomy experiments.

## A Nobel Prize

Over a period between 1922 – 1923, Edmund Hubble observed that several celestial objects were actually *outside* the Milky Way, eventually overturning the previously held scientific orthodox that the universe was constant and limited to the Milky Way. Further observations by Hubble and others showed that the universe consisted of countless numbers of galaxies all moving away from each other, some at near-light speed. This observation led to the theory of the Big Bang. If all objects are moving away from each other, then they must have had a common point of origin. And, if that much matter was in one "place" at one "time," then the universe must have begun with a Big Bang, much like a bursting fireworks explosion expels its fragments away from the point of explosion in three-dimensional space. Observing the rate of expansion and the size of the universe, the age of the universe was/is calculated to be about 13.7 billion years.

One of the implications of such a violent event was that residual radiation should remain in the universe in the form of black body electromagnetic radiation (including RF). Through the mid-20th century, many predictions and spirited debates took place regarding this topic. Most of the physicists concluded that this residual noise would have a temperature in the range of 2 – 5 K with the most likely range being 2 – 3 K. If this residual temperature of the universe could be found it would all but prove the Big Bang theory to be valid.

Then in 1964, two engineers, Arno Penzias and Robert Wilson (**Figure 10.12**) were working on a project at Bell Labs to reduce noise in microwave radio links. They became puzzled by a minimum 3 K noise temperature no matter which way they pointed their 6.2 meter horn-reflector antenna at 4 GHz. They would later find that they had discovered the residual energy from the Big Bang and received the Nobel Prize for their discovery. We now know that the actual temperature is 2.7 K and the noise is in near-perfect conformance to the theoretical black body frequency distribution predicted by Planck. Thus an antenna was the indispensable tool for observing the first direct evidence of the beginning of the universe: microwave radiation from the Big Bang.

**Figure 10.12 — Arno Penzias and Robert Wilson in front of the horn antenna used to first measure the electromagnetic residue of the Big Bang. [Courtesy of Alcatel-Lucent USA]**

Penzias and Wilson found that the absolute minimum noise an antenna can detect is 2.7 K, which is about –194 dBm/Hz. This will be true of any antenna located anywhere in the universe and pointing in any direction! It also sets a *fundamental* limit to background noise.

This measurement may seem a bit puzzling. Since the horn antenna used by Penzias and Wilson was Earth-bound, the antenna was pointed at the sky through the atmosphere. Since the atmosphere is considerably warmer than absolute zero, one might expect that they would have observed a much higher value of thermal noise. In fact, the atmosphere is a "thin gas" that doesn't radiate (or absorb) black body radiation in the same manner as "true" black bodies. Once again, the answer to this dilemma lies in the laws of quantum mechanics. For our purposes, the atmosphere does not emit black body radiation over the frequencies investigated by Penzias and Wilson, allowing the above measurements to be made. In contrast the atmosphere has an extremely high black body temperature at much lower frequencies.

## Atmospheric and Galactic Noise

There are countless objects in the observable sky that emit much higher temperatures. If the antenna is pointed at these objects, higher noise powers will result. **Figure 10.13** is provided to illustrate Earth-bound sky noise as a function of frequency. By using Boltzmann's constant the actual dBm/Hz values can be easily computed.

As shown in Figure 10.13, atmospheric and galactic noise pose a basic limitation to communication systems at frequencies below about 1 GHz. The air molecules in the atmosphere are constantly being bombarded by electromagnetic radiation and particles from the Sun. The effects are twofold: ionization of the atmosphere (particularly at higher atmospheric elevations) and electrical charging. Furthermore, rapid vertical movement of liquid particles (particularly water droplets) in the atmosphere adds to electrical charging. The most dramatic electromagnetic noise generation is lightning caused by sudden discharging of highly charged clouds.

The Milky Way Galaxy (our home) consists of billions of stars and a myriad of other fascinating celestial structures. Many of these structures emit radiation at radio frequencies as well. Detailed study of these sources of radiation began after WWII and became known as *radio astronomy,* which has clearly added very significant data to the understanding of a wide variety of celestial objects as well as fundamental discoveries in cosmology.

Figure 10.13 also shows a minimum of natural noise occurring between about 1 and 10 GHz. This is the range of frequencies where most SETI (search for extraterrestrial intelligence) activity has been conducted, since this band provides the minimum sky noise for Earth-based observations.

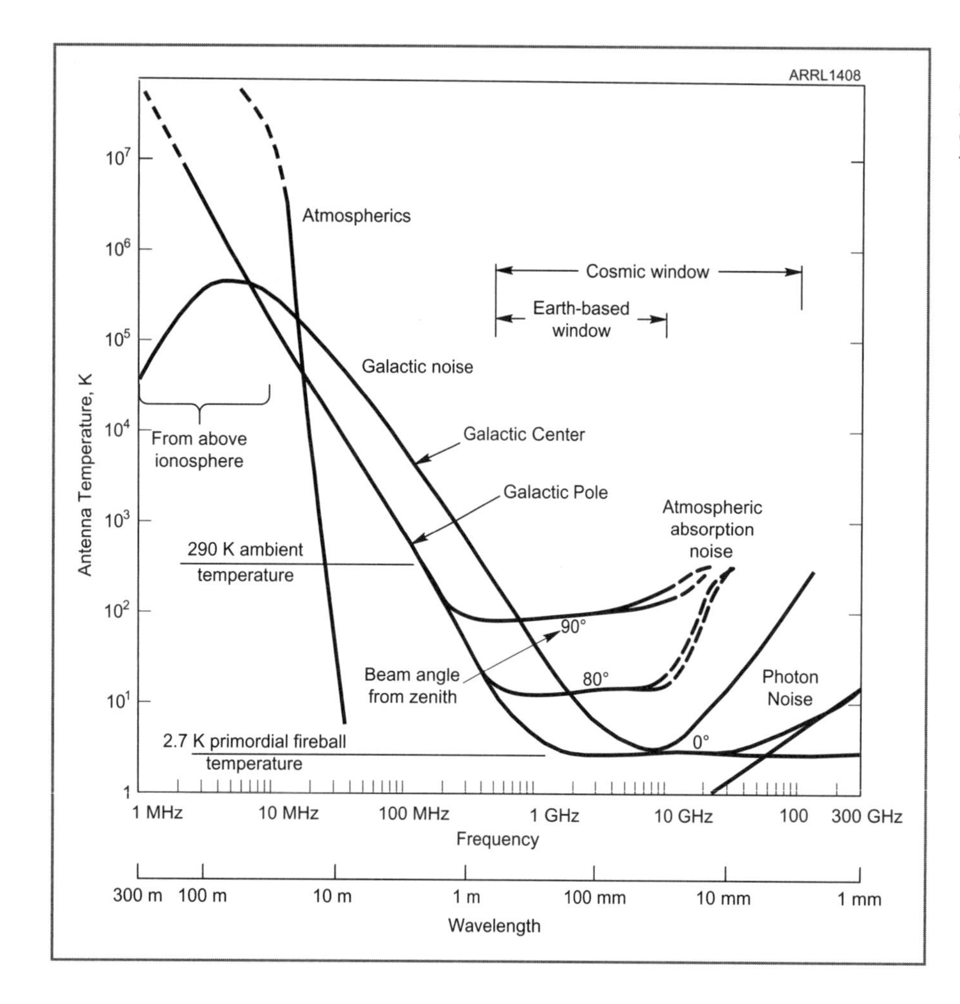


**Figure 10.13 — This graph shows not only noise from beyond the Earth, but also the noise due to atmospheric generation in terms of a noise temperature.**

Therefore if "they" were interested in interstellar communication they might use these same wavelengths below about 10 GHz.

## Photons in Radio Waves

We have already stated that electromagnetic waves are both "waves" and "particles." The particle is called a photon. When the photon has enough momentum, dislodging electrons becomes possible — thus creating an electric current as in a photoelectric or "solar cell." Although defining the photon relative to radio waves is usually not of practical value, a brief description is presented for completeness. Again Planck's constant is used to calculate both the momentum and the energy of the photon.

The energy (in Joules) of a photon is linearly proportional to the frequency of the wave.

$$J = hf \qquad \text{(Equation 10.44)}$$

where $h$ is Planck's constant as in Equation 10.1 and $f$ is the frequency in Hertz. This is the amount of energy contained by one photon. If we should ever need to know how many photons we are emitting per second, we simply divide our output power by the power/photon. Keep in mind 1 Joule = 1 watt-second (1 J = 1 W-s), and to keep terms in order, remember that $h = Js$ and $f = 1/s$, therefore we end up with Joules, an energy term.

$$photons\,/\,second = \frac{watts}{hf} \qquad \text{(Equation 10.45)}$$

For example if we are transmitting 100 W at 144 MHz, we are transmitting about $10^{27}$ photons/second. Notice that as we raise the frequency but keep the power constant, we transmit fewer photons.

When we consider the momentum of a photon, classical Newtonian physics once again breaks down. "Momentum" in classical physics *must* have mass, but a photon has no mass. However, a photon must travel at the speed of light or it loses its energy. Any "thing" that has no mass and no energy is a "nothing." The value of a photon's momentum is the same relationship for its energy ($hf$). However, since momentum is usually stated as a direction as well as a function of energy, it is usually expressed in vector notation.

This result is also a bit peculiar in that the power of a radio *wave* can be calculated from its field strength *or* by the number of *particles/second.* The answer, in watts, will be identical. This is the essence of Einstein's wave-particle duality of electromagnetic radiation. The power of electromagnetic radiation is not "shared" between the "wave" and the "particle," but coexists in both simultaneously.

## Atmospheric Noise Considerations at MF and Low HF

Under most circumstances, RF power at these relatively long wavelengths does not penetrate the atmosphere. This is true for both extraterrestrial galactic noise intersecting the Earth and terrestrial noise propagating toward the sky. In both cases the power is either absorbed and/or reflected. On the Earth we observe this noise as thermal noise coming from a huge equivalent noisy resistor (the atmosphere) under a huge reflector (the ionosphere). However, the noise problem at these frequencies is considerably worse than the Earth's 290 K thermal noise.

Figure 10.13 dramatizes the noise problem at these frequencies. **Figure 10.14** provides a more detailed plot of atmospheric noise at MF and HF frequencies with the "noise

**Table 10.1**
**Noise Relationships**

This table shows the relationships among noise power levels, equivalent noise temperature, and the S unit equivalent (assuming –73 dBm = S-9 and the assumed difference between each S unit is 6 dB). $T_e$ represents the equivalent noise being generated by a thermal black body source. Clearly atmospheric lightning is not a black body source, otherwise the atmospheric temperature would be orders of magnitude hotter than the Sun! Lightning strikes can produce thousands of degrees of heat, but not billions of degrees.

| *dB reference to 290 K* | *Equivalent dBm/Hz from Figure 10.14* | *$T_e$ K* | *Equivalent S units in 1 kHz bandwidth* |
|---|---|---|---|
| 0 dB | –174 dBm/Hz | 290 | S-0 –17 dB |
| 20 dB | –154 dBm/Hz | $290 \times 10^2$ | S-0 + 3 dB |
| 40 dB | –134 dBm/Hz | $290 \times 10^4$ | S-4 – 1dB |
| 60 dB | –114 dBm/Hz | $290 \times 10^6$ | S-7 + 1 dB |
| 80 dB | –94 dBm/Hz | $290 \times 10^8$ | S-9 + 9 dB |
| 100 dB | –74 dBm/Hz | $290 \times 10^{10}$ | S-9 + 29 dB |

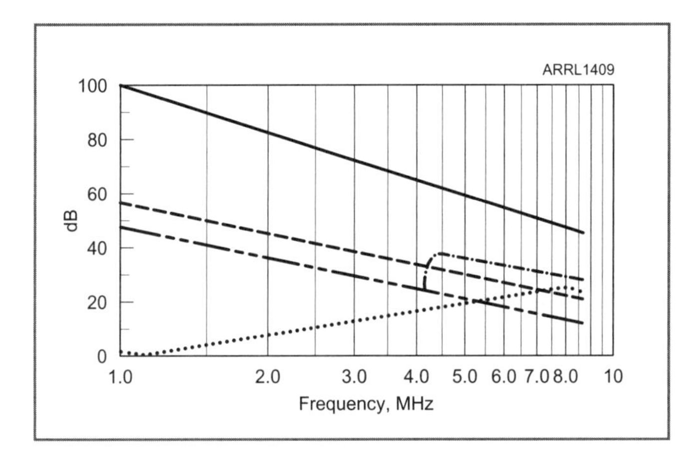


**Figure 10.14 —This figure plots the worst and best case presence of noise in the 1 – 10 MHz spectrum. The Y-axis shows the dB value above thermal noise at 290 K. Table 10.1 is provided to show the equivalent power and temperature to the Y-axis**

floor" on the graph being the noise temperature at 290 K. **Table 10.1** shows the relationships among noise power levels, equivalent noise temperature, and the S unit equivalent.

The solid black line in Figure 10.14 shows the maximum natural atmospheric noise. The natural atmospheric noise level dominates the over-all noise power when it is at its maximum and may be taken as the total contributor to the noise level. Lightning is the major contributor, the equivalent of hundreds of ultra-high-power spark gap transmitters keying at hundreds of times per second. When propagation permits, this massive power will appear at a receiver hundreds or thousands of kilometers away. In effect, the actual nighttime noise level at a given location at MF will be the grand sum of lightning occurring over the Earth's dark hemisphere and the quality of propagation to these areas. This is the mechanism that creates a constant "static" from the sum of thousands of strikes/second.

The dotted line shows the minimum natural noise level. Before the spread of electrical grids worldwide, this would have been the lower threshold for noise. Notice the huge variation of natural noise possible at 1 MHz, about 100 dB. Again, lightning is the main source of atmospheric noise at these frequencies and varies widely depending upon the season, time of day, and geographic location.

The line with alternating dots and dashes shows the galactic noise. In a quiet location where man-made noise is low and during periods of simultaneous low natural noise, galactic noise may predominate the total noise power at frequencies above about 5 MHz. The reason for the low frequency roll-off is the fact that the source is extraterrestrial and thus the ionosphere will reflect these frequencies back into space and not affect terrestrial receivers. The actual roll-off frequency is determined by the MUF and D-layer absorption.

The line with dashes shows the mean man-made noise level at a quiet rural location. When natural noise falls to low levels, man-made noise becomes the dominating factor. Like natural noise, nighttime noise is typically much higher than daytime noise because of propagation effects. The usual areas producing the most noise are highly populated regions such as China, Japan, North America and Europe. Consequently, pointing a directional antenna in these directions will result in a higher noise response, if, of course the band is "open" to these areas.

Statistically we can add to this noise level to adjust for additional local noise. For these general locations simply add the following expected power (dB) to this line:

1) "Normal" rural area, +11 dB
2) Residential area, +18 dB
3) Business/commercial area, +24 dB

Of course these are average difference values. Actual additional noise due to proximity to populated areas can vary widely. Marconi certainly had a much better noise environment when he completed the first transatlantic radio contact! Also, man-made noise will almost certainly increase in the future. This will result in ever increasing difficulty for working DX on the low bands.

The line with two short dashes and a long dash alternating shows the "best" case of man-made noise at a quiet location.

Here, "quiet" location might mean Antarctica, a remote island, or Greenland — but only when propagation to noisy areas is not good.

## Receive Antenna Parameters for MF and Low HF

In the previous section we described the various noise sources in this frequency range. The problem at these frequencies is similar to the problem of radio astronomy in one important way. Both applications present the challenge of detecting a point source on the hemispheric sky with that hemisphere filled with noise. In contrast, the problem for MF and low HF receivers is dealing with very high noise temperatures, while Penzias and Wilson faced measuring temperatures near absolute zero. The general idea of point source versus hemispheric source points to some clear objectives for MF and low HF receive antenna design.

If we assume that the sky has a uniform noise temperature over its entire hemisphere (which it usually does not have!) and we increase the gain of our receive antenna, the noise power we receive will be the same as the hemispheric pattern antenna, as in the radio astronomy explanation. However, if we increase the gain of our receive antenna, a desired signal will increase as a function of the higher gain since it is a point source. Suppose we use an isotropic receive antenna with the resulting total received power including both noise and the desired signal,

$$P_{r(iso)} = P_n + P_s \quad \text{(Equation 10.46)}$$

Now if we assume the noise to be constant over the hemispheric sky, and add gain (3 dB for this example) and, more importantly directivity, then

$$P_{r(3\ dBi)} = P_n + 2P_s \quad \text{(Equation 10.47)}$$

Notice that the noise power has remained constant but the signal power has doubled. Thus we have increased our SNR by a factor of two. Adding more directivity will further increase the SNR linearly as the numerical directivity increases.

$$\text{SNR} \propto \text{directivity} \quad \text{(Equation 10.48)}$$

The receive antenna efficiency can be relatively low since we are mainly concerned with the received SNR and the received noise is likely to be very high. Again the relationship is

$$\text{Gain} = \text{Directivity} \times \text{Efficiency} \quad \text{(Equation 10.49)}$$

If our receive antenna is 100% efficient, then our antenna noise temperature will be equal to the noise temperature of the sky. For example, by Figure 10.14, a very quiet condition for 1.8 MHz implies an antenna noise temperature of about 290 × $10^4$ K, which places the antenna thermal noise floor at about –134 dBm/Hz or –104 dBm for a 1 kHz bandwidth. Therefore the received desired signal strength in dBm is used to compute the SNR against –104 dBm noise.

Now, what happens when we use a small receive antenna such as a 20 foot vertical, or a small loop? How much inefficiency can we tolerate? For this calculation we use the same equation for radio astronomy equivalent noise temperatures, Equation 10.43. For example, let us assume we are using a K9AY loop. From Figure 9.9 in Chapter 9 we see the maximum gain is –26.6 dBi, which is very good for such a small loop. We will first calculate the noise powers assuming that the loop is isotropic in the hemisphere (+3 dBi). The efficiency of such a loop is about 0.2%.

$$T_{ant} = 290 \times 10^4 (0.002) = 6344 \text{ K} \quad \text{(Equation 10.50)}$$

We will also assume a weak signal condition, where the SNR = 1. Assume 99.8% of the noise source for this antenna is attributed to the actual temperature of the antenna $R_l$, and 0.2% will originate in the antenna's $R_r$. We do not care what the actual values of $R_l$ and $R_r$ are, since our given efficiency defines the ratio. $R_l$ is simply the ohmic value of the antenna, and thus the noise power will be –174 dB/Hz. Notice that the equivalent sky noise temperature is only about 13 dB higher than the thermal noise emanating from the antenna's thermal resistance. Again, this is due to the antenna's inefficiency, thus lowering the "received" noise power.

The sensitivity equation (Equation 10.32) further lowers the SNR due to the receiver's noise figure. A typical noise figure for MF receivers is 15 dB, placing the receiver noise temperature at 9170 K. Often the receiver's attenuator will be used in an effort to lower intermodulation and other nonlinear responses in the receiver proper. These nonlinear receiver responses have the effect of raising the equivalent noise response and/or creating spurious signals in the receiver. In-band intermodulation products are reduced a minimum of 3 dB for every 1 dB of attenuation added. Therefore finding the "sweet spot" is a balance of improving the intermodulation performance versus not affecting the sensitivity limit. The "sweet spot" will change for different noise power (and signal) inputs.

Thus for the given efficiency, a hemispherical response, and a very quiet atmospheric noise condition, our weak signal will not be detected. However, the K9AY loop also has directivity, and by Equation 10.46 we gain back some of the SNR. It is critical that the desired point source is aligned with the maximum directivity of the antenna, or as close as possible. The object is to maximize directivity in the direction of the desired point source. Sky noise can be assumed to be arriving from all elevation angles, while some distant weak desired signal will be arriving at a relatively low angle. A small vertical antenna will naturally reject the noise arriving from higher angles, where a small loop will not. A low dipole will be a very poor receive antenna under these circumstances since its directivity maximum is "straight up." The loop will respond to low-angle vertical polarized waves, but it will also respond to noise from "straight up." Consequently the small vertical has an inherent SNR advantage over the loop. Small loops that also offer azimuth gain, such as the K9AY loop, will also benefit from some high-angle rejection but not as well as a vertical.

Starting with the vertical and then adding elements results in the additional advantage of azimuth as well as elevation gain. The Beverage also has a low angle response and gain in the desired direction. The Beverage also exhibits higher efficiency than smaller loops, allowing more flexibility with receiver controls. This explains why Beverages, vertical 4-squares and 8-circle antennas are the usual preference for low frequency receiver antennas, not to mention a quiet antenna location. Multiple plots of these antenna configurations are presented in Chapters 7 and 9. When considering the all-important SNR figure of merit, the total steradian pattern must be used to compare expected SNRs, especially high angle responses. Unfortunately, these very effective receive antenna arrays tend to be very large, compared with small loops. Therefore the small loop might be the best (or only) choice when limited by real estate.

## Shannon's Limit

There is one more expression to define the *absolute* physical limits to *any* communication link, including radio communication. Although not usually considered in an antenna text, the Shannon Limit is the *fundamental* limitation of *any* communication link, including all radio links. Since it is a *physical* limit that also involves noise, we include at least a basic discussion in Antenna *Physics*. Also by reviewing the Shannon Limit, we find that noise becomes a fundamental *theoretical* limitation to any radio link regarding the rate of information conveyed as well as a fundamental *theoretical* limitation to the dynamics of the radio link. Therefore, antennas play a crucial role in determining the fundamental *theoretical* limits of any radio communication link from *both* physical limitations.

The receiver must interpret the received waveform as containing some desired *information.* To this end the waveform, or *signal*, must contain information that is *unknown* to the receiver or there is no point in transmitting the information. The format of the signal must be known by the receiver, but the information carried by the signal is usually unknown. For example, when we tune an SSB receiver we *assume* a certain type of waveform, but the information we receive may be unpredictable (for better or worse).

Claude Shannon (**Figure 10.15**) was working at Bell Labs in the late 1940s when he provided the world with a remarkable and profound insight: Information, in whatever form, can be directly equated with a ratio of a desired "signal" containing power or energy to that of the noise power or energy in a "channel." We usually think of *"information communication"* as being some abstract phenomena. Shannon actually successfully equated it to *physical* terms. His work directly led to the formation of an entirely new discipline, Information Science.

Now we return to the concept of SNR, but now include it in an equation of fundamental importance:

**Figure 10.15 — Claude Shannon in 1952 at Bell Labs. [Courtesy of Alcatel-Lucent USA]**

The Shannon Limit

$$C = B \times \log_2\left(1 + \frac{S}{N}\right) \quad \text{(Equation 10.51)}$$

where

$C$ is the capacity in information bits/second
$B$ is the bandwidth of the "channel" in Hertz
S/N is the signal-to-noise ratio

In this equation the *capacity (C)* term needs some explanation. We usually think of a digital communication system specified in bits per second. This says nothing about the *information per bit. "Information,"* in turn, is only conveyed when the receiver learns something from the transmitter it did not know before the transmission.

A simple analogy is when someone tells you something that you already know. No new *information* has been sent and received. In effect the *energy* and time taken to transmit the data was wasted. Now imagine the speaker knows *exactly* what you *know* and *don't know.* The speaker then cleverly crafts what he/she says so that each word and inflection conveys the maximum possible amount of new *information* without repeating anything that you already know. The speaker also knows exactly how much information you are capable of receiving, the characteristics of your hearing, and so on. This is an example of maximizing the information contained for a limited acoustic energy expended, constrained by the audio bandwidth in the presence of undesirable noise, and "noise" within the listener's hearing, nervous system, neural receptors and whatever else affects the reception of the desired *information*.

This example is closely analogous to a digital communication system, including radio links, which, of course, include antennas. Now imagine a digital communication system sending some text written in English. If the transmitter sends a "q" then there is no information conveyed by also following the "q" with a "u," assuming, of course, all English words that use a "q" are *always* followed by a "u." In effect, the time and energy spent sending the "u" are wasted since the receiver already *knows* English.

The next leap is to imagine that some *code* is developed that permits each *bit* of information conveyed to contain the maximum amount of information. Both the transmitter and receiver know English, and both also have the ability to encode (transmitter) and decode (receiver) the 100% efficient coded message. Therefore, one of the implications of Shannon's Limit is to convey the maximum amount of information per bit possible. This has resulted is a multi-billion dollar industry developing and implementing codes, all to overcome the effects of *noise* and the resulting *bit errors*.

The type of modulation used in a radio system must also be optimized to provide maximum spectral efficiency. Of course this topic also fills volumes. However, we must consider a theoretically perfect modulation scheme together with some ideal code to reach the Shannon Limit.

There is another intriguing implication of the Shannon Limit equation. Notice that the equation does not have an error term. The "limit" is quite literal. If the information is coded with theoretical precision, there will be zero errors if the equation is satisfied. Below the limit, no information can be communicated. Above the limit the error rate is zero (again assuming a "perfect code").

As an aside, the equations that define the amount of information in a bit are quite similar to the equations defining entropy in thermodynamic systems. Despite the fact that there is no relationship between information theory and classical "entropy" (except the equations look similar), the amount of information in a bit is termed its "entropy"! See **Figure 10.16**.

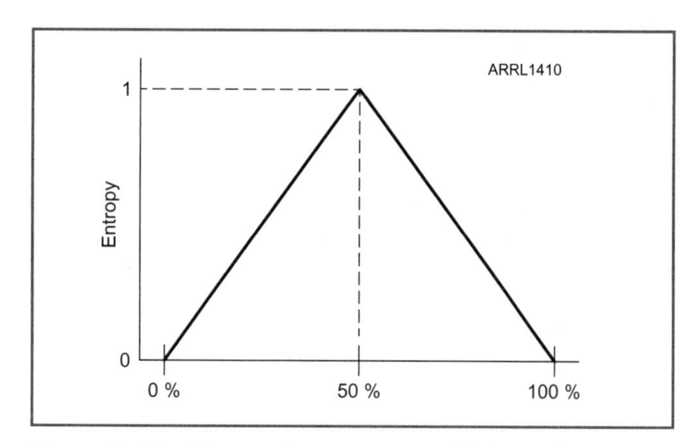


**Figure 10.16 — The maximum amount of information (maximum entropy) is sent when the receiver perceives a 50% probability that a bit will be a 1 or a 0. There is no entropy when there is either a 0% or a 100% chance that a bit will be a 1, as the receiver already knows the outcome. To reach the Shannon Limit, the challenge is for each bit sent to contain an entropy of 1.**

Finally, let us assume we could build a communication system with *zero noise*. Using Equation 10.51 we find that the channel capacity becomes *infinite* even for an *arbitrarily narrow bandwidth*. With zero noise, we could run error-free Gb/sec links through 1 Hz bandwidth channels. Also, any radio link could have *infinite* range if all noise (and interference) could be eliminated. Indeed the universe would be a very different place without the existence of random radio noise being present everywhere.

Even today many people, including many engineers, are not aware of this incredible and counterintuitive understanding of information and noise. The reader is encouraged to pursue Shannon and information science further. However, from a *physical* standpoint, Shannon's Limit represents the starting-point for defining limitations on any communication system. If you know the noise level and bandwidth, you can quickly calculate a *physical limitation to the channel's information capacity*. Finally, because this limit is a *physical limit*, it applies to all information systems, radio, fiber, and even biological systems, including human beings. Shannon's Limit has found direct applications in diverse areas as neuroscience and electronic communication systems.

## Extreme System Design

In this section we will apply fundamental limits from antenna physics and information science to define a system limited only by these physical constraints. This analysis will use the concepts shown in Figure 10.7 combined with the Shannon Limit.

Sometime in the future we may wish to communicate with another civilization in a solar system in the direction of an otherwise empty sky. We can use a very slow data rate since it may take many years to reach the target receiver. We have already received signals from this civilization thus we know what frequency and coding to use. Our parameters are an equivalent 1 Hz bandwidth and a minimum SNR of 1/1. What is the maximum capacity of this system? By Equation 10.49 we find the capacity is 0.69 bits/second. Now what is the sensitivity of the receiver assuming a background noise of 2.7 K and a receiver noise temperature of 1 K?

From Equation 10.32 we rewrite the appropriate terms

$$\text{Sensitivity} = -194 \text{ dBm} + 10 \log_{10}(1) + NF\ (-198 \text{ dBm}) + 0(NF)$$

where –194 dBm/Hz is the noise power from the Big Bang (since we are using only 1 Hz bandwidth there is no correction for bandwidth), the receiver noise temperature is 4 dB below the antenna noise temperature, and the F = 1 which equals 0 dB. Therefore the sensitivity is 1.5 dB above the antenna temperature, or –192.5 dBm. Since the assumed SNR is 1, then the desired signal must be at least –192.5 dBm.

We assume equal gain antennas at each end of the link and a carrier frequency of 6 GHz ($\lambda = 0.05$ m). What is the EIRP necessary to deliver this power at a distance of 100 light years (LY)?

$$1 \text{ light year} \cong 9.47^{15} \text{ meters} \quad \text{(Equation 10.52)}$$

We recall from Equation 2.5 the isotropic loss is

$$\left(\frac{\lambda}{4\pi R}\right)^2$$

In this case

$$\left(\frac{\lambda}{4\pi R}\right)^2 = \left(\frac{0.05 \text{ m}}{4\pi(100)9.47^{15} \text{ m}}\right)^2$$

Based on Equation 2.7 in Chapter 2, we can now set up the necessary equation:

$$-192.5 \text{ dBm} \cong 63^{-24} \text{ W} = \left(P_t G_t G_r \left(\frac{0.05}{4\pi(100)9.47^{15}}\right)^2 \text{ W}\right) \quad \text{(Equation 10.53)}$$

or

$$P_t G_t G_r = \frac{63 \times 10^{-24}}{\left(\frac{0.05}{4\pi(100)9.47 \times 10^{15}}\right)^2} \text{ W} \quad \text{(Equation 10.54)}$$

or

$$P_t G_t G_r = \frac{63 \times 10^{-24}}{176 \times 10^{-42}} = 357 \times 10^{18} \text{ W} \quad \text{(Equation 10.55)}$$

This is the amount of EIRP and receiver antenna gain to satisfy the minimum link budget. Now we can derive the necessary antenna gains and transmit power.

At 6 GHz the aperture of an isotropic antenna is

$$A_{iso} = \frac{0.05^2}{4\pi} \cong 200 \times 10^{-6} \text{m}^2 \quad \text{(Equation 10.56)}$$

Now let us assume that at each end are 200 meter diameter parabolic antennas that are 100% efficient (or very close to it). Therefore the antenna aperture at both ends is $31.4 \times 10^3$ m$^2 = A_{parabola}$. This represents a power gain of $15.7 \times 10^6$ or about 72 dBi for *each* antenna. Therefore from Equation 10.52 we derive

$$P_t = \frac{357 \times 10^{18}}{\left(15.7 \times 10^6\right)^2} = 1.45 \times 10^6 \quad \text{(Equation 10.57)}$$

The value of the linear power gains of the two antennas (here equal) in the link budget analysis are multiplied. This is identical to adding the dB gain numbers. The necessary transmitter output power is 1.45 million watts assuming zero transmission line losses to communicate a maximum of 0.69 bits/second of information a distance of 100 light years. Thus the equations defined by the physics of the antenna, radio performance, and information theory apply to wireless garage door openers the same way they apply to interstellar communication links.

## Physics Defines the Ultimate Limits to Radio Communication

In this chapter we have defined the *fundamental physical* limitations to a radio communication link. First, even if we point our antenna with a background of "quiet" sky we will still have an antenna noise temperature of 2.7 K (assuming a 100% efficient antenna with no sidelobe response). Second, even if we achieve the Shannon Limit by maximizing the information sent with each bit, we will still be limited by how much information we can send per unit time since we will have some finite noise from the Big Bang. Third, we are forced to conform to the physical realities of link budgets, the inverse-square law, antenna aperture, receiver sensitivity, and transmit power. These limitations are due to the fundamental *physical* characteristics of free space itself.

In addition to these limitations defined by this *physics,* there are further *engineering* limitations mainly constrained by the state-of-the-art in all relevant areas: transmission lines, amplifiers, mixers, data converters, and, of course, even antennas are never perfect. They all have losses and a host of other undesirable effects. Also, the myriad of matter's physical characteristics and the resulting effects upon antennas and radio links must be considered.

From these daunting realities, we actually described a system that is limited only by physics, not current engineering state-of-the-art. This text attempts to touch on all the basic variables pertaining to the antenna and the antenna's place in a system. Further advances in theoretical limits will have to wait for the inevitable evolution of human knowledge and the resulting upgrading of our understanding of the limits of *physics.*

Appendix

# You Already Know Calculus

Back when I was a student studying mathematics, "calculus" seemed like a daunting and mysterious subject. My teachers never explained to me what it is or why it is so valuable. I was only told to master algebra, geometry, trigonometry, and then "pre-calculus." Had I simply known what it was — and more importantly, its great usefulness — I would have been much more excited to prepare for it. Why not simply tell students what calculus is and, more important, how powerful a technique it is?

The intention of this Appendix is not to *teach* calculus. The purpose is to explain what it is, and convince you that you already understand the basic principles. Everyone who has ever driven a car understands the relationships among acceleration, velocity, and distance driven. Looking at this relationship in a slightly different manner will lead to the first realization of what calculus is.

## A Simple Example

See **Figures A.1** and **A.2** for the following discussion. Imagine a motionless car at a starting line. The car is not accelerating, has no velocity and is at point 0 (starting line, X = 0) along a long straight road. Now let us consider these three variables:

1) Acceleration
2) Velocity
3) Position (along the road (X-axis))

These three variables will be plotted against *time.* At *time* = 0 seconds, the car begins to move, and therefore has acceleration. Acceleration is defined with two functions of *time*, for example 5 *miles/hour/second.* In other words, the car has a *velocity* of 5 *miles/hour* after 1 *second*, 10 miles/hour after 2 seconds, and so on. Also notice that velocity is defined with only one term of time (*miles/hour* or *mi/h*). *Position,* however is defined with *no* function of time. A mile is simply a mile along the X-axis; it does not require time to define it.

Now, if we accelerate 5 miles/hour/second at a constant rate, it is a simple matter to multiply this acceleration rate to determine the velocity. For example, if we accelerate at this

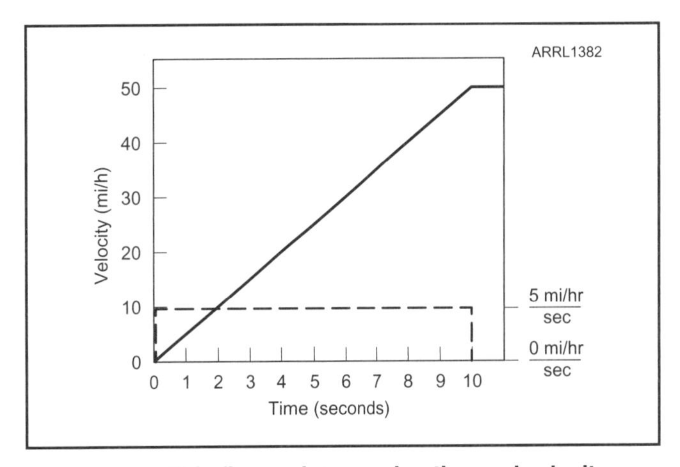


**Figure A.1 — This figure plots acceleration and velocity against time. At time 0 seconds we begin 10 seconds of acceleration. At the end of 10 seconds our acceleration drops to zero, so our velocity remains constant at 50 mi/h.**

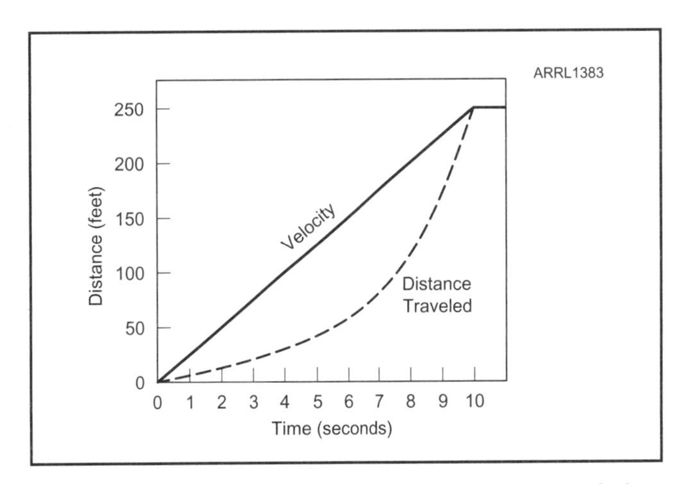


**Figure A.2 — This figure shows that the distance traveled for an increasing velocity is a nonlinear function (exponential). When the acceleration stops after 10 seconds, the velocity remains constant at 250 feet/second, and the distance traveled increases at a linear rate of 250 feet/second, as the exponential curve becomes a straight line.**

rate for 10 seconds, our velocity will be 50 miles/hour. At any point in time, we know *exactly* how fast we are going.

In this simple example, all we need to do to calculate velocity is multiply acceleration by time where

$$\frac{\left(mi/h\right)}{s}(s) = \left(mi/h\right) \qquad \text{(Equation A.1)}$$

The seconds cancel and we are left with mi/h (or mph).

But what about distance traveled (*X*)? If our velocity is a constant 50 mi/h, then the calculation is similarly trivial, we multiply velocity by time to get miles.

$$\left(mi/h\right)(h) = mi \qquad \text{(Equation A.2)}$$

However, in our above example, velocity is *changing* for the first 10 seconds, so we cannot simply multiply by time to get distance traveled. Arithmetic breaks down and we need something more sophisticated to "add up" the velocities over time to get distance traveled. For this we need to "integrate" the velocity over time.

For this operation it is more convenient to use *feet/second* as the velocity, since miles will be a small fraction after only 10 seconds of acceleration. So let's redefine acceleration as 5 feet/second/second.

We can now define the velocity as $5t$. That is, velocity increases at a *rate* of 5 feet/second/second; in this case we'll use time ($t$). So after, say, 4 seconds, our velocity will be 20 mi/h.

If we want to find how far we have traveled during the first 10 seconds, we need to set up an integral from $t = 0$ to $t = 10$. The equation we use is a *definite integral* with the start and end points, and we define the velocity as a function of time ($5t$) since the velocity is simply 5 feet/second/second (the acceleration) multiplied by time. Now to determine the distance traveled in the same 10 seconds, we cannot simply multiply the velocity × time because the velocity is gradually changing with time. We need to use integral calculus to calculate the *function* of velocity over time. Again the distance traveled (*X*) is equal to the *function* of velocity *integrated* over the proper amount of time (10 seconds).

$$X = \int_0^{10} 5t\,dt \qquad \text{(Equation A.3)}$$

The actual calculation of the time to solve this problem *does* involve knowledge of integral calculus. However, understanding that it is the method of determining the distance traveled at the end of 10 seconds is a giant leap to understanding what integral calculus *is*. We will cheat here and present the actual solution:

$$X = \int_0^{10} 5t\,dt = (10)\frac{5}{2}t^2 - (0)\frac{5}{2}t^2 \qquad \text{(Equation A.4)}$$

Now we take two points, where $t = 0$ and $t = 10$. Finding the difference will provide the answer. When $t = 10$, the result is 250 feet. When $t = 0$ the result is 0 feet. Therefore

$$X = 250 - 0 = 250 \text{ feet} \qquad \text{(Equation A.5)}$$

Therefore, starting from zero velocity, with an acceleration of 5 feet/second/second, after 10 seconds the car will have traveled 250 feet. Similarly, we can calculate the total distance traveled for any time. After 5 seconds, the car has traveled 62.5 feet. Without calculus an exact calculation is impossible.

## Integral and Differential Calculus

We can think of integral calculus as a very powerful tool for "summing up" functions, including functions that cannot be "summed" by simpler mathematical tools (addition, multiplication, and so on). Distance traveled is simply a sum of all the velocities (including changing velocities) over a portion of time. Velocity is simply a sum of all accelerations (and changes of accelerations) over time. Thus if you know the function that defines acceleration over time, it is easy (using calculus) to find the velocity, and also the distance traveled.

But what about calculating velocity from distance traveled? If we know the distance traveled in 1 hour (for example, 50 miles), we can easily calculate the *average* velocity (50 mi/h). However, we cannot calculate the velocity at any given point in time. We might have driven 100 mi/h in a rural area, and then slowed down through a town. However, if we know the *function of* the distance traveled over time, then we can calculate the velocity at any given time by taking the *derivative* of the function.

For example, if the function of distance traveled is

$X = 50t$ miles

Then our velocity has indeed been constant and the calculation is simple. However, if our distance traveled is $5/2t^2$ feet/second, then we can substitute any value of seconds and calculate *velocity* by *differentiating* $5/2t^2$. The differential is written as

$$\frac{dx}{dt}$$

meaning we are finding the *rate of change* of $x$ as $t$ changes.

Again we will cheat and present the answer for the derivative. This is simply finding the rate for change in velocity over time. In this case, as above, the velocity is increasing for the duration of the 10 seconds (as in a drag race). Thus the velocity is simply 5 mi/h/sec ($t$) or

$$\frac{d\frac{5t^2}{2}}{dt} = 5t$$

You now know the two basic divisions of calculus: *integral* and *differential.*

*Integral calculus* is "summing up" even difficult functions to provide "the integral of" function.

1) The *(time)* integral of acceleration is velocity.

2) The *(time)* integral of velocity is distance traveled since we are integrating (summing) over time.

*Differential calculus* is "finding the rate of change," or the *derivative* of a function.

1) The time derivative of distance traveled is velocity.

2) The time derivative of velocity is acceleration.

If acceleration is zero, velocity is constant. If acceleration is positive (going faster), velocity increases. If acceleration is negative (braking), velocity slows. If velocity is changing, there is some non-zero value for acceleration. The derivative of a constant function (not changing) is always zero.

We can also say that the *second derivative* of distance traveled is acceleration, since we have to take the derivative of a derivative. The form is:

$$A = \frac{d^2X}{dt^2}$$

and the distance traveled (*X*) is the integral of the integral of acceleration, the form being

$$\iint Adt$$

It's really that simple!

Now we can explore other examples. Energy is the time integral of power. For example, if we run a 1 kW machine for 1 hour, we have used 1 kW-hour of energy, another simple calculation. However, if we are constantly changing the power being used according to "smooth" functions, we will need calculus. For example, suppose we have a transistor amplifier that exponentially increases its power until there is a large surge of power, the transistor is destroyed and all power consumption suddenly stops. If you know the *function* that defines the destruction sequence you can calculate the energy consumed by the sequence. In this way, your power meter is performing a power integration to tell the power company how much your bill should be. It has integrated all the power you have used, no matter how nonlinear the consumption has been, to yield the integral of the power over time: the energy used.

## Not Limited to Time

Time is a very frequently used variable for a wide variety of real-world calculations with or without calculus. However, it's not the only possible variable. Many dielectrics' constants will change as a function of frequency. Such a differential term would take the form:

$$\partial \varepsilon_r / \partial \omega$$

This equation signals we are defining how the dielectric constant of a material ($\varepsilon_r$) is changing with changing frequency ($\omega$). In this text we also use a *partial derivative*. Let's assume that our material's ($\varepsilon_r$) dielectric constant changes not over just frequency, but also over temperature. Thus $\varepsilon_r$ is *partially* dependent upon frequency and *partially* dependent upon temperature. Therefore

$$\varepsilon_r(total) = \frac{\partial \varepsilon_r}{\partial \omega} + \frac{\partial \varepsilon_r}{\partial T} + reference\ function$$

This is known as a partial differential equation. This equation informs us how the dielectric constant of a material changes with frequency and with temperature. Of course we need some starting point value here represented by a reference function. By plugging in the actual numbers, we can perform an actual real calculation. The above equation only illustrates the form. Again, solving for an actual solution usually requires several years of training in advanced mathematics, but understanding *what it means* does not.

The applications are endless: the rate of growth of trees over different temperatures, humidity, soil types, sunlight, and other factors. Or we could analyze the rate of weight gain in humans as a function of exercise, carbohydrate intake, genetics, toxic exposures, and so on.

In Chapter 3 we presented Maxwell's equations using a special form of an integral

$$\oint x$$

This simply informs us that we are summing (in Maxwell's case magnetic or electric fields, here simplified as *"x"*) over a closed surface, usually a sphere. This looks like a difficult operator, but in principle it is quite simple to understand.

We all know how to find the average of a number of values: add up the values (sum) and divide by the number. In Chapter 7 we used an integral of a sin function to find the sum of the area under a sin curve, and then divided by the maximum value. This is a necessary step to determine the radiation resistance along an antenna element with a sinusoidal current distribution, easily accomplished with integral calculus, impossible otherwise.

As a final example, in Chapter 8 we set up a *matrix* of differentials to describe how *five* differential variables are all *interdependent* in order to set up an equation that defines antenna loss due to soil resistance. This equation uses techniques at least one year past calculus, but again, it's just one more step in complexity for precisely describing multiple variables all interdependent with respect to a final desired term, in this case loss resistance.

We live calculus functions all the time. They define almost everything that changes as a result of something else changing. We solve them all the time without knowing that we are doing so. Many relationships (such as driving a car) are second nature and become instinctive. Understanding calculus permits insights into a very wide breath of problems impossible without it.

To conclude, many if not most of the complicated looking equations presented in advanced science and engineering texts are just a more sophisticated way of defining something that simpler techniques cannot. That's what calculus is all about, but you already knew that.

Appendix

# Summary of $R_r$ Terms Described by Kraus

Most authors of advanced antenna texts discuss radiation resistance. Of all the authors, Kraus offers the most comprehensive treatment and for the most part is in agreement with the other authors. Reference: John D. Kraus, *Antennas*, 2nd ed, McGraw-Hill, 1988.

In particular, the discussion of $R_r$ for ground-mounted vertical antennas is a common source of confusion. Kraus provides several approaches to solve for $R_r$ under different circumstances. I have introduced a new term ($R_{rp}$) to distinguish between the $R_r$ of the antenna versus the $R_r$ at a particular point along the antenna ($R_{rp}$). Following is a summary of these terms.

1) $R_r = \frac{P_r}{I^2}$ (from Kraus, page 216)

This conceptual definition has several weaknesses. The term $P_r$ is difficult to measure directly in the best of circumstances. In the case of ground-mounted verticals it is nearly impossible.

Furthermore, attempts to use computer modeling become extremely problematic. In the free space hemisphere above ground, a far field power integration includes the effects of power loss *and* losses due to propagation effects. In the dielectric hemisphere below ground, any specific surface of integration will yield a false value, since you are only integrating power at a specific distance from the feed point and inherently ignore power loss beyond that distance. Worst of all, integration of power in the ground usually cannot distinguish between power lost to heat and power actually *radiated* underground (which also ends up being dissipated as heat!).

Using computer modeling techniques, power integration is typically performed over a surface, such as a sphere. Within a lossy dielectric power "radiated" into the dielectric and "lost" to heat are both distributed through the three-dimensional dielectric volume. Consequently an accurate differentiation between radiated and absorbed power is nearly impossible. We rightly consider underground radiation to be "lost," but a calculation of $R_r$ must account for all radiated power, "desired" or not.

The portion of power that can be directly attributed to antenna loss and not propagation loss is best calculated by a direct measurement of $R_r$ at a point of maximum current and then compared with a modeled antenna over lossless ground. Only loss in the return current can provide the proper term to calculate losses in ground-mounted verticals. This value simply appears as an equivalent resistance at the feed point. Theoretically it is possible to actually perform a power integration that will correctly differentiate radiated and lost power, but this would likely be an exceedingly difficult model.

For direct measurements we must somehow *know* that the antenna is lossless, since there is no compensation for loss in this equation. This equation also does not provide a term for the $A_e$ or the $h_e$ of the antenna, which additionally limits its usefulness. Although this equation does automatically compensate for a change in $R_r$ for $A_e$, it does not provide the necessary terms to tell us why.

The reader should be wary of any attempt to directly calculate $R_r$ using this equation, particularly for ground-mounted verticals.

2) $R_r = \frac{S(\theta,\phi)_{max} r^2 \mathbf{\Omega}_A}{I^2}$ (from Kraus, page 847)

This is Kraus' general definition of $R_r$. It represents the simplest closed form equation that takes the two critical terms ($A_e$ and $h_e$) into account. The numerator contains the equivalent of the power input term from the simpler equation. However, the radiated power term is defined in terms of gain, which in turn, can also be easily converted to aperture, or in this case, to a power density concentrated into a steradian value. The actual definition of this general equation is the 3 dB beamwidth, again stated in steradian value. A 3 dB beamwidth corresponding to a steradian value *approximately* contains all of the power radiated. However, if we hypothetically create an area on the sphere that contains all the power (defined by a "solid" steradian angle) normalized to the maximum gain, this equation becomes precise.

From a practical view, the 3 dB beamwidth is usually easier to measure than the Poynting vector over the entire sphere, so this equation is not only closed form but is also much more useful. Since we now know the gain, we also know what the lossless feed impedance will be, so all the terms become deterministic, including loss.

3) $R_r = \frac{h_e^2 Z_0}{4A_e}$ (from Kraus, page 42)

It is instructive to note that this equation is the first definition Kraus offers for $R_r$ in the introductory Chapter 2. It is also instructive to note that this equation can be directly derived from Equation 2 above. As Kraus further notes, this equation's form is particularly applicable to ground mounted verticals and small antennas (relative to wavelength). Indeed, in these applications, $R_r$ becomes a much more critical term than in most other antenna configurations. The reason for the great utility of this equation is that it takes into account *both* antenna loss and antenna gain. Most importantly, the radiated power term is calculated by a distributed and maximum current *on the* antenna and not a far more difficult radiated power calculation.

Antenna loss results in both a reduction in the antenna's average current and maximum current ($h_e$) and gain (represented by $A_e$). As long as these terms remain linearly proportional there can be no change in $R_r$ in the presence of loss, unlike the implication from Equation 1. Also, this equation provides the accurate effect gain has on $R_r$, an inverse proportion. This equation represents the proper definition for the ground mounted vertical since it includes all relevant variables in a closed-form.

In the case of very low conductivity dielectrics, it is exceedingly difficult to separate radiation into the dielectric and actual heat loss in the dielectric. Eventually the radiated power will indeed be lost to heat, but in the far field where it can be considered as a propagation loss, not an antenna loss. From a practical sense, any power lost in the dielectric is *loss*, radiated or lost to heat in the near field since it is not radiated into free space. However, the value of $R_r$ "doesn't care" it only responds to power lost in the antenna proper (including through mutual impedances) or power reflected back to the antenna. Again, extreme care must be taken in the actual calculations of "loss" and "radiation."

Using relatively simple antenna modeling programs such as *NEC*, it is possible to set up a model that can force return losses into an equivalent single resistor and eliminate the effects of far field loss simultaneously. The equivalent loss resistor is placed just above the ground, with the feed point located just above it, both very close to the ground. This simple technique will be suitably accurate for most "real" grounds. Antenna loss proper will converge upon a simple single resistance. The antenna is modeled using a perfect ground to eliminate propagation losses. A decrease in $h_e$ and thus efficiency will track perfectly with a reduction in $A_e$ (gain) just as Equation 3 states.

4) $R_r = 94.25\frac{h_e^2}{A_e} = 94.25\frac{h_e^2}{3\lambda^2/4\pi} = 395\left(\frac{h_e}{\lambda}\right)^2$

This equation provides a direct and abbreviated method to calculate the radiation resistance of a short vertical: Kraus does not present this special equation, but does derive a similar expression for very short dipoles. This equation is published by many sources, and can be seen to be a direct derivation of Equation 3. For ground mounted verticals less than about ⅛ λ, the gain versus physical height becomes nearly constant at 4.77 dBi, or a power gain of 3. Thus the aperture simplifies to

$$\frac{3\lambda^2}{4\pi}$$

for any vertical height below about ⅛ λ. The other simplification involves $h_e^2$. For short verticals, the current distribution can be assumed to be linear (a small portion of a sine curve). Therefore, a linear short vertical's $h_e$ value will be ½ its physical height multiplied by 2 (to include the image), or about 6 Ω for a ⅛ λ physical height. For a perfectly top loaded ⅛ λ vertical, $h_e$ becomes 1× its physical height × 2 (to include the image), or about 24 Ω. Therefore a properly top-loaded short vertical has a fourfold increase in $R_r$ when compared to an unloaded simple vertical of the same physical height.

5) Radiation Resistance at a Point which is Not a Current Maximum (Kraus, page 227)

This section make two simple points: First, with linear antennas greater than ½ λ, the choice of feed point location will change the current distribution along the antenna, and thus the location of the current maximum and directivity. Second, for a lossless antenna, the real portion of the feed point impedance can be taken to be the radiation resistance at that point. Therefore for radiated power calculations, the simple formula $I^2R_{feed} = P_r$ applies. An example of this is in AM broadcast antennas, where it assumed that the antenna is lossless. That is usually close enough for radiated power calculations, but never precise. In effect, Kraus is describing here what I have defined as $R_{rp}$ for clarification.

Appendix

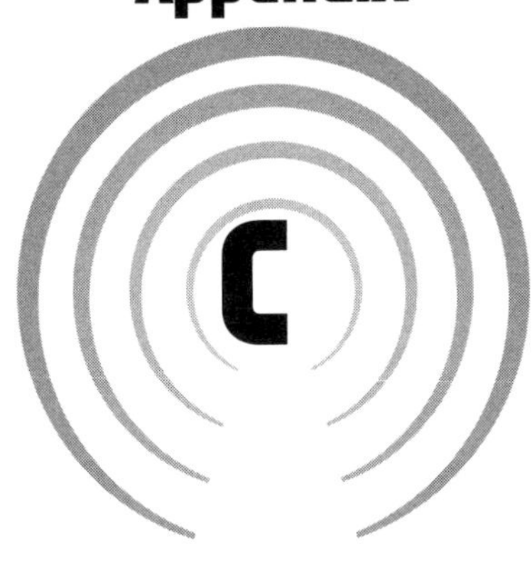


# References

The following articles and texts were valuable in the course of preparing this book.

International Telecommunication Union Recommendation ITU-R P.372-6, Radio Noise, **www.itu.int/dms_pubrec/itu-r/rec/p/R-REC-P.372-6-199408-S!!PDF-E.pdf**

"Project Cyclops: A Design Study of a System for Detecting Extraterrestrial Intelligent Life," NASA, Ames Research Center, 1971.

Bianchi, Cesidio, and Meloni, Antonio, "Natural and Man-made Terrestrial Electromagnetic Noise: An Outlook," Instituto Nazionale di Geofisica e Vulcanologia, Roma, Italy, *Annals of Geophysics*, Vol 50, No 3, June 2007, p 435.

Balanis, Constantine A., *Antenna Theory*, 3rd Edition (Wiley-Interscience, 2005).

Devoldere, John, *ON4UN's Low-Band DXing*, 5th Edition (ARRL, 2010).

Drentea, Cornell, *Modern Communications Receiver Design and Technology* (Artech House, 2010).

Fujimoto, K., Henderson, A., Hirasawa, K. and James, J. R., *Small Antennas* (Research Studies Press, 1987)

Johnson, Richard C. and Jasik, Henry, *Antenna Engineering Handbook*, 2nd Edition (McGraw-Hill, 1984).

Jordan, Edward C., *Electromagnetic Waves and Radiating Systems* (Prentice-Hall, 1950; also in paperback by Pearson Education, 2003).

Kraus, John D., *Antennas*, 2nd Edition (McGraw-Hill, 1988).

Lawson, Dr. James L., *Yagi Antenna Design* (ARRL, 1986).

Maxwell, James Clerk, "A Dynamical Theory of the Electromagnetic Field," *Philosophical Transactions of the Royal Society of London*, Vol 155, Jan 1, 1865, pp 459 – 512.

McLean, James S., "A Re-Examination of the Fundamental Limits on the Radiation Q of Electrically Small Antennas," *IEEE Transactions on Antennas and Propagation*, Vol 44, No. 5, May 1996, pp 672 – 675.

Silver, H. Ward, Ed., *The ARRL Antenna Book*, 23rd Edition (ARRL, 2015).

Taub, Herbert and Schilling, Donald, *Principles of Communication Systems*, 2nd Edition (McGraw-Hill, 1986).

Uda, S., "On the Wireless Beam of Short Electric Waves," Institute of Electrical Engineers of (JIEE), Mar 1926

Wangsness, Roald K., *Electromagnetic Fields* (Wiley & Sons, 1979).

Wheeler, H. A., "Fundamental Limitations of Small Antennas," *Proceedings of the I.R.E.*, Vol 35, No. 12, Dec 1947, pp 1479 – 1484.

Wheeler, H. A., "The Radiansphere Around a Small Antenna," *Proceedings of the I.R.E.*, Vol 47, No. 8, Aug 1959, pp 1325 – 1331.

Yagi, H., "Beam Transmission of Ultra Short Waves," *Proceedings of the I.R.E.*, Vol 16, No. 6, Jun 1928, pp 715 – 740.

Zavrel Jr., Robert J., "How Antenna Aperture Relates to Gain and Directivity," *QEX*, May/Jun 2004, pp 35 – 39.

Zivkovic, Z., Senic, D., Bodendorf, C., Skrzypczynski, J., & Sarolic, A. (Faculty of Electrical Engineering, Mechanical Engineering and Naval Architecture, Split, Croatia), "Radiation Pattern and Impedance of a Quarter Wavelength Monopole Antenna Above a Finite Ground Plane," *20th International Conference on Software, Telecommunications and Computer Networks (SoftCOM),* IEEE, Sep 2012, (pp 1 – 5).

# Appendix D

# Symbols and Abbreviations

| | |
|---|---|
| ***A*** | area (vector notation) |
| *A* | *area* |
| *A* | ampere |
| $A_c$ | collecting aperture |
| $A_e$ | effective aperture |
| $A_{em}$ | maximum effective aperture |
| $A_{er}$ | effective aperture of the receive antenna |
| $A_{et}$ | effective aperture of the transmit antenna |
| $A_l$ | loss aperture |
| $A_p$ | physical aperture |
| $A_s$ | scattering aperture |
| B | bandwidth |
| ***B*** | magnetic flux (vector) |
| BER | bit-error ratio |
| *c* | velocity of electromagnetic radiation in free space |
| C | Coulomb |
| *C* | capacity bits/second |
| *C* | capacitance (farads) |
| *C* | circumference of a loop antenna |
| *D* | directivity |
| ***D*** | electric flux density (vector) |
| *DF* | dielectric dissipation factor |
| *dB* | decibel |
| *dBi* | antenna gain relative to an isotropic antenna |
| *dBil* | linear polarized gain relative to a linear polarized isotropic antenna |
| *dBic* | circular polarized gain relative to a circular polarized isotropic antenna |
| *dBd* | antenna gain relative to a ½ wave dipole (0 dBd = 2.15 dBi) |
| *dBm* | power expressed referenced to 1 milliwatt (1 mW) |
| *dBW* | power referenced to 1 watt (1 W) |
| *E* | energy |
| $E_b$ | energy per bit |
| ***E*** | electric field (vector) |
| EIRP | effective radiated power relative to an isotropic antenna |
| ERP | effective radiated power relative to a ½ wave dipole |
| *f* | frequency, Hz |
| *F* | noise factor |
| *G* | gain |
| $G_r$ | gain of the receiver antenna |
| $G_t$ | gain of the transmitter antenna |
| ***H*** | magnetic field (vector) |
| *h* | Plank's constant |
| $h_e$ | effective height |
| $h_p$ | physical height |
| *I* | current |
| $I_{ave}$ | average current along an antenna element |
| $I_0$ | maximum current point on an antenna |
| *J* | Joules |
| *j* | designates a complex function |
| *k* | Boltzmann's constant |
| $k_e$ | dielectric constant |
| *l* | length |
| *L* | inductance (Henrys) |
| *m* | meters |
| *N* | Newton |

| | |
|---|---|
| $NF$ | noise figure |
| $N_0$ | noise energy |
| | |
| $P$ | power |
| PADL | phase, amplitude, direction, location |
| $P_i$ | power input to an antenna |
| $P_n$ | noise power |
| $P_r$ | radiated power |
| $P_r$ | received power |
| $P_{rr}$ | re-radiated power |
| $P_t$ | transmitted power |
| | |
| Q | quality factor |
| $q$ | electrical charge |
| | |
| $R$ | resistance in ohms |
| $R_b$ | bit rate |
| $R_l$ | loss resistance |
| $R_{load}$ | resistive part of a load impedance |
| $R_r$ | radiation resistance of an antenna |
| $R_{rp}$ | radiation resistance at a point on an antenna |
| $R_t$ | terminal resistance |
| $r$ | radius |
| | |
| $\mathbf{S}$ | Poynting vector, power flux in W/m$^2$ |
| $S$ | power flux density |
| $s$ | second |
| SNR | signal-to-noise ratio |
| | |
| $T$ | temperature in Kelvins |
| $T_{ant}$ | antenna noise temperature |
| $T_e$ | effective noise temperature |
| $T_0$ | 290 Kelvins |
| | |
| $V$ | voltage |
| $V$ | velocity |
| $V_n$ | noise voltage |
| $VF$ | velocity factor |
| | |
| $Z$ | impedance |
| $Z_{feed}$ | impedance at an antenna's feed point |
| $Z_0$ | impedance of free space |
| $Z_0$ | characteristic impedance of a transmission line |
| | |
| $\Gamma$ | reflection coefficient |
| $\varepsilon_0$ | permittivity of free space |
| $\mu_0$ | permeability of free space |
| $\omega$ | angular frequency (2p*f*) |
| $\lambda$ | wavelength |
| $\theta$ | a degree or radian value, corresponding to latitude in spherical coordinates |
| $\phi$ | a degree or radian value, corresponding to longitude in spherical coordinates |
| $\Omega$ | ohms |
| $\mathbf{\Omega}$ | steradian solid angle |
| $\Omega_a$ | beam area |

Appendix

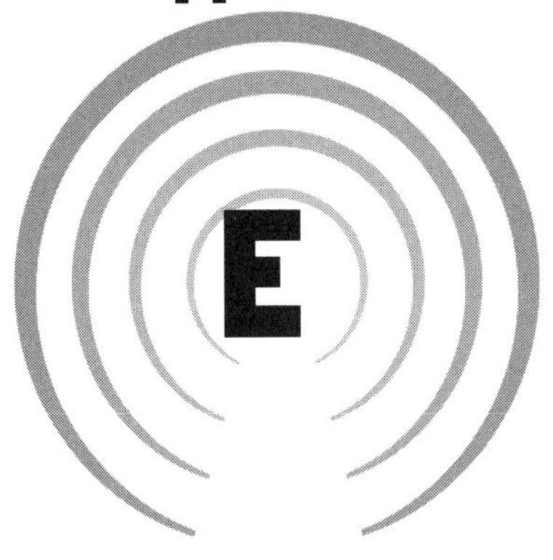


# Comparison of Elevation Plots of Vertical-Polarized Ground-Mounted 160 Meter Antenna Arrays

These three plots (**Figures E.1** to **E.3**) are presented as a simple visual comparison among various configurations using the same gain scales. All data assumes lossless ground systems over average earth. The relative shapes of these plots are valid for higher frequencies as well, so this may serve as a basic comparison for higher frequencies. However, the absolute gain numbers will change with different frequencies. ARRL plotting is used to offer more detailed graphing for small differences among the gain figures. All three graphs include a ¼ λ vertical for reference comparisons.

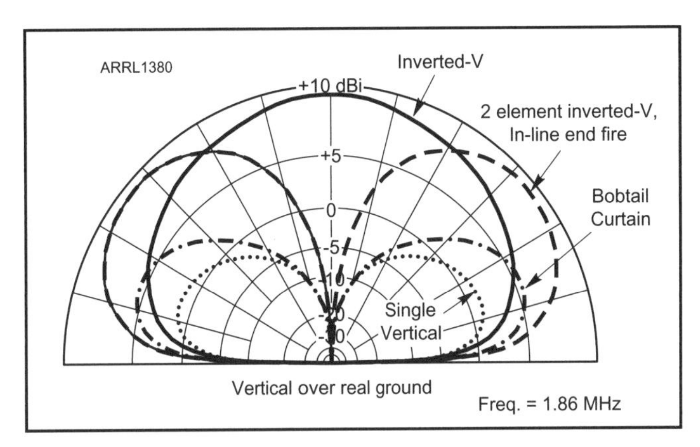


**Figure E.2 — Bidirectional vertical arrays. This figure shows four plots: a reference ¼ λ single vertical; a Bobtail curtain; a single inverted V (off the end); and two in-line inverted Vs fed in end-fire configuration. The inverted V ends are 0.25 meters above ground, and the apex angles are 90 degrees (apex height is 28.3 meters).**

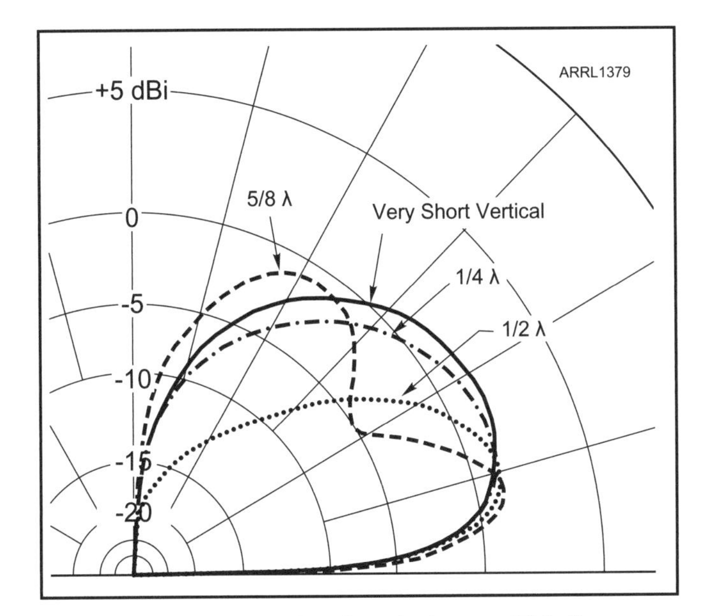


**Figure E.1 — Omnidirectional vertical arrays. This figure plots the elevation gain for four vertical heights: very short (5 meters), ¼ λ, ½ λ, and ⅝ λ heights. Notice the high degree of similarity among all the arrays. This plot assumes zero ground losses. When a non-optimum ground system is used, careful attention must be given to the antenna's radiation resistance which is a direct correlation of the antenna height.**

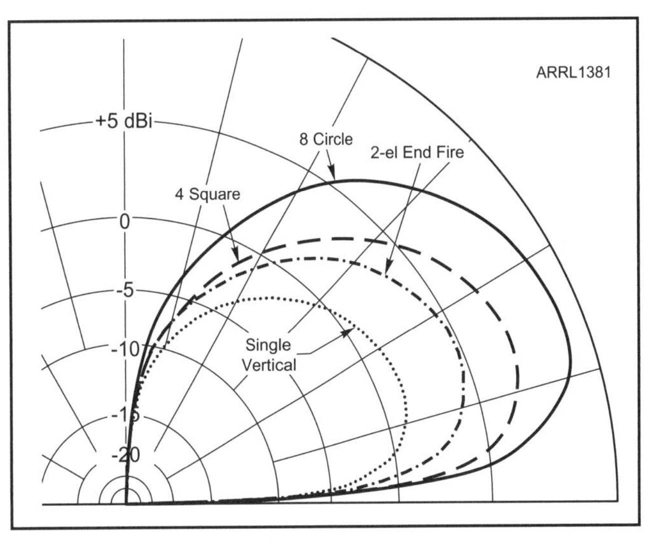


**Figure E.3 — Mono-directional vertical arrays. This figure shows only the maximum forward gain azimuth of the patterns. The four arrays are: single vertical; 2-element end-fire fed in quadrature; 4-square; and 8-circle. Separation of elements is ¼ λ on all multielement arrays. This is an excellent example of a near 3 dB gain for each doubling of the number of elements and thus also the aperture.**

# Notes

# Notes

# Notes

# Notes

# Notes

# Notes

# Notes

# Notes

# Notes

# Notes

# INDEX

Note: The letters "ff" after a page number indicate coverage of the indexed topic on succeeding pages.

## S



## T



## U



## V



## W



## Y